TROPICAL FORESTS

Ecosystem

TROPICAL FORESTS

Peter D. Moore

Illustrations by Richard Garratt

Facts On File
An imprint of Infobase Publishing

TROPICAL FORESTS

Facts On File, Inc.
An imprint of Infobase Publishing
132 West 31st Street
New York NY 10001

ISBN-10: 0-8160-5934-9
ISBN-13: 978-0-8160-5934-8

Library of Congress Cataloging-in-Publication Data
Moore, Peter D.
    Tropical forests / Peter D. Moore; illustrations by Richard Garratt.
        p. cm. — (Ecosystem)
    Includes bibliographical references.
    ISBN 0-8160-5934-9
    1. Forest ecology—Tropics—Juvenile literature. 2. Forests and forestry—Juvenile literature. I. Title.
    QH84.5.M66 2008
    577.3—dc22   2006037441

Facts On File books are available at special discounts when purchased in bulk quantities for businesses, associations, institutions, or sales promotions. Please call our Special Sales Department in New York at (212) 967-8800 or (800) 322-8755.

You can find Facts On File on the World Wide Web at http://www.factsonfile.com

Text design by Erika K. Arroyo
Illustrations by Richard Garratt
Photo research by Elizabeth H. Oakes

Printed in the United States of America

Bang Hermitage 10 9 8 7 6 5 4 3 2 1

This book is printed on acid-free paper.

To Amelia, Amanda, Madeleine, and Michael.
Your hands may be very small but in them lies the future.

◆ ◆ ◆ ◆ ◆

Contents

Preface

Increasingly, scientists, environmentalists, engineers, and land use planners have come to understand the living planet in a more interdisciplinary way. The boundaries between traditional disciplines have blurred as ideas, methods, and findings from one discipline inform and influence those in another. This cross-fertilization is vital if professionals are going to evaluate and tackle the environmental challenges the world faces at the beginning of the 21st century.

There is also a need for the new generation of adults, currently students in high schools and colleges, to appreciate the interconnections between human actions and environmental responses if they are going to make informed decisions later, whether as concerned citizens or as interested professionals. Providing this balanced interdisciplinary overview—for students and for general readers as well as professionals requiring an introduction to the Earth's major environments—is the main aim of the Ecosystem set of volumes.

The Earth is a patchwork of environments. The equatorial regions have warm seas with rich assemblages of corals and marine life, while the land is covered by tall forests, humid and fecund, and containing perhaps half of all Earth's living species. Beyond this are the dry tropical woodlands and grassland and then the deserts, where plants and animals face the rigors of heat and drought. The grasslands and forests of the temperate zone grow because of the increasing moisture in these higher latitudes, but grade into coniferous forests and eventually scrub tundra as the colder conditions of the polar regions become increasingly severe. The complexity of diverse landscapes and seascapes can nevertheless be simplified by considering them as the great global ecosystems that make up our patchwork planet. Each global ecosystem, or biome, is an assemblage of plants, animals, and microbes that have adapted to the prevailing climate and the associated physical, chemical, and biological conditions.

The six volumes in the set—*Deserts, Revised Edition; Tundra; Oceans, Revised Edition; Tropical Forests; Temperate Forests, Revised Edition;* and *Wetlands, Revised Edition*—span the breadth of land-based and aquatic ecosystems on Earth. Each volume considers a specific global ecosystem from many viewpoints: geographical, geological, climatic, biological, historical, and economic. Such broad coverage is vital if people are to understand how the various ecosystems came to be, how they are changing, and, if they are being modified in ways that seem detrimental to humankind and the wider world, what might be done about it.

Many factors are responsible for the creation of Earth's living mosaic. Climate varies greatly between the Tropics and poles, depending on the input of solar energy and the movements of atmospheric air masses and ocean currents. The general trend of climate from the equator to poles has resulted in a zoned pattern of vegetation types, together with their associated animals. Climate is also strongly affected by the interaction between oceans and landmasses, resulting in the ecosystem patterns from east to west across continents. During the course of geological time even the distribution of the continents has altered, so the patterns of life currently found on Earth are the outcome of dynamic processes and constant change. The Ecosystem set examines the great ecosystems of the world as they have developed during this long history of climatic change, continental wandering, and the recent, meteoric growth of human populations.

Each of the great global ecosystems has its own story to tell: its characteristic geographical distribution, its pattern of energy flow and nutrient cycling, its distinctive soils or bottom sediments, vegetation cover, and animal inhabitants, and its history of interaction with humanity. The books in the Ecosystem set are structured so that the different global ecosystems can be analyzed and compared, and the relevant information relating to any specific topic quickly located and extracted.

The study of global ecosystems involves an examination of the conditions that support the planet's diversity. But environmental conditions are currently changing rapidly. Human beings have eroded many of the great global ecosystems as they have reclaimed land for agriculture and urban settlement and built roads that cut ecosystems into ever smaller units. The fragmentation of Earth's ecosystems is proving to be a serious problem, especially during times of rapid climate change, itself the outcome of intensive industrial activities on the surface of the planet. The next generation of ecologists will have to deal with the control of global climate as well as the conservation and protection of the residue of Earth's biodiversity. The starting point in approaching these problems is to understand how the great ecosystems of the world function and how the species of animals and plants within them interact to form stable and productive assemblages. If these great natural systems are to survive, then humanity needs to develop greater respect and concern for them, by understanding better the remarkable properties of our patchwork planet. Such is the aim of the Ecosystem set.

Acknowledgments

I should like to express my gratitude to the editorial staff at Facts On File for their untiring support, assistance, and encouragement during the preparation of this book. Frank K. Darmstadt, executive editor, has been a constant source of advice and information and has been meticulous in his checking the text and coordinating the final assembling of materials. My gratitude also extends to Ms. Alana Braithwaite, his assistant, and to the production department. I should also like to thank Richard Garratt for his excellent illustrations and Elizabeth Oakes for her perceptive selection of photographs. I owe particular thanks to my wife, who has displayed a remarkable degree of patience and support during the writing of this book, together with much needed critical appraisal. I must also acknowledge the contribution of many generations of students in the Life Sciences Department of the University of London, King's College, who have been a constant source of stimulation, critical comment, and new ideas. I also acknowledge a considerable debt to my colleagues in teaching and research at King's College, especially those who have accompanied me on field courses and research visits to many parts of the world. Their work, together with that of countless other dedicated ecologists, underlies the science presented in this book.

Introduction

Tropical forests are the most diverse and complex of all the world's ecosystems. They are also among the most vulnerable and threatened habitats on Earth. It is particularly important, therefore, that everyone be aware of their remarkable biological properties and their immense value to humanity through their influence on the atmosphere and global climate. The goal of this book is to inform students in a way that combines factual accuracy with a sense of wonder at the delicacy of interactions between the living and nonliving worlds that is so well displayed in the forest ecosystem.

Tropical Forests is structured such that information is both accessible and easy to assimilate. The chapters are arranged in a logical structure, beginning with the physical setting, including the geography, geology, and climatic conditions within which the tropical forests are found. The great biodiversity of the forests, from microbes to mammals, is covered in some detail, and emphasis is placed on the adaptations of organisms to their environment and to the other species that surround them. In order to encourage students to develop a holistic concept of the forest, the interactions between organisms and their physical surroundings are considered in some detail, as are the processes linking the two into an integrated ecosystem. An appreciation of the complexity of food webs and nutrient cycles will help students determine what underlies the innate stability of the forest. Finally, the rapid changes of land use and climate change are examined, and the dubious future of the forests is debated. Throughout the book, diagrams, photographic illustrations, and informative sidebars are used to assist the student in finding out more about the forests and in understanding the functioning, value, and sheer beauty of this remarkable ecosystem.

In the year 1493, Christopher Columbus became the first European to write a description of a tropical forest. He had arrived at the islands of the West Indies and discovered a New World that clearly impressed him greatly.

> Its lands are high and there are in it very many sierras and very lofty mountains. . . . All are most beautiful, of a thousand shapes and all are accessible and filled with trees of a thousand kinds and tall, and they seem to touch the sky. And I am told that they never lose their foliage, for I saw them as green and lovely as they are in Spain in May and some of them were flowering, some bearing fruit.

Columbus had discovered the tropical rain forest.

The features of the forest that so impressed Columbus continue to enthrall today's visitors of tropical forests. The sheer diversity of life there is most clearly illustrated by the trees, which provide the forest with its massive structure, resembling the architecture of a cathedral. Columbus hit upon an important botanical fact in his reference to tree diversity: the vast majority of plant species within the tropical forests are trees. Within the arching forest canopy but hidden from Columbus's eye as he floated in his ship offshore is a multitude of living creatures occupying each layer of branches and leaves. Even the roof of the forest is tangled and entwined with climbing vines and suspended plants at vertiginous heights, their roots never touching the forest floor (see illustration on page xviii). Birds, bats, and beetles spend their entire lives high in the canopy, feeding upon the flowers and fruits that Columbus could see blossoming at heights of 100 feet (30 m) and more in the expansive sea of evergreen leaves. Beneath that canopy and hidden from the eyes of the distant observer lies the dark, damp interior of the forest's depth. Descending from the brilliantly lit surface of the emergent branches is like diving into the ocean, as light levels fall and a stillness envelops the observer. Layers of branches and leaves become more shaded until the observer's eye reaches the forest floor, where the supportive columns of tree trunks convey a cloistered air. Although the floor itself is moist and thick with decaying leaves, ground-dwelling plants are less abundant than might be expected. The idea of the impenetrable jungle, so popular in adventure novels, is generated at the forest's edge, where disturbances by rivers or trails give access to the light, and the light stimulates plant growth close to the ground. In the midst of the forest, far from any disturbance, the floor is relatively clear and the struggling bustle of life takes place overhead.

In the 500 years since Columbus and his followers first encountered the New World tropical forests, much has been learned about their ecology and diversity. They occupy about 6 percent of the world's land surface but contain approximately half of the Earth's living species. They are thus the focus of global biodiversity. These figures, however, are vague approximations because the variety of life is so great and the numbers of scientists so small that no one can do more than hazard a guess at how many species the tropical forests contain—it could run into the tens of millions. Scientists can land a machine on the moons of Saturn, but they still cannot count all the creatures of the forest.

WHAT ARE THE TROPICAL FORESTS?

There are many different kinds of tropical forests. Perhaps most famous are the equatorial rain forests that lie in a belt around the equator in three main regions of the world—South America, Africa, and Southeast Asia. These are the forests that contain the highest *diversity* of plants and animals, support the greatest mass of living material per unit area, and maintain an uninterrupted pattern of growth throughout the year. Their canopies are evergreen, constantly supporting flowers and fruits from the forest's variety of plant components and supplying the needs of myriads of vertebrates, insects, and microbes. It is this type of forest that is often referred to as "jungle." The word *jungle* actually comes from the ancient Sanskrit language of India, where the word *jangala* means a wilderness. The word was taken up by the English, corrupted in both its pronunciation and its meaning, and came to be applied specifically to the tropical forests, particularly to the impenetrable growth of vegetation associated with forest edges.

Moving north and south from the equator, the forest becomes somewhat less massive in its structure and evergreen trees gradually give way to deciduous ones as the climate becomes increasingly dry. Eventually the complex forest is replaced by thick, thorny woodland savannah as the increasing length of a dry season leads to prolonged drought. In some tropical locations, such as northern India, intense rainfall comes in a single period of the year, the monsoon, and a distinctive monsoon forest develops as a result. On high mountains in the Tropics, the forest is subjected to very different environmental conditions and takes on yet another form. Montane rain forest is sometimes called "elfin woodland" because its small stature and lichen-draped boughs give it a dwarfed appearance.

Tropical Forests will describe all of these different forest types and examine the climates in which they develop, the plants and animals that inhabit them, and the complexity of

The upper canopy of the Amazon rain forest, showing emergent trees and the dense cover of the layered canopy below. *(Large-Scale Biosphere-Atmosphere Scientific Conference)*

the interactions between all of the different components that compose the tropical forest ecosystem. The book will also tackle some more general questions that the tropical forests pose, including why these habitats are so much richer in living things than any other type of ecosystem, and how the forests survived during the extreme changes in global climate over the past two million years. Questions will be raised regarding the complexity of their ecological interactions and whether complexity makes the forest more or less stable when faced with stress. These are important questions that need to be answered if the tropical forests are to survive as human populations continue to expand and environmental changes accelerate.

WHY ARE THE TROPICAL FORESTS IMPORTANT?

In these times there are many problems facing humanity—social, political, and ecological. It is important to consider whether the tropical forests are of particular significance in a world with so many difficulties to face and whether they are of greater value to humanity than other biomes.

There are many reasons why the tropical forests require such close attention, an important one being that they are so diverse in species. Ecologists still do not fully understand why one type of ecosystem should support more species than another or how so many species can live alongside one another. Perhaps each species has its own unique way of making a living, avoiding having to compete too intensively with other species. Whatever the mechanisms supporting this diversity, the sheer numbers of plants and animals living in the tropical forests means that they are a vast, potential resource for human welfare. The loss of an area of tropical forest puts more species in danger of extinction than does the loss of the same area in any other habitat, and losing species is something we cannot afford. People need other species to support them. Plants are required to trap solar energy and provide food. The numbers of plants used in global food production is extremely small and there is great potential for recruiting new plants in the service of humankind. Timber is a vital building material; many tropical forest trees are highly valued for the properties of their woods. To lose a tree species or even a particular genetic strain of a species is regrettable and is a loss that can never be remedied. Conservation of species and genetic resources is thus a wise policy for the future when changing human needs and knowledge may place new demands on food and timber as well as other valuable materials, such as rubber.

About half of the world's drugs are based on plant products, many of these being from tropical forest species. In a complex environment like the forest, every plant is a potential food resource for the many animals in the community, and adaptation to such a competitive situation has led to the development of a range of chemical defenses in plants to deter the voracious grazers. Many of these chemicals have a direct impact on animal (including human) physiology. Some affect the function of the heart, others the kidney or the liver; many are highly toxic and can be used to kill unwanted cells in the body, such as cancer cells, if used in small quantities. The great arsenal of plant defense compounds found in tropical forest vegetation represents an immense resource for the pharmaceutical industry. Natural defense mechanisms may also provide new pesticides for agriculture that can be used to control insects and other organisms that attack crops but that are less persistent or harmful to the environment than artificial pesticides.

Plants are not the only source of toxins and possible drugs for the future. Some predatory animals of the tropical forests have developed poisons that they use in disabling their prey. Many reptiles and spiders have venoms that affect mammalian nervous systems. Although these are harmful when applied in large quantities, small doses can prove beneficial in the treatment of nervous disorders. Even vampire bats have their uses. The saliva of the vampire bat contains chemicals that prevent blood clotting, and these are proving useful in the treatment of strokes in humans. Once such compounds have been discovered and their mode of action studied, chemical modifications can be tried to see if the natural compounds can be improved upon. Nature, in other words, is a great resource of ideas that can be taken up by pharmacists and chemists for human applications.

Quite apart from the value of individual species in agriculture, forestry, and the pharmaceutical industry, as an interactive ecosystem the entire forest is important in maintaining global balance of the atmosphere and climate. The forest recycles water, intercepting rainfall, taking up water from the soil, and returning water vapor to the atmosphere. The rain forest not only absorbs but also produces rainfall through its capacity to recycle the water on which it depends. The loss of forest means that water must move over and through soils to the oceans, eroding landscapes and destroying agricultural potential as it moves. In the mountains the water movement can cause landslides and ensuing loss of lives and livelihoods; in the lowlands it often results in devastating floods. Forests are thus important protectors of the tropical landscape, so their conservation, maintenance, and management are vital for the development of human populations.

Not only is the local well-being of humans affected when forests are lost, but also there is a global impact. Tropical forests contain large reserves of carbon, locked in the organic materials that compose their woody trunks and their leaf canopies. These reserves of carbon can be transformed into the gas carbon dioxide by tree felling and burning. The gas then enters the atmosphere and becomes an agent of climatic change. In the course of this book, the role of the tropical forests in the global economy of carbon will be examined. Scientists are very eager to understand whether the tropical forests act as a sink for the excess carbon entering the air as a result of industrial combustion of fossil fuels. It is also important to discover whether the current changes in atmospheric chemistry and the resulting climatic changes will influence the growth, structure, and composition of the forests. This book will face these questions and weigh the evidence.

The importance of the rain forest thus lies in its resources of wood, fiber, latex, toxins, and drugs, and in its contribution to the world's hydrological cycle of water and atmospheric cycle of carbon. But this is by no means the sole importance of the forest to humanity. People require more than food, building materials, and drugs for their survival; they also need mental and spiritual stimulation, education, and recreation. The tropical forests are indeed a jungle, in the true sense of a vast wilderness. For residents of the concrete jungle of the city it is important to be aware that there are still places where nature operates free from the excesses of human manipulation. The diversity of life in forests is a source of wonder and pleasure to countless people, whether it is the bizarre complexity of orchid flowers or the lavish colors of a macaw's feathers or a toucan's bill. The forests and their inhabitants, therefore, are important because they bring enjoyment to many people. One could also say that they are important because the animals and plants themselves have a right to existence on this planet, and their survival cannot be judged solely in terms of their value to the human species. But this is an ethical question rather than a scientific one.

In *Tropical Forests* we will thus analyze the forest and its composition, looking in detail at some of its component parts. We will also investigate the whole system, to see how the various parts are integrated and how the forest behaves as a part of the landscape and the world order. In recent years, the tropical forests have come to the fore in being recognized as vital components of the Earth's eco-logical system, but they are poorly known and understood in comparison to other great biomes of the world. This is in part due to their relative inaccessibility: It is easier for a member of the public to visit the ice sheet of Antarctica than the interior of a tropical forest. It is also difficult to display visually the nature and the composition of a forest. Apart from aerial surveys, forests are just not easy to portray on film for television consumption. It is easier to film a cheetah chasing and slaying a wildebeest on the African savannah than a jaguar killing a capybara in the Amazon swamps. This is a challenge that photographers and filmmakers are rapidly taking up, but the forest has certainly had less exposure to the general public than many other habitats, where filming is less frustrating and where panoramic shots can be more impressive.

The fact that tropical forests are so important in many different ways, together with the need for greater public appreciation of the ecology and function of the forests, is justification for the writing of this book.

1

Climate and Tropical Forests

The general impression of tropical forest climate is heat and humidity. These characteristics are found in the great rain forests of the equatorial regions, but the picture is not quite as simple and uniform as it may seem. Lowland tropical forests are certainly warm, but in the highlands of the Tropics this is not always the case. Equatorial forests are usually wet throughout the year, but the forests that lie at any distance from the equator may experience periodic drought. The relationship between the tropical forests and climate, therefore, is more complex than might be expected.

■ PATTERNS OF TROPICAL FOREST DISTRIBUTION

The map shows the broad pattern of tropical forest distribution over the Earth's land surface. True rain forest lies in a band around the equator and falls into three main geographical blocks. In the New World there is the American rain forest formation that has its main distribution in the Amazon Basin in Brazil, but also has outliers in Central America, the West Indies, and along the Atlantic coast of Brazil. The African rain forest has its main distribution in the west of Africa in the coastal regions and extending east through the Congo Basin as far as Lake Victoria in the east. An outlier of this forest is situated in eastern Madagascar. In Southeast Asia lies the Indo-Malayan rain forest, occupying western Thailand and Malaysia, together with the islands of Sumatra, Borneo, Java, the Philippines, and New Guinea. Outliers and extensions of this forest are found along the western coast of India, Sri Lanka, Burma, Vietnam, eastern Australia, and some of the South Pacific islands.

Beyond the limits of rain forest lie moist deciduous forests, which are mainly found in South America, Africa, and parts of southern Asia. The remaining regions of the Tropics are taken up with dry deciduous forests, savannah woodlands and grasslands, thorn scrub, and desert. The bulk of the tropical rain forest areas of the world lie close to the equator, mainly between latitudes 15°N and 15°S (see sidebar on latitude on page 3). The drier deciduous forests lie mainly between latitudes 20°N and 20°S, and beyond them, especially on the continental landmasses, are regions of dry scrubland and desert.

■ GLOBAL CLIMATE PATTERNS

To understand the distribution pattern of tropical forests, one needs to study closely the climate patterns over the Earth. There are several obvious questions that need to be answered, such as why the Tropics are hotter than the Arctic and Antarctic regions, why the equatorial regions are so wet, and why conditions generally become drier as one moves away from the equator. To answer these questions it is necessary to consider the distribution of energy around the Earth. This requires information about the movements of air masses in the atmosphere and water masses in the oceans.

The first point to examine is the way in which energy arrives at the surface of the planet. Energy on Earth ultimately comes almost entirely from the Sun. There are some exceptions to this statement: the hot interior of the Earth releases some geothermal energy, natural radioactive decay also produces some energy, and the gravitational pull of the moon creates tidal energy. But the vast bulk of the energy arriving on Earth is in the form of solar radiant energy. Energy comes in a variety of forms, which physicists differentiate in terms of *wavelength*. Light itself, for example, consists of a spectrum of energy of different wavelengths, from the very short violet end to the longer-wavelength red end (see the diagram on page 28). This accounts only for the visible forms of energy; there are many shorter-wavelength forms, such as ultraviolet, and even shorter gamma rays. There are also longer-wavelength forms, such as infrared, heat energy that can be felt but not seen. Although the Sun emits energy over a very broad spectrum, most of the solar radiation that reaches the surface of the Earth lies between 0.000001 and 0.00016 inches. Such minute numbers are difficult to work with, so the measurements used by scientists are metric, and the wavelength is expressed in micrometers (1 μm is one millionth of a meter) or in nanometers (1 nm = one billionth of a meter); the Earth receives energy mainly in 0.2-μm to 4-μm wavebands. The Earth's atmosphere does not absorb energy within this range, so it passes directly to the surface of the planet. Shorter wavelengths, however, such as some ultraviolet radiation, is absorbed by certain components of the atmosphere, such as high-altitude ozone. In addition, dust in the atmosphere, together with cloud cover, can result in the absorption or reflection of some of the incoming energy.

When solar energy strikes the surface of the Earth, much of it is likely to be absorbed. An object placed in sunlight becomes warm as a result of the absorption of energy and as its temperature rises it radiates some energy back into the atmosphere. In a similar way, a land surface heated by sunlight radiates heat energy, and heat energy has a longer wavelength than light energy. Some of the heat energy radiated by land surfaces is taken up by the atmosphere, because the atmosphere is less transparent to longer-wavelength heat than it is to light. By absorbing the reradiated heat energy in this way, the atmosphere acts like a thermal blanket around the Earth, retaining heat that would otherwise be rapidly lost. This absorption, known as the *greenhouse effect,* is due to the presence of certain gases in the atmosphere, including water vapor, carbon dioxide, oxides of nitrogen, and several others, which have become known as *greenhouse gases.* It is called the greenhouse effect because the entire Earth system acts like an enormous greenhouse. Light passes through glass panes of a greenhouse without being absorbed to any great extent by the glass. The glass itself does not become hot. When light strikes the soil within the greenhouse it causes the soil to become hot, and the reradiated heat does not pass out through the glass roof because long-wave, heat energy does not easily penetrate glass but remains within the greenhouse, warming the air it contains. Without the greenhouse effect the Earth would rapidly cool as soon as the Sun set in the evening. Nights would become unbearably cold and life on the planet would be much more difficult. As will be discussed later (see "The

The global distribution of evergreen and deciduous tropical forests.

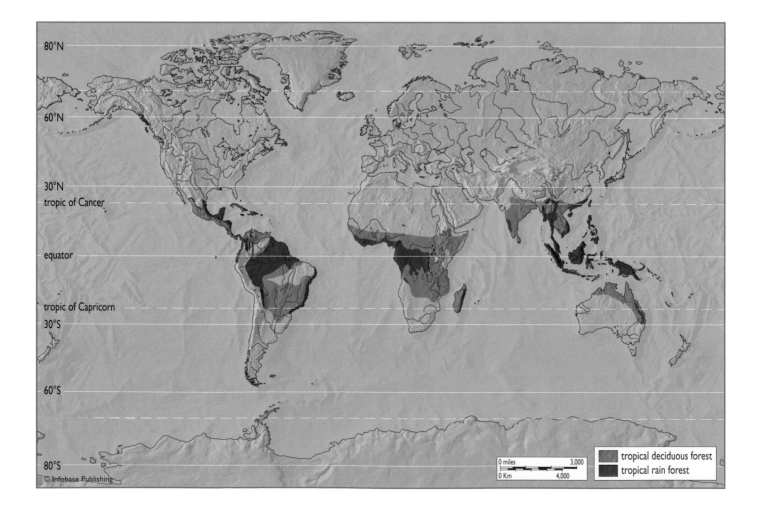

Latitude

The Earth is a sphere, which rotates around an axis. This axis passes from the North Pole through the center of the Earth to the South Pole. The equator is an imaginary line that passes around the Earth and is equidistant from both of the poles. In cross section, therefore, the equator forms a right angle to the Earth's axis, as shown in the diagram. The equator is regarded as having an angle of zero degrees (0°), which means that the angle of the North Pole is 90°N and that of the South Pole is 90°S. Between these extremes it is possible to denote the position north or south of any point on the Earth's surface by reference to the angle it lies at in relation to the equator. This is called its *latitude*. Lines of latitude run around the Earth north and south of the equator, ranging from 0° to 90°; those with high values (close to the poles) are termed the high latitudes, and those close to the equator have low values and are called low latitudes. Tropical forest is restricted to low latitudes. These lines are, of course, purely conceptual but are an important means of defining any location on the surface of the Earth. A second coordinate is required to fix a precise position, and this is given by its *longitude*. Lines of longitude run from pole to pole over the Earth's surface and are numbered from 0° to 359°, with the base line (0°) running through Greenwich in London, England. This choice of the zero line of longitude is a consequence of history, but it has proved convenient because the 180° line (the International Date Line) runs through the Pacific Ocean

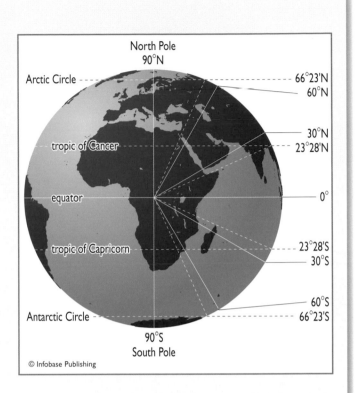

Latitude is measured as the angle at the center of the Earth for each location on the surface, taking the equator as 0° and the poles as 90° north and south.

rather than over land surface, where changing date with location on either side of a line would be extremely confusing for the local inhabitants.

Forest as a Carbon Sink," pages 209–211), changes in the intensity of the greenhouse effect in the atmosphere have a profound influence on global climate.

There is a complication in this process because the amount of energy absorbed by a surface varies. A black surface absorbs sunlight and becomes heated much more effectively than a white surface, and a dull, matte surface absorbs more efficiently than a shiny surface. The efficiency of a surface in absorbing light and becoming heated in the process is dependent on how much of the incoming energy is reflected; the reflectivity of a surface is expressed as its *albedo* (see sidebar).

As can be seen from the figures in the sidebar, different types of land surface and vegetation have a strong effect on how much of the Sun's energy is absorbed. The figures for the surface of water also show that albedo depends quite strongly on the angle of the Sun. These aquatic data apply to a water body with a calm and smooth surface. Waves can

create very complex patterns of reflection. These variations in albedo, depending on the nature of the surface, have an important impact on the world's climate. Oceans and land surfaces, forests and deserts, ice sheets and wetlands all have different capacities to absorb and reflect incoming radiation, which influences the degree to which they become warm in the sunshine and the way they radiate heat back into the atmosphere. The angle at which radiation strikes the Earth's surface also influences absorption and reflection. The diagram shows the way in which the Earth's curvature results in energy being received at different angles according to latitude. The Sun is overhead at noon at the equator twice a year, when the length of day and night is equal over the whole planet. The Earth's tilt on its axis of 23.5°, however, means that this situation is not maintained. The Sun is overhead at the tropic of Cancer (66.5°N) during the Northern Hemisphere summer and at the tropic of Capricorn (66.5°S) during the Southern Hemisphere summer. Throughout the

Albedo

When light falls upon a surface, some of the energy is reflected and some is absorbed. The absorbed energy is converted into heat and results in a rise in the temperature of the absorbing surface. The proportion of light that is absorbed and the proportion reflected vary with the nature of the surface. A dark-colored surface absorbs a greater proportion of the light energy and reflects less than a light-colored surface; it therefore becomes warmer more rapidly. A dull, matte-textured surface absorbs energy more effectively than a shiny one. A mirror reflects most of the light that falls upon it, which is why mirrors are usually cold to the touch. The angle at which light falls upon a surface also affects the proportions of light absorbed and reflected. Light arriving vertically is reflected less than light arriving from a low angle. Scientists express the degree of reflectivity of a surface as its *albedo*, stated as the amount of light reflected divided by the total incident light.

In order of efficiency of reflectivity, the following list gives examples of the albedo of various surfaces:

SURFACE	PROPORTION REFLECTED	ALBEDO
Snow	75–95%	0.75–0.95
Sand	35–45%	0.35–0.45
Concrete	17–27%	0.17–0.27
Deciduous forest canopy	10–20%	0.10–0.20
Road surface	5–17%	0.05–0.17
Evergreen forest canopy	5–15%	0.05–0.15
Water (overhead sun)	5%	0.05
Overall average for Earth	39%	0.39

year, however, the angle of the Sun within the low-latitude tropics is high when compared with that of the higher latitudes. In the polar regions, the angle of the Sun above the horizon is never high. This difference in solar angle between the poles and the Tropics means that energy is spread over a wide area in the high latitudes, whereas it is concentrated in a smaller area in the low latitudes. The heating potential of radiation in the polar regions is thus lower than that found in the Tropics, as shown in the diagram below.

Seasonal differences in daylength are also more pronounced in the high latitudes. In the Tropics there is some variation in daylength depending on the position of the overhead Sun at noon, but the variation is very small and days are roughly equal to nights in their length at all times

Light from the Sun arrives at the Earth as parallel beams, but the light heats different amounts of surface area of the Earth according to the angle at which it arrives. Tropical areas receive more intense energy than the polar regions. The tilt of the Earth upon its axis means that the position of the overhead noonday Sun varies with season.

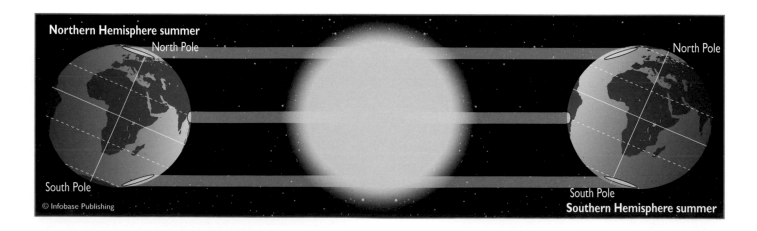

of the year. At higher latitudes, however, there are increasingly wide disparities in daylength with season. When the three factors of solar angle, area of energy dispersal, and seasonal differences are put together, it is evident that the polar regions will have a smaller input of energy per unit area of Earth's surface than that of the Tropics. Much of the energy absorbed by the Earth is radiated back into space, but the ratio between absorption and radiation varies with latitude. The accompanying diagram shows the quantities of energy absorbed and radiated at different latitudes. It can be seen that energy absorption at the equator is about five times that at the poles but radiation is only fractionally greater. More energy is being absorbed at the equator than is being radiated, while the reverse is true for the poles.

The average temperature at the equator is 79°F (26°C), which varies very little with season, while at 80°N the January average is -22°F (-30°C) and the July average is 37°F (5°C). As the diagram shows, at this high latitude, the Earth is actually losing more energy than it is receiving, so the Arctic can be regarded as an *energy sink*. But the Arctic is not becoming colder year by year, and the equatorial regions are not becoming hotter. There must be energy movement from tropical areas with an abundant supply to the energy sinks at high latitude. There are two means by which the Earth's energy can be redistributed in this way, the atmosphere and the oceans.

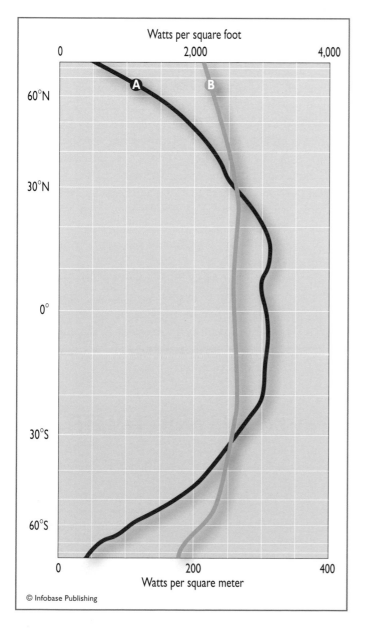

Curve A represents the distribution of solar energy absorbed by the Earth's surface. The polar regions absorb considerably less energy than the equatorial regions. Curve B shows the radiation of energy from the Earth's surface at different latitudes. The high latitudes radiate more energy than they receive, while the Tropics absorb more energy than they radiate. Stability is maintained as a result of the redistribution of energy by the atmosphere and the oceans.

© Infobase Publishing

■ THE ATMOSPHERE AND CLIMATE

Heating in the Tropics and cooling at the poles results in a global energy imbalance, which in turn gives rise to atmospheric turbulence. Warm air has a lower density than cold air, so there is a tendency for cool, dense air to move toward the equator, pushing the warm, low-density air upward. Air masses approach the equatorial regions from both north and south of the equator, creating the *intertropical convergence zone* where the two air masses meet. The warm, moisture-laden air rises and cools in the process (see sidebar "Lapse Rate," page 10). Cooler air is less able to hold moisture, so water condenses into raindrops and the equatorial regions receive high levels of precipitation, resulting in development of the tropical rain forests. In the upper atmosphere, around 10 miles (16 km) above the ground, the air that has been forced upward moves out toward the poles (see diagram [left] of energy at different latitudes), and since the air has now cooled and become denser, it falls toward the Earth's surface once more, creating a high-pressure belt between about 25° and 30° north and south of the equator. These falling air masses have lost their moisture over the equatorial regions, so they cause dry conditions when they arrive back at the Earth's surface. These belts are where the world's deserts are mainly found. Between the equator and these dry latitudes precipitation gradually declines, and this, as will be seen (see chapter 3, "Types of Tropical Forest," pages 53–75), has a profound impact on the structure and composition of the tropical forest.

The hot, dry air of the desert belts may be deflected back toward the equator, or it may travel toward the poles.

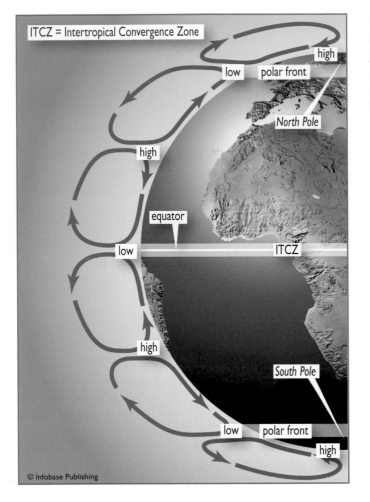

The circulation of the atmosphere occurs in a series of cells, creating areas of low pressure near the equator and in the region of the polar fronts, and areas of high pressure close to the tropics of Cancer and Capricorn and at the North and South Poles.

If it takes the latter route, then these air masses meet cold, polar air moving in the opposite direction. Where the two air masses collide they create a boundary zone, called the *polar front,* and this region is characterized by unstable weather conditions. The precise position of the polar front is extremely variable in space and time, but the mix of air masses results in the cyclonic depressions and accompanying precipitation typical of latitudes roughly between 40° and 60° north and south of the equator. The polar front is also associated with a strong wind, the *jet stream,* which moves from west to east at an altitude of around 35,000 feet (10,000 m) at the boundary between the lower part of the atmosphere, the *troposphere,* and the upper part of the atmosphere, the *stratosphere.* This boundary is called the *tropopause.* The jet stream causes cyclones (low-pressure systems) and anticyclones (high-pressure systems) to move around the world in an easterly direction. The lati-

tude of the polar front and the jet stream varies with the season. In summer in North America the polar front and jet stream lie roughly along the Canadian border, dipping south to Washington, D.C., in the east. In winter it runs approximately from Los Angeles through northern Texas to South Carolina.

In the Southern Hemisphere, the polar front encircles Antarctica and lies within the Southern Ocean. For much of its length, where the Southern Ocean meets the South Atlantic Ocean and the Indian Ocean, it lies approximately along the 50°S latitude. In the region where the Southern Ocean meets the South Pacific, however, the southern polar front is located along latitude 60°S. Some of the air forced upward in the turbulent regions around the polar fronts moves on toward the poles at high altitude. Here the air becomes very cold and dense, so it begins to lose altitude and its descent creates another high-pressure zone over the poles, as shown in the diagram on this page. As in the high-pressure zone of the outer Tropics and subtropics, the outcome is the production of low-precipitation, desert conditions.

The pattern of winds over the Earth's surface, shown in the diagram on page 7, is generated by the circulation of these atmospheric cells of air mass movements, but they rarely move in a simple north–south direction. The reason for this is the spin of the Earth on its axis. Any free-moving object in the Northern Hemisphere, whether in the oceans or in the atmosphere, tends to be deflected to the right, hence motions become clockwise. In the Southern Hemisphere free objects are deflected to the left, so the motions become counterclockwise. This tendency was first described by a French engineer, Gaspard-Gustave de Coriolis (1792–1843) in 1835, and has come to be called the *Coriolis effect.*

There are also seasonal differences in the pattern of air movements over the Earth's surface, and these have an important effect on seasonal climatic variations in the Tropics. They are caused by movement of the intertropical convergence zone north and south of the equator with changes in the season (see sidebar on page 8).

■ THE OCEANS AND CLIMATE

The atmosphere thus provides a mechanism for the redistribution of tropical heat, but it is not the only means by which

(opposite page) The pattern of air movement over the Earth's surface and the distribution of high- and low-pressure systems in January and in July. In the Northern Hemisphere, the northward movement of the intertropical convergence zone causes a reversal of the winds in the Indian Ocean, resulting in monsoon conditions over India.

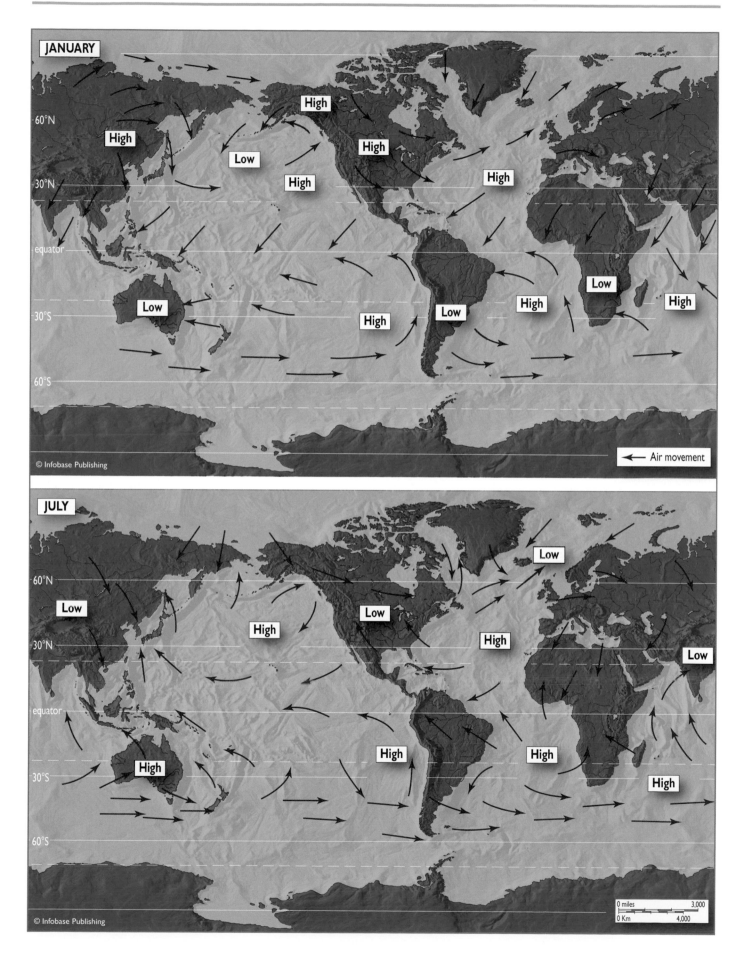

The Intertropical Convergence Zone

The abundant supply of energy to the equatorial regions heats the Earth's surface in that area causing the initiation of convection currents in the atmosphere. Hot, low-density air is forced upwards by cooler, denser air moving in from higher latitudes both north and south of the equator. These air masses thus converge on the equatorial regions creating the intertropical convergence zone (ITCZ). But the location of the ITCZ varies with season because the angle of tilt of the Earth means that different latitudes experience an overhead Sun at midday at different times of the year. In the Northern Hemisphere summer the ITCZ moves north of the equator and in the Southern Hemisphere summer it moves south. This has an impact on the pattern of winds in the Tropics, as shown in the diagram on page 6. In the Northern Hemisphere winter the ITCZ is deflected south and as a consequence winds over northern Africa and the Indian Ocean are southerly. The tropical forests of equatorial Africa thus receive dry air from the Sahara Desert, and the winds over India, Burma, and Thailand are offshore. In the Northern Hemisphere summer, by contrast, the ITCZ is located farther north and this brings winds off the oceans into West Africa, East Africa, India, and parts of Southeast Asia. Onshore winds in the Tropics are rich in water content, and they bring summer precipitation into lands that were dry in winter. This wet season is called the *monsoon*. The monsoon permits the growth of forest even in some dry tropical areas, such as the Himalaya foothills of northern India.

energy is taken from the Tropics and dispersed into the higher latitudes; the oceans also have a profound influence. Surface water moves around and between the great oceans in the form of currents and, in general, the pattern follows that of the winds and is driven by them. In the Southern Hemisphere the main surface circulatory pattern of the oceans is in the form of counterclockwise motions, whereas in the Northern Hemisphere the movements are clockwise. In the North Pacific Ocean and the North Atlantic Ocean there are west–east currents bringing warm tropical water into the higher latitudes.

Oceanic circulation occurs in deep water as well as at the surface. There is a global movement of water around the earth that is driven by changes in the density of seawater, itself a product of salinity and temperature. Warm water with relatively low levels of salt is less dense than cold water with higher salt concentrations. As water moves around the world it changes in its temperature and its salinity. This circulation pattern is called *thermohaline circulation* (*thermo-* refers to the water temperature, and *-haline* relates to its saltiness), or the *oceanic conveyor belt*. The overall pattern

The Oceanic Conveyor Belt

In addition to the surface currents in the oceans, which are driven mainly by winds in the atmosphere, there is a global circulation of waters driven by the changing density of seawater, as shown in the diagram on page 9. The surface water of the oceans tends to be warmer, less salty, and hence less dense than the deep water. When these warm surface waters move into the high latitudes, they give up some of their heat to their surroundings and consequently cool, become denser, and sink. This chilling of surface water occurs most strongly in the North Atlantic Ocean, where the warm tropical surface waters of the Gulf Stream pass northward between Iceland and the British Isles and meet the cold waters of the Arctic Ocean. As the warm waters lose their heat to the waters of the Arctic Ocean, they cool, sink, and begin to make their way back south through the Atlantic and eventually into the Southern Ocean surrounding Antarctica. Passing eastward into the Pacific Ocean, these cold, deep waters eventually surface either in the Indian Ocean or in the North Pacific. There they become warm and drift westward once more around the Cape of Good Hope in South Africa and into the Atlantic to begin their circulation over again.

This thermohaline circulation of water through the oceans of the world plays an important part in the redistribution of energy between the Tropics and the high latitudes, particularly the North Atlantic.

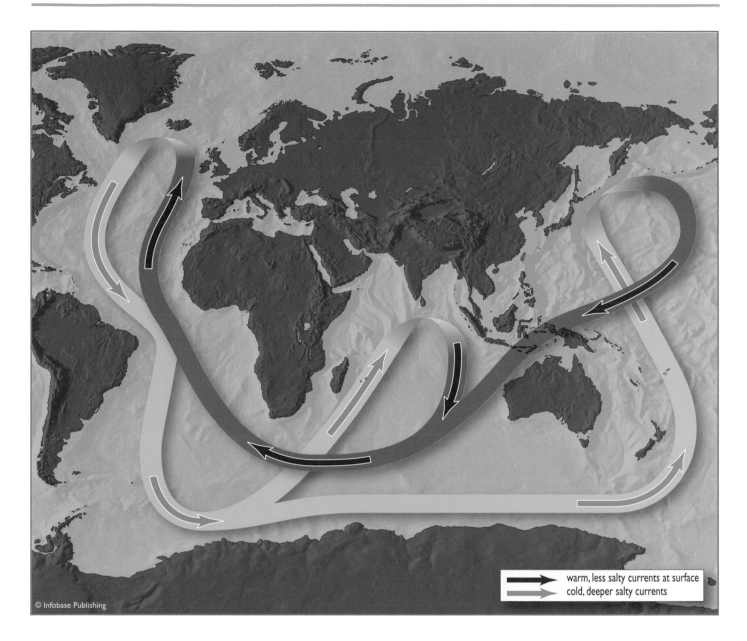

warm, less salty currents at surface
cold, deeper salty currents

© Infobase Publishing

The global circulation of oceanic waters, called the oceanic conveyor belt. Warm, low salinity and low-density water moves along the surface of the ocean. On arrival in the North Atlantic Ocean the water cools and forms a deep water current that flows in the opposite direction.

of global thermohaline circulation is shown in the diagram above and is explained in the sidebar on page 8.

Between them, the atmosphere and the oceans thus redistribute energy around the surface of the Earth. Their currents and movements ensure that the Tropics, although hot, do not become totally roasted, and the polar regions, although cold, are less frigid than one might expect from their overall energy budgets, as shown in the diagram on page 5. Lands fringing the world's oceans have an *oceanic climate*.

■ TROPICAL CLIMATES

A study of the energy distribution and redistribution over the Earth's surface explains why the Tropics are warmer than other parts of the globe, but it does not fully explain why the equatorial regions are also generally wet. Rain forests are present around the equator because of large quantities of precipitation, mainly in the form of rainfall, which is spread throughout the year. The atmosphere of the region is thus constantly humid (see sidebar on page 10), which is favorable to the growth of many types of plants.

Warm air is able to hold more water than cold air, so the solar heating of the equatorial regions creates conditions that are suitable for rapid evaporation of water. The air takes up water from the oceans, wetlands, rivers, and soils, as well as from the leaves of the plants that clothe the ground so that it becomes saturated with water vapor. Its relative humid-

Humidity

Humidity is an expression of the quantity of water vapor in the air. This can be measured in absolute terms, such as the mass of water vapor in a given volume of air. Humidity may also be measured in terms of vapor pressure. All of the gases in the atmosphere (largely nitrogen and oxygen) contribute to the air pressure, and water vapor also plays a part. The pressure of water vapor in the atmosphere is an expression of the air's humidity or wetness. The amount of water vapor that can be contained within a packet of air depends on the temperature of that air. Air at low temperature is able to hold less water than air at high temperature; so, if air cools, some of the water vapor condenses into water droplets that may fall as rain. The temperature at which the air becomes saturated with water, and condensation occurs, is called the dew point. By definition, there-fore, any packet of air at a temperature above its dew point is not fully saturated with water. The difference between the pressure of water vapor present in this air and the vapor pressure needed to saturate it (the *saturated vapor pressure*) is called the vapor pressure deficit. This is a measure of the air's capacity to absorb more water vapor. If the vapor pressure deficit is large, water in the environment is more likely to evaporate to fill the gap. An alternative way of expressing this deficit is the relative humidity of the air. To compute this term, the quantity of water present in the air is calculated as a percentage of the amount of water needed to saturate the air under the given conditions of temperature and pressure. A high relative humidity means that the air is close to saturation and is therefore unable to take up much more water vapor.

ity, in other words, approaches 100 percent (see sidebar "Humidity" above). As has been explained, warm air is less dense than cold air, so the hot, saturated air becomes forced upward by cooler, denser air moving into the intertropical convergence zone. As it is forced upward, the air begins to cool (see sidebar "Lapse Rate" below) and its capacity to hold water is reduced. Since the air is at or close to saturation, the dew point is quickly reached and water condenses into

Lapse Rate

Solar energy passes through the atmosphere and is absorbed by the surface of the ground. As the ground becomes warmer, some of the heat is transferred to the air in contact with it, so the atmosphere becomes heated at its base. A fixed amount of air, call it a "package," warmed in this way expands, which means that the molecules of gas move around faster and separate from one another, so they occupy a larger volume. The package of air thus becomes less dense, and when it is less dense than the air immediately above, it is displaced by the cooler, denser air and is forced upward. As it rises, the warmer air is subjected to less pressure because the weight of air above it is smaller. When it experiences lower pressure, the package of air expands again, but this expansion results in the molecules of the gases losing energy as they have to push other molecules out of the way. As they lose energy the molecules slow down and the temperature of the gas falls. This process is called adiabatic cooling. The word adiabatic means that there is no energy exchange between the gas and its surroundings. Adiabatic cooling, therefore, is caused by the expansion of atmospheric gases with altitude, and the outcome is a gradient of falling temperature with height above sea level. The steepness of this gradient, or the rate at which temperature falls with altitude, is called the *lapse rate*. If the air is not saturated with water, that is, if it is relatively warm and its relative humidity is less than 100 percent (see sidebar "Humidity," above) then the term used for the fall in temperature with altitude is dry adiabatic lapse rate (DALR). At higher altitudes the air may become so cool that the water vapor it contains is sufficient to saturate it. At this stage condensation occurs and water droplets form, resulting in rain. The air now has a relative humidity of 100 percent and further cooling with altitude then follows the saturated adiabatic lapse rate (SALR), which is usually less steep than the DALR. This is because the condensation of water releases latent heat. Lapse rates vary with many factors, but on average the DALR is about 5.4°F with every 1,000 feet in altitude (9.8°C per 1,000 m), while the SALR is about 3°F with every 1,000 feet (5.5°C per 1,000 m).

liquid droplets. These may drift as cloud, but often coalesce into larger drops too heavy to remain suspended and they fall as rain.

The equatorial zone thus receives much rainfall, as is shown in the diagram. If this map is compared with the map of tropical forest distribution on page 2, then it can be seen that the rain forest is found within the high-rainfall areas. Regions to the north and south of these receive progressively less rainfall corresponding mainly to the distance from the intertropical convergence zone, and these areas are characterized by forests with an increasing proportion of deciduous trees. The evergreen rain forest is largely restricted to areas receiving more than 40 inches (100 cm) of rain per year, while the drier tropical forests can be found in regions between about 20 and 40 inches (50–100 cm). But annual figures for rainfall can be misleading because the rain may all fall within a short period of time and the remainder of the year may be dry, which would not be suitable for rain forest growth. It is perhaps more meaningful to define the climate required by rain forest in a way that expresses the consistency of high rainfall and temperature. Rain forest needs a minimum of 2.4 inches (6 cm) of rain each month of the year and a mean annual temperature of 77°F (25°C) with no more than 9°F (5°C) variation between monthly means. If some months have substantially less rainfall than this or if the temperature varies beyond the required levels, then the vegetation that develops will not be rain forest. Dry seasons will produce mixed or deciduous forest depending on the severity of the drought.

Climatologists and biogeographers often use diagrams to express seasonal variation in climate in such a way that they can be rapidly understood and interpreted. These are called climate diagrams; their composition is explained in the sidebar on page 127.

The climate diagram for a region can immediately give a clue as to whether rain forest growth is possible. The precipitation and the temperature curves need to be consistently high throughout the year and the precipitation curve should always be above the temperature curve, indicated by vertical shading. Tropical forest growth is possible even where

---

The global pattern of precipitation, showing the high level of rainfall in equatorial regions. The distribution of tropical forests (see page 2) has a very similar pattern to the equatorial high-precipitation zone.

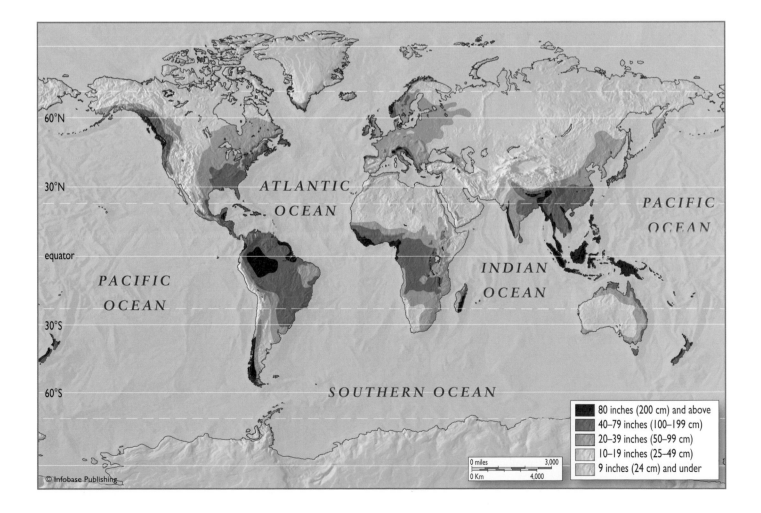

Climate Diagrams

Climate is the average of weather conditions over a long period of time for any given region. It is possible to describe climate in very broad terms, such as the average temperature at a location through the entire year, or the total of precipitation in a year. But this does not express the variation in climate through the year, which often displays a distinct pattern. To illustrate such seasonal variation, climatologists construct climate diagrams, an example of which is shown in the diagram, which represents the climate of Calabozo in Venezuela. The horizontal axis (abscissa) shows the months of the year, running from January to December for Northern Hemisphere sites and July to June for Southern Hemisphere sites. In this way the warmest, or summer, period always occurs in the center of the graph. Venezuela is in the Northern Hemisphere, so its time scale runs from January to December. Two climatic features are displayed on the graph, temperature and precipitation, and the scales are arranged in such a way that 18°F (10°C) on the left side is graduated to match 0.8 inches (2 cm) of precipitation on the right. Monthly means of temperature and precipitation are then plotted on the graph and the scales are calibrated in such a way that if the precipitation graph rises above the temperature graph, then there is an excess of water at the site. When precipitation falls below temperature on the scales employed, it indicates that there will be a shortage of water in that month. Whether the shortage represents a stress to the vegetation at the site depends, of course, on the nature of the plants and their drought resistance, but the diagram provides an indication of dry and wet seasons. To make these contrasts even stronger, a convention of shading the diagram is employed, with stippled symbols representing potential drought, vertical shading meaning water abundance, and black shading representing very considerable excess of water. As can be seen in the diagram, the climate of lowland Calabozo shows a dry winter period, from November to April, then a very wet summer, from May to October.

Although rainfall varies considerably, the temperature changes little throughout the year, and this is characteristic of the Tropics. In the higher latitudes, of course, seasonal change in temperature is much more marked and the curve may even fall below the base line of freezing point during certain months. Climate diagrams can be modified to show months in which there is a very real possibility of frost, which is usually displayed by a bar

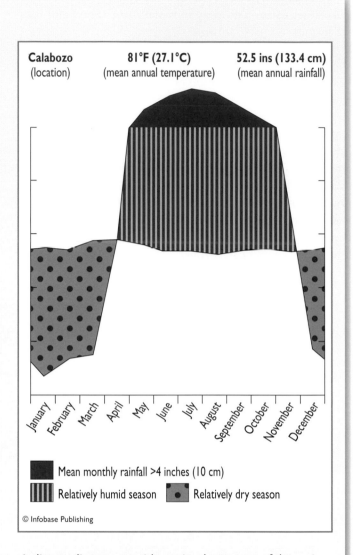

| Calabozo | 81°F (27.1°C) | 52.5 ins (133.4 cm) |
| (location) | (mean annual temperature) | (mean annual rainfall) |

■ Mean monthly rainfall >4 inches (10 cm)

||| Relatively humid season   • Relatively dry season

A climate diagram provides a visual summary of the main climatic features of a site. These diagrams are devised in such a way that the monthly rainfall (1 unit is 0.8 inches; 20 mm) and the monthly mean temperature above freezing point (1 unit is 18°F; 10°C) have the same scale representation on the vertical axis. As a consequence, when the monthly mean temperature and the monthly precipitation curves are drawn, humid and dry periods become apparent. For monthly rainfall data in excess of 4 inches (10 cm), the scale is reduced to one 10th. The horizontal axis represents the months of the year, which are arranged to retain the summer months in the center. For Southern Hemisphere sites, therefore, the scale runs from July to June.

running along the abscissa covering those months. In the Tropics this is only likely on high mountains or deep within continental regions, far from the winter warming effect of the oceans.

these conditions are not met, but the forest that develops is unlikely to be true evergreen rain forest; it is more likely to consist of mixed evergreen and deciduous forest, or, in conditions of more extreme drought, fully deciduous forest or savanna woodland.

Some typical climate diagrams for the wet Tropics are shown here. The first three come from evergreen rain forest sites situated in the Congo River basin of central Africa (Stanleyville), the Amazon Basin of South America (Uapés), and New Guinea in Southeast Asia (Suva), respectively. All are close to the equator and show very little variation in the monthly average temperature throughout the year, having annual average temperatures of about 78–80°F (25–27°C). Rainfall at all three sites is consistently high throughout the year, with total annual input ranging from 72 inches (184 cm) in the Congo to 115 inches (292 cm) in New Guinea.

Although there are some fluctuations in rainfall from month to month, especially in the Congo, monthly rainfall at all sites is generally in the range of 6–10 inches (15–25 cm). This is well in excess of the minimum required to support evergreen rain forest.

Climate diagrams from a range of tropical forest regions. The upper three sites of the diagrams are: Stanleyville in the Congo, Africa; Sao Gabriel in the Amazon Basin, South America; and Suva in New Guinea, Southeast Asia. All three of these sites are situated in evergreen tropical rain forests. The lower two sites are Calabozo in Venezuela, South America, and Mapoon in Queensland, Australia. Both of these sites lie in regions with a pronounced drought season and their forests are largely deciduous.

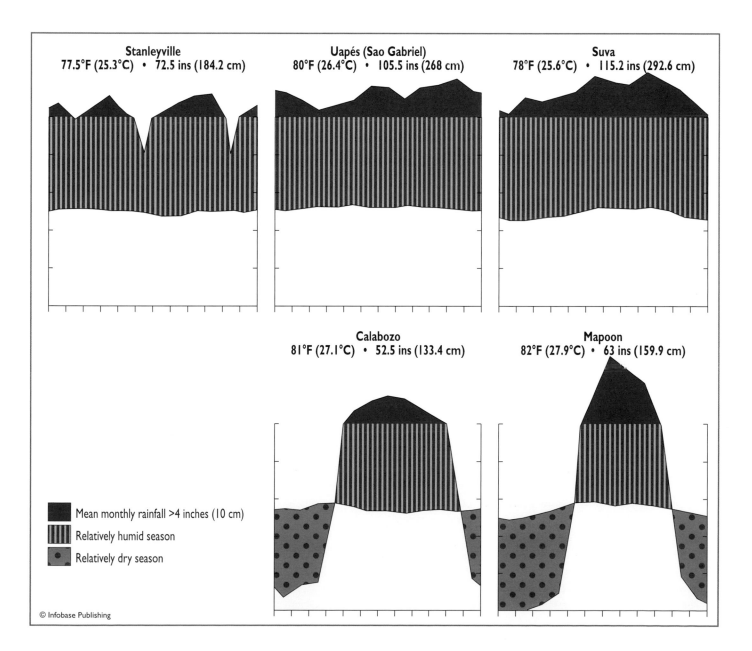

Stanleyville
77.5°F (25.3°C) • 72.5 ins (184.2 cm)

Uapés (Sao Gabriel)
80°F (26.4°C) • 105.5 ins (268 cm)

Suva
78°F (25.6°C) • 115.2 ins (292.6 cm)

Calabozo
81°F (27.1°C) • 52.5 ins (133.4 cm)

Mapoon
82°F (27.9°C) • 63 ins (159.9 cm)

Mean monthly rainfall >4 inches (10 cm)
Relatively humid season
Relatively dry season

The remaining two sites differ from the typical evergreen rain forest climate diagrams in having dry periods during the year. The first diagram, from Calabozo, shows a site south of Caracas in Venezuela, which has an abundance of summer rainfall but suffers drought during the winter months of December through March. This variation is due to the seasonal changes in wind patterns as the intertropical convergence zone alters its position. In winter the winds over northern South America are mainly offshore and dry, but in the Northern Hemisphere summer, when the ITCZ has shifted northward, the trade winds blow in off the Atlantic Ocean, bringing moist air and high levels of precipitation. The winter drought, however, is sufficient to affect the nature of the forest; deciduous trees that can avoid drought stress by going dormant are at an advantage. Queensland in northern Australia has a similar climate, as shown in the last of the diagrams. This site lies in the Southern Hemisphere, so the arrangement of months is different from that in the other diagrams, but the summer conditions still lie in the center of the diagram. As in the case of the Venezuelan site, there is winter drought and then a very high input of rainfall in December through March (the local summer) as the winds from the Pacific regions to the north strengthen and bring high levels of precipitation to the area. These are the winds that keep India dry during those Northern Hemisphere winter months, as shown in the climate diagrams on page 6. In the Indian sites it is the coming of the northern summer and the northward movement of the ITCZ that brings southerly winds from the Indian Ocean and monsoon conditions over India. In all of these sites with dry seasons, however, the forest needs to be composed of drought-resistant trees.

■ TROPICAL STORMS

Although climate diagrams describe the average range of conditions that occur throughout the year, month by month, tropical regions also experience extremes of climatic conditions that can have a considerable impact on natural communities and human populations. The atmosphere in equatorial regions is highly unstable (see "The Atmosphere and Climate," pages 5–6). The Earth's surface is heated by the Sun and transfers some of this heat to the atmosphere immediately above, and as the atmosphere is heated from below, convection currents are generated that lead to rising masses of air and the inevitable precipitation that results from falling temperature with height (see sidebar "Lapse Rate," page 10) and the condensation of water vapor. Regular, sometimes daily storms develop as the heat builds up and the convection currents strengthen. Electrical storms are thus frequent in the Tropics.

More severe storms can result because of the denser air that moves into the space vacated by the rising air currents. The converging air masses do not move directly toward a focal point, but swirl around in a circular motion, converging in a spiral form around the region of low pressure created by the rising air. This type of movement in the atmosphere is called a *cyclone,* and its motion is a consequence of the deflection of moving objects by the spin of the Earth, the Coriolis effect. Cyclones spin in a counterclockwise direction in the Northern Hemisphere and a clockwise direction in the Southern Hemisphere. Cyclones occur in the temperate regions as well as in the Tropics, where they are called depressions, and they usually bring wind and rain. Temperate cyclones are quite large features, often having a diameter of 1,250–2,000 miles (2,000–3,000 km) and are clearly visible on satellite pictures of the Earth. In the temperate zone cyclones form in the region of the polar front, where cold air from the polar regions collides with tropical air masses, forcing the warmer air upward. They then track in an easterly direction.

In the Tropics, cyclones develop mainly over the oceans where warm water (usually over 80°F; 26.5°C) heats the atmosphere and creates the necessary conditions to generate a tropical depression. The spiral of winds around the warm low-pressure system at this stage can reach speeds of 35 miles per hour (56 km/h). Within about 10 days of its initiation, this spinning air mass can develop through a tropical-storm stage into a full-scale tropical cyclone, with wind speeds of 66 miles per hour (100 km/h). In the center of the cyclone there is relatively little air movement, called the eye of the storm. This calm eye is usually about 16–31 miles (25–50 km) in diameter and is surrounded by a rapid transition into the rapidly circulating winds, where air movement is at its fastest. Tropical cyclones are smaller than temperate ones and are usually no more than 400 miles (650 km) in diameter, but they are also fiercer and produce more torrential rain. A tropical downpour from a cyclone can produce 20 inches (50 cm) of rain in a single day.

Tropical cyclones are most common in the summer and fall, when the sea temperature is at its highest, and they usually form at some distance from the equator, usually around latitudes 5°N and 5°S. In the Pacific Ocean they are usually referred to as *typhoons,* while in the Atlantic they are called hurricanes, and in Australia, "willie willies." Although they begin their development at sea, they often hit land, and it is then that they can cause damage to tropical forests, especially along coastal regions. Unlike temperate depressions that move from west to east, tropical cyclones tend to move from east to west, so a cyclone may begin as a thunderstorm off the coast of West Africa and develop into a hurricane when it hits the Caribbean. Madagascar, the island off the east coast of Africa and lying on the western edge of the Indian Ocean, also receives tropical cyclones, as does the Bay of Bengal in eastern India and the South China Sea on the western edge of the Pacific Ocean. Many climatologists believe that the intensity and frequency of tropical cyclones is increasing as global temperatures rise (see "Climate Change and Tropical Forests," pages 217–218).

The high winds associated with cyclones can uproot trees, especially when they grow on shallow soils, and the intense downpours of rain can cause flooding and landslides as a result of soil erosion. These are important factors in the disturbance of tropical forests, but they may also create opportunities for a wide diversity of living organisms as they cause constant change in the forest structure.

■ MONSOON

Many tropical regions receive their rainfall in a distinct seasonal pattern with alternating wet and dry seasons (see "Tropical Climates," pages 9–14). The rains are caused by alterations in wind patterns as the intertropical convergence zone (ITCZ) migrates north and south with the changes in the solar angle and the heating pattern of the Earth. The rains have been given the name *monsoon* from the Arabic word *mausim,* meaning a season.

Four main regions in the Tropics receive monsoon rains. India and the western part of Southeast Asia, includ-

ing Burma and northwest Thailand, have monsoons in the summer as the ITCZ moves north and winds sweep into the Asian continent from the Indian Ocean. The Himalaya Mountain chain blocks these rain-bearing winds and prevents them from moving on into Tibet and western China. As a consequence, much of the moisture from the winds falls on the southern slopes of the mountains and drains back southward, through the Himalayan foothills (see illustration below), into the Indian Ocean via the Ganges River, passing through Bangladesh. The eastern fringe of Southeast Asia has a winter monsoon as the ITCZ moves south and the northeasterly winds come in from the Pacific. Northern Australia also has its monsoon rains at this time.

Africa receives summer monsoon rains both in the west and in the east. In the west, the northward movement of the ITCZ brings southerly, moisture-laden winds from the Atlantic Ocean onto the coast of Sierra Leone, Liberia, and the Ivory Coast. In the east, southerly winds from the Indian

This river bed near Dehra Dun in northern India is seen here in January, when the dry season is well advanced and no water flows down from the Himalaya Mountains. In the summer monsoon this will be transformed into a torrent, and the forests will receive their input of water. *(Peter D. Moore)*

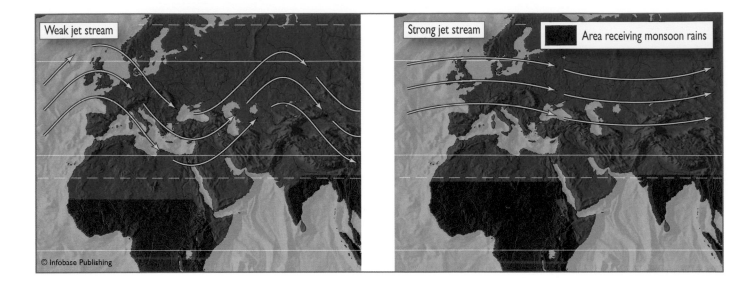

The jet stream is a high-altitude movement of air masses from west to east around the Earth. When the air mass movement is relatively weak and slow, it meanders widely, and this holds back the monsoon rains in Africa south of the Sahara (the Sahel) and in northern India. When the jet stream is strong, it moves in a relatively straight line, and the monsoon rains are not held back.

Ocean also penetrate deeply into the Horn of Africa into Somalia, Ethiopia, and the Sudan, bringing monsoon rains. The entire strip of land to the south of the Sahara Desert, called the Sahel, is dependent on these rains for its productivity, and forests extend farther north in Africa than would otherwise be possible because of the summer monsoon.

The monsoon rains are critical for the survival of many human populations, especially in India and Africa. About 55 percent of the world's population lives in lands that depend on monsoon rains, so any alteration in the pattern or the strength of the monsoon is of deep concern. In the 1970s the monsoon failed for several consecutive years in the Sahel, causing agricultural failure and severe food shortages. It seems likely that the monsoon failed to penetrate far north because of a meandering pattern in the jet stream, the easterly flow of air around the planet in the temperate latitudes, as shown in the diagram above. When the jet stream is weak and slow moving, it meanders and holds back the northward penetration of the monsoon in both Africa and India. When the jet stream is fast moving, it flows in a more direct manner, allowing a more northerly extension of the monsoon rains into these dry regions.

Both the regular delivery of the monsoon rains and the less predictable and more catastrophic arrival of tropical cyclones bring a supply of water that is vital for the survival of tropical forests. But when the water arrives in too great a concentration it can cause floods.

TROPICAL FLOODS

Flooding in tropical forests may occur as a result of the sudden arrival of too much freshwater or saltwater. A river basin may drain a catchment in which tributary streams arrive from other regions, and these may receive water seasonally because of the pattern of precipitation or, in the case of mountainous regions, snowmelt at certain times of year. The drainage capacity of the river basin is exceeded and water overflows the banks, encroaching on surrounding forest. Or water may arrive less predictably as a result of tropical storms of such severity that a valley cannot cope with the abundance of water and flooding results. This type of flood is often more damaging to the forest and can result in a landslide, erosion, and the physical destruction of forest areas. Saltwater flooding is confined to coastal forests, where exceptionally high tides or surges of salt water associated with typhoons may lead to flooding. Coastal mangrove forests are often subject to this kind of flooding. More severe flooding of coastal forests can result from tidal waves, or tsunamis, which are produced by major geological events such as underwater earthquakes. The Indian Ocean tsunami of December 2004 led to extensive flooding of coastal forest in Indonesia, Sri Lanka, and Thailand.

The forests of the Amazon Basin in Brazil are subject to annual flooding that is prolonged because of the very extensive system of tributary rivers, as shown in the diagram on page 17. There are many tributary rivers, including the blackwater rivers, such as the Rio Negro from Colombia the north, the clearwater rivers that drain the highlands to the east of the Andes Mountains, and the sediment-rich rivers, including the Rio Madeira, which enter the Amazon from the south, draining from the Andes themselves. Increased rainfall arrives in the northern part of the watershed from April through August, while rainfall in the south of the catchment is greatest from November through March. Parts

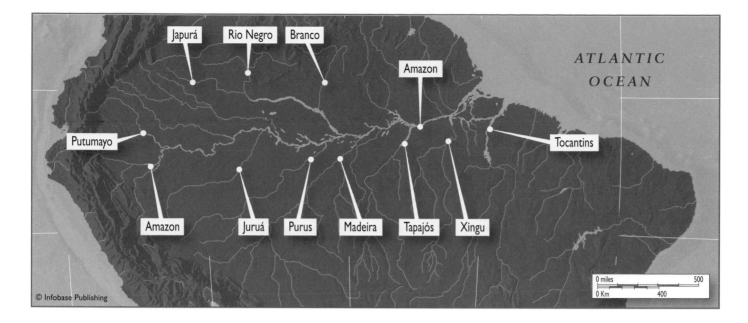

The Amazon River and the lower parts of many of its tributaries flood when different regions of the catchment have their peak rainy seasons. The major tributaries, named on this map, drain much of northern South America.

of the Amazon forest are thus flooded at almost any time of year, but on average any given location will be flooded for between four and seven months of the year. The regions of the forest most subject to flood are those lying alongside the river and its tributaries (see illustration below), an area

of *floodplain* that covers approximately 96,500 square miles (250,000 km²), considerably larger than Florida.

The Amazon forests are thus regularly flooded, but even when they are not subject to floods they still receive water from consistent rainfall. This is not the case for tropical forests within areas affected by monsoon rainfall, such as those in northern India. These receive all of their water input in a single wet season, and this can cause damaging floods. River valleys are dry for much of the year and then subject to large volumes of torrential water, eroding banks, trees, and sometimes human settlements. This can have an impact in the

The low-lying forests of the Amazon Basin are regularly flooded. This flooding provides fish with an opportunity to forage widely and to eat the fruits of many trees, dispersing them to new areas. *(Leeman)*

Trees, Water, and Climate

The study of water movements is called *hydrology*. Trees are extremely influential in the movement of water between the soil and the atmosphere. As they photosynthesize, they lose water vapor to the atmosphere from their leaves, termed *transpiration*. *Evaporation* is an inevitable consequence of having open pores in the leaf surface, through which carbon dioxide is absorbed, and it would lead to the desiccation of a plant if there were no reliable source to replenish the water loss. Plants have roots that tap the water resources of the soil, and the loss of water from the leaves creates a pull on the column of water present in the conducting tissues, the *vessels* within the wood of the plant stem, ultimately creating sufficient force for the root to take up water from the soil. Trees can thus be regarded as transport mechanisms that take water from the soil and carry it to the atmosphere. In many respects they are passive transport agents. They do not need to expend energy in pumping water because the evaporative process, coupled with the strong adhesive forces between water molecules, caused by hydrogen bonding, that make a water column hard to disrupt, creates the pull needed to maintain the transpiration stream. There are some ways in which transpiration benefits the plant; the leaves are cooled as water evaporates, preventing overheating; and the flow of water from the soil also brings a supply of inorganic elements, such as nitrogen, phosphorus, and potassium, needed to build new tissues.

Trees also influence the water cycle by *intercepting* rainfall. Standing beneath a tree in a rainstorm can provide an alternative to an umbrella as a protection from the worst of the deluge. The leaves intercept the falling rain, which creates a film of moisture over their surface, and only when the leaves are saturated with water does the drip of penetrating rain, or *throughfall,* commence. Some of the intercepted rain may trickle along leaf stalks, twigs, and branches, eventually running down the main trunk into the ground. This is called *stem flow*. When the rainfall ceases, the water on leaves and stems evaporates directly back into the atmosphere, so this component of the precipitation never reaches the ground. A rain gauge situated beneath a forest canopy will record considerably less precipitation than one placed above the canopy, because the forest intercepts and recycles water before it ever reaches the soil. On the other hand, in humid, misty conditions when there is no actual rain, but the air is very damp, condensation may occur on the leaves and can create an input of throughfall and stem flow water that would not be measured in a rain gauge outside the forest. This additional input of water to the soil, created by the forest canopy, is called *occult precipitation, occult* meaning unseen. The movement of water vapor into the atmosphere by the combined effects of evaporation and transpiration is given the term *evapotranspiration*.

Forest canopies also shade the forest floor, keeping them cool (see "Microclimate," pages 25–33) and thus reducing direct evaporation from the soil surface. The overall effect of the presence of forest cover on soil moisture is thus very complex. The diagram summarizes all the effects of a forest canopy on the abundance of water in the soil. The presence of trees effectively reduces soil moisture because of rain interception and evaporation and because of transpiration. Water is added to the soil, however, by occult precipitation, and soil surface evaporation is reduced because of canopy shading. In experiments where ecologists have deforested *watersheds* and then monitored the effect on the flow of water from exit streams, the impact of deforestation is to increase water losses in the streams, sometimes by as much as 40 percent. So trees remove more water than they add to a soil, and the loss of forest causes more water to move over the land surface and through soils, creating erosion and floods.

(opposite page) The interactions between tropical forest trees and water are complex. Leaf canopies intercept rainfall and carry it down the outer part of their trunks to the ground as stem flow, or allow some of the water to permeate the canopy and drop to the ground as throughfall. But some of the intercepted rain evaporates back into the atmosphere. Mist may also condense on leaves as occult precipitation. The transpiration of water by trees carries water in the opposite direction, from the soil into the atmosphere.

lower reaches of rivers such as the Ganges, which enters the Bay of Bengal through the coastal forests of the Sundarbans in Bangladesh. Monsoon waters from the entire length of the Himalaya foothills and mountains drain through this estuary, and highly damaging floods often result. These floods have become more severe as a result of forest clearance in the mountains, which results in more water passing through the river system (see sidebar above).

DROUGHT AND FIRE

As has been seen, most tropical forests have an abundant supply of water. Some may even experience regular flooding of the forest floor, as in the case of the Amazon floodplain. The monsoon forests, however, and the forests that merge into the savannah woodlands and grasslands, experience a regular seasonal drought. In general, the duration and the severity of the drought increase as one moves away from the equator. The total amount of rainfall decreases with distance from the equator, and the drought season is associated with the slightly cooler conditions that occur when the overhead noonday Sun and hence the ITCZ has moved to the other hemisphere. As will be seen (see chapter 3, "Types of Tropical Forest," pages 53–75), the evergreen forest passes through a semievergreen forest to deciduous forest along this latitudinal and precipitation gradient. Changes in temperature along the gradient are relatively small and are unlikely to be of significance for most plants and animals, but the drought has a strong impact.

Drought is first experienced in the upper canopy because the lower levels are protected by the upper layers and remain relatively humid (see "Microclimate," pages 25–33). This means that the move from evergreen to deciduous leaves occurs first among the top canopy trees. The forest begins to change in appearance with the passage of seasons as some these emergent trees lose their leaves in the drought. The occurrence of seasonal drought provides plants and animals with a means of measuring time, however, which is difficult in an environment where daylength is relatively unchanging. Seasonal changes provide a stimulus to processes such as migration and reproduction.

The series of climate diagrams from India on page 21 illustrates the differences between the climates of these various forms of tropical forest. The coastal site of Marmagao is the wettest, with a total annual rainfall of 100 inches (254 cm). There is a winter dry season, but this is relatively short (about three to four months) and the very high summer rainfall is sufficient to support evergreen rain forest. Inland the total rainfall declines quite rapidly with distance from the ocean, even though the altitude rises to 2,400 feet (780 m), and the annual precipitation at Belgaum is just 51 inches (130 cm). The length of the dry season has also increased to four to five months, and the vegetation in this area is semievergreen forest. Traveling north to the inland town of Bidar, the annual precipitation has fallen to 34 inches (86 cm) and the dry season extends for seven months of the year. The monsoon rains, however, are heavy and are still prolonged over about five months, so a moist monsoon forest that is largely deciduous is found here. The more northern site of Jaipur has an even longer dry season, lasting about eight months, and the summer monsoon brings just 24 inches (61 cm) of rain in a short but intense wet season. Monsoon forest survives here, but is dry and deciduous.

An important biogeographical question is whether the nature of the forest is determined by the overall total rainfall or by the length of the dry season. In fact, both factors are important, and neither can be considered in isolation from the other. High total precipitation can in part compensate for a lengthy dry season, but it is not a complete substitute for more prolonged wetness. Evergreen tropical forests generally need at least 80 inches (200 cm) of rainfall per year if there is any significant dry season during the course of the year. They can withstand a dry season of four or five months, but only if the total rainfall exceeds 120 inches (300 cm). So the wet season needs to be very wet if a true evergreen rain forest is to grow under conditions with a substantial dry season. Semi-evergreen forest, in which a proportion of the trees are deciduous, requires a minimum annual precipitation of about 40 inches (100 cm) to survive if the dry season lasts for less than six months. In situations where the dry season lasts longer, the wet season has to be more humid, giving an annual rainfall of 80 inches (200 cm) or more. Much of northern India has less rainfall than this, and the dry season can last nine months or more, so the only forest that can be supported is deciduous monsoon forest or savannah woodland. These two types of woodland blend into one another, but the boundary occurs at the point where rainfall is below about 20 inches (50 cm) per year and the dry season lasts longer than six to nine months.

The total amount of precipitation and the pattern in which it is distributed through the year thus combine to determine the type of forest that develops in an area. On the basis of a climate diagram, therefore, it is possible to predict the kind of tropical forest that should occur at a site. In practice, of course, there are some additional factors that can influence the distribution of forests. Low-lying wet areas, fed by rivers draining from extensive *catchments,* may support forest vegetation even when the precipitation seems too low or the length of the dry season in the region appears too long. There are also situations where human activity, such as grazing and burning or a combination of the two, prevents the growth of forest in a location where the climate appears appropriate. The boundary between forest and savannah is often determined by human management of tropical woodlands and grasslands by means of fire. Climate, therefore, is not the only factor controlling the pattern of forest development.

Even climate is not always predictable from year to year. Climate is described as the average weather conditions over a period of many years, but a wide range of variation may be found from one year to the next. Dry episodes sometimes occur even in regions occupied by the wet evergreen rain

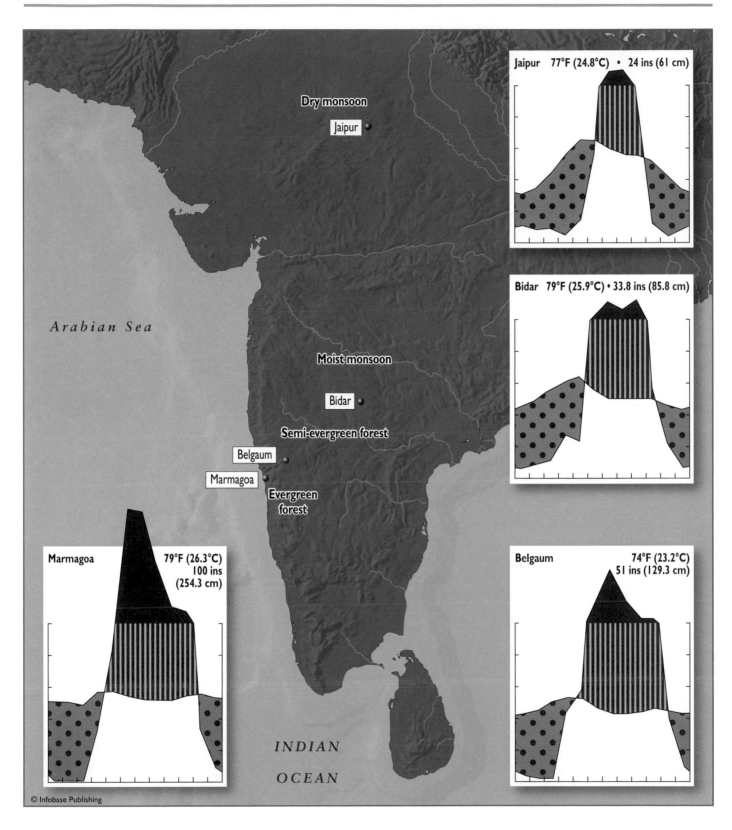

Jaipur 77°F (24.8°C) • 24 ins (61 cm)

Bidar 79°F (25.9°C) • 33.8 ins (85.8 cm)

Marmagoa 79°F (26.3°C) 100 ins (254.3 cm)

Belgaum 74°F (23.2°C) 51 ins (129.3 cm)

Dry monsoon
Jaipur

Arabian Sea

Moist monsoon
Bidar

Semi-evergreen forest
Belgaum
Marmagoa
Evergreen forest

INDIAN
OCEAN

© Infobase Publishing

forest, and the impact of occasional extreme climatic events can alter the face of the forest.

The island of Borneo lies in the South China Sea, and its southern part, Kalimantan, belongs to the state of Indonesia. Much of the lowland region of this island has a consistently

Climate diagrams from different parts of India show how the monsoon rains decrease in quantity and in duration as one moves from the coast inland. The decline in precipitation is accompanied by a change in tropical forests from evergreen to deciduous.

wet tropical climate, with an average annual rainfall of 110 inches (280 cm) and a minimum of four inches (10 cm) in each month. This climate lies well within the requirements of evergreen rain forest, and this is the characteristic vegetation of the region. There are occasions, however, when monthly rainfall falls below four inches for two consecutive months, which constitutes a dry period, but this rarely happens more than once a year. A drought lasting more than three months is very rare, having occurred only twice in half a century of records. In 1997 and 1998 two droughts of two to three months each took place in which there was almost no rain for the entire dry period, and the impact on the forest was devastating. Approximately 25 percent of the trees in the forest died as a direct result of the drought, and the impact proved greatest on the largest of the trees. But worse was to come. The hot, dry forest proved extremely flammable and soon after the end of the second dry spell a fire broke out that swept through the entire region. Most of the trees that had survived the drought were killed by the fire and the overall death rate among the trees rose to almost 80 percent.

Drought in the tropical forest is thus an extremely important aspect of its ecology, whether caused by a predictable dry season or by erratic and unpredictable fluctuations in climate. Drought is often accompanied by fire, sometimes caused by electrical storms, and sometimes generated deliberately or accidentally by human agencies. When this happens, the entire structure and composition of a forest can be altered in a matter of hours. Regular burning may destroy rain forests or reduce them to dry savannah woodlands, in which case they may regenerate when the management regime is changed. In northeast Queensland, Australia, for example, savannah woodland has been maintained in some areas by a long history of burning by native aboriginal cultures. In the Iron Range National Park, annual rainfall is over 67 inches (170 cm) with a dry period lasting for four months, between July and October. Burning has been prevented in the park since the 1920s and rain forest is rapidly expanding into the former savannah woodland. This large-scale experiment demonstrates that vegetation patterns in the Tropics can be deflected from their natural development in response to climate. But it also illustrates the way in which climate takes charge once more when the deflecting factor (in this case, human use of fire) is removed.

■ TROPICAL MOUNTAINS

The climate of the tropical regions is not uniform along belts of latitude. It is affected by the proximity of the ocean, which may bring additional precipitation, depending on the local ocean currents and the prevailing winds. It is also affected by topography, the physical form of the landscape.

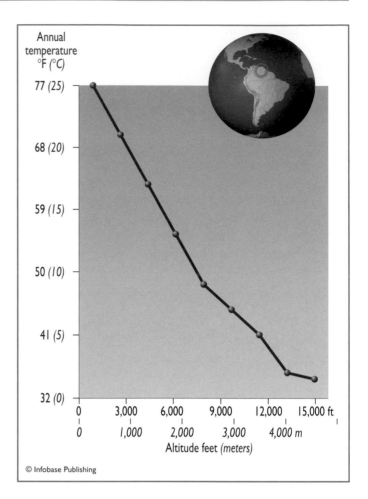

The relationship between mean annual temperature and altitude in Venezuela. Compare with the vegetation zones shown in the diagram on page 23.

Mountains are landscape features that generate their own climates and thus have a considerable impact on the general latitudinal climate belts and consequently on the vegetation of a region.

The temperature falls with increasing altitude (see sidebar "Lapse Rate," page 10) and this drop in temperature has an effect on vegetation. The above diagram shows the relationship between annual average temperature and altitude in the Andes Mountains of Venezuela. There is a steady decline in temperature for the first 7,000 feet (2,100 m) and then the curve becomes steeper.

Changes in precipitation with altitude are complex, however, because falling temperature of air as it is forced upward over a mountain leads to condensation and precipitation. Many tropical mountains accumulate layers of clouds around them, so although temperature falls with altitude the precipitation does not; it may even increase. Much depends on the direction of the wind. In the Venezuelan Andes, for example, the trade winds from the north of the mountains push moist oceanic air upward and precipitation increases with altitude, becoming greatest between 7,000 and 10,000 feet (2,100 to 3,000 m). In this zone a broad band of cloud forms, maintaining wet conditions, and the altitude of this cloud belt changes with the season. Above this zone the precipitation drops so that the cooler upper parts of the mountain are also relatively dry. On the protected, southern side of the mountains, this effect does not occur and precipitation remains high until the alpine zone is reached where conditions become drier.

The zonation of vegetation over the mountain is influenced by these differences in climate, as shown in the next diagram, below. On the northern side of the mountains, conditions in the lowlands are dry, leading to the development of desert scrub, but the increasing precipitation with altitude allows first deciduous and then semi-evergreen forest to grow. In the cloudy belt, a distinctive forest, logically called cloud forest, forms (see "Montane Forests," pages 65–67). Above this is alpine bush and scrub vegetation, which gradually become more dwarfed and sparse until the permanent snows are reached. On the south side of the

mountains, the lowlands are part of the evergreen rain forest of the Amazon Basin. This forest gradually changes its composition and structure with altitude; plant geographers use the terms submontane and montane forest to describe these vegetation types. Eventually, as in the north, the forests become increasingly small in stature and merge into the scrub of the subalpine and alpine zones.

This pattern of climatic and vegetation zonation is common to all tropical mountains, although the precise altitude of the zones and the nature of the vegetation depends on local conditions and the various types of plants available. The tropical mountains of East Africa, for example, show

---

The relationship between altitude and vegetation zonation in the mountains of northern Venezuela. On the north side of the mountains, adjacent to the Caribbean Sea, humid cloud forests develop between 7,000 and 10,000 feet (2,000 to 3,000 m). Compare with the temperature change with altitude shown in the diagram on page 22.

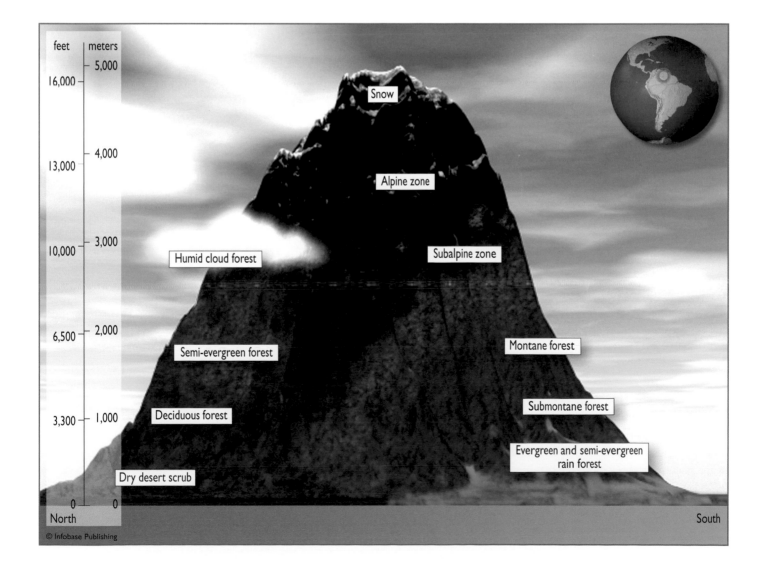

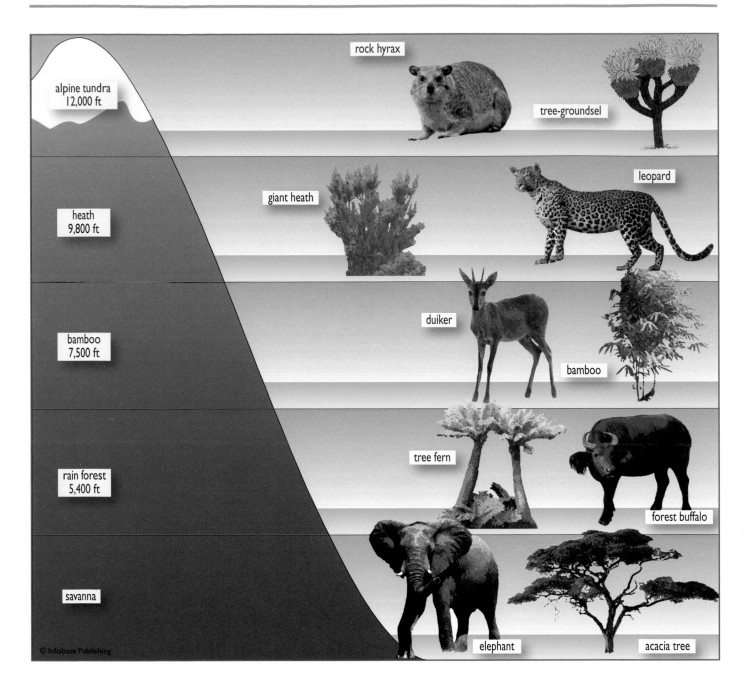

The zonation of vegetation on Mount Kenya, East Africa. The montane forest vegetation is represented here by dense bamboo forests.

less variation between their southern and northern aspects than those of the Venezuelan Andes. Mount Kenya, which lies close to the equator, is shown in the diagram above. The base of this mountain stands in a dry climatic region and bears savannah vegetation. Increasing altitude brings lower temperature but also increased precipitation, so the lower slopes of the mountain, between 5,400 and 7,500 feet (1,600 to 2,300 m), carry a rain forest cover. Dense bamboo forest lies above the rain forest zone and is equivalent to the

Venezuelan cloud forest in its altitude and climate, occupying the wettest part of the mountain. At its upper altitudinal limit the bamboo gives way to subalpine scrub (see illustration on page 25) and open woodland of strangely shaped trees (see "Montane Forests," pages 65–67) below the snows and glaciers of the mountain peak.

The zonation of climate and vegetation on tropical mountains is thus not a simple altitudinal reflection of the latitudinal zones of the lowlands. The climate change with latitude as one proceeds north and south of the equator generally brings increased seasonality and drought, leading to the replacement of evergreen rain forest with deciduous forest and then open dry savannah woodlands. Altitude may bring increased wetness, especially in the

At the timberline on Mount Elgon in Uganda, bamboo forest is replaced by tall heathland dominated by the tree heather (*Erica arborea*). *(Peter D. Moore)*

drier Tropics, and produces the cool, moist conditions that support distinctive forest types, such as the cloud forests and the montane forests.

■ MICROCLIMATE

Climate is the average set of weather conditions over a long period of time for a particular geographic region. It is an important determinant of vegetation and therefore the type of ecosystem that develops in any particular part of the world. But when one considers individual organisms, such as a single tree, an herbaceous plant, a reptile, or a flying insect, their experience of climate is a very local one. They are concerned with such factors as the supply of light and water at their particular location, the availability of radiant heat, or the humidity of the atmosphere. These local climatic factors that operate on a very small scale in a habitat contribute to the *microclimate*. Forest habitats create complex microclimates because of their elaborate

structure. The word *structure* in ecology is used in a very specific way and denotes the architecture of an environment; it should not be confused with the *composition* of a habitat, which refers to the various species of organism that make up the living community.

Of all the forests of the world, the evergreen rain forest has the most complicated structure (see sidebar "Forest Structure," page 54), with the tree canopy arranged in several layers above the forest floor. This intricate architecture in the canopy has an important impact on the microclimate within the forest, affecting light penetration, temperature, and humidity at all elevations. The microclimate, in turn, determines the types of plants and animals that can survive and make a living at the different canopy levels.

Light represents the supply of energy to the ecosystem (see sidebar "Energy," page 77). Sunlight arrives at the uppermost canopy of the forest during the daytime and it penetrates through the layers of the canopy, eventually reaching the forest floor. Light is available to the forest in three different forms: direct sunlight, light from blue sky, and light that has been filtered through cloud. As can be seen from the diagram on page 26, light from these three sources may penetrate down through the forest in three different ways. Light may pass directly and unhindered through to the forest floor via small gaps in the leaf canopy. Direct sunlight passing in this manner creates a dappled effect on the forest floor, with patches of brightly lit ground within a dark matrix. These well-lit patches are called *sunflecks*. Light may also strike leaf surfaces, in which case some light is reflected, some absorbed, and some transmitted through the leaf to the next layer down. Energy is removed by reflection and absorption, so the intensity of the light penetrating a leaf is considerably reduced. The third possible path for light is one of reflection. Some of the reflected light from leaves, together with reflected light from tree trunks, continues on

	❶ Transmitted	❷ Direct	❸ Reflected
Sun	✓	✓	✓
Blue sky	✓	✓	✓
Cloud	✓	✓	✓

its path toward the forest floor. As the diagram illustrates, the overall light climate of a forest is therefore very complicated, with the possibility of nine different types of light reaching the forest floor.

Of these various sources of light, only the sunfleck component contains the full intensity of sunlight; all of the others have lost energy in the course of their passage through the canopy. Measurements of the light intensity on the floor of evergreen tropical forests usually show that the shade light contains only 0.1–1 percent of the energy of full sunlight. In other words, the leaf canopy has extracted almost all the energy available by the time it reaches the floor, so life on the ground and in continuous shade for a photosynthetic plant in such a forest is almost impossible. The sunfleck environment therefore assumes very considerable importance for plant life on the ground. A sunfleck comes directly from the Sun, so it has the full intensity that would be experienced at the top of the forest canopy, but it may not last for very long. Clouds may pass over the Sun and obscure it, and the Sun is constantly altering its position because of the rotation of the Earth, so sunflecks move around on the forest floor, illuminating different patches in the course of a day. Nevertheless, the short burst of high-intensity light within the sunfleck may supply a plant with the bulk of its energy needs for a day. Basking animals, such as reptiles or butterflies, may also rely on the sunfleck for energy, but they have the advantage of mobility, so they can seek out the well-lit patches and follow them when in need of radiant energy.

The complex light climate of the forest affects not only the intensity of light patterns on the floor but also the *quality* of the light. Light *quality* is a term applied to its spectral composition, and this is important to living organisms because it influences the process of photosynthesis (see sidebar on pages 28–29).

Light transmitted through leaves or reflected from their surfaces is depleted in the wavelengths absorbed by the photosynthetic pigments, the chlorophylls and carotenoids. The remaining light is proportionally richer in light of green wavelength, which cannot be used by plants. This means that as light passes through successive layers of leaves or is reflected from plant surfaces, it becomes increasingly unsuitable for photosynthesis (see illustration on this page). Plants living on the forest floor, therefore, are not only deprived of light with adequate intensity for photosynthesis but also supplied with light of poor quality for the job. This means that the light arriving as sunflecks, which is unadulterated by the

Light penetrates through the upper canopy of the forest to illuminate the lower layers of trees. The tree shown here is a banyan tree. *(Commander John Borniak, NOAA/Department of Commerce)*

upper layers of leaves, is even more important for plants living deep in the shade. In order to use it, however, they must be able to cope with sudden and considerable changes in light intensity, which is quite a big physiological problem for them. Sunlight contains a high proportion of light within the wavebands 400 to 700 nm. Light from the blue sky and clouds lies largely within the 300- to 500-nm wavebands, so it lacks the red component needed for activation of the second peak in chlorophyll absorption but provides energy input at the lower wavelength that is suitable for photosynthesis.

Light penetration in the forest is not only vital for animal vision and plant photosynthesis but is also important as a source of heat. As has been discussed (see "Tropical Climates," pages 000–000), the average temperature of tropical air remains quite uniform through the changing months of the year. But daily changes in temperature take place as

(opposite page) Light can penetrate to the floor of a tropical forest in many different ways. Light may pass directly to the floor without interruption, or it may be transmitted through the leaf canopy, or it may be reflected. There are three sources of light: direct sun, blue sky, and cloud, so there are nine possible ways that light can reach the forest floor.

Light, Vision, and Photosynthesis

The British physicist Sir Isaac Newton (1642–1727) first demonstrated that the light perceived by human eyes consists of a wide spectrum of different wavelengths of energy. He passed light through a glass prism and split the "white" light into a series of different colors. Another British scientist, James Clerk Maxwell (1831–1879), showed that the light seen by the human eye is only a very small band in a much wider spectrum of radiant energy, called the *electromagnetic spectrum* (see illustration). Although the physics of light is very complex, it can be simply regarded as a wave motion in which the wavelength (the distance between one wave crest and the next) is extremely variable. The wavelength of light is measured in units called *nanometers* (nm), where 1 nanometer is 10^{-7} centimeters, or 2.54×10^{-7} inches—a very small unit. The radiation from radioactive sources, gamma rays, has a wavelength of only 10^{-5}–10^{-3} nm, while radio waves used for broadcasting are measured in miles (10^{9}–10^{15} nm). In the middle of this great spectrum lies visible light, ranging in wavelength from 400 to 700 nanometers. The shorter the wavelength, the higher the energy content of the radiation, so the radiation from radioactive materials, as well as the short-wave ultraviolet radiation from the Sun, can damage the molecules in living cells, including the DNA that contains genetic information. Long-wavelength radiation, such as radio waves, by contrast, are very low in their energy content so do not cause molecular damage. Between the two, within the wavebands of visible light, the energy content is sufficient to cause minor changes in molecular configurations, so living organisms can employ chemical systems for detecting them, as in the retina of the eye, but they are not so energetic that they cause serious damage. For this reason, the relatively narrow visible light spectrum has proved the most useful for animals

The electromagnetic spectrum showing the different types of radiation associated with different wavelengths. Note that the wavelength is expressed on a logarithmic scale. Visible light occupies a very narrow band of the total spectrum. The wavelength of light is extremely short, so physicists measure it in nanometers (nm). One nanometer is 10^{-9} (one billionth) of a meter and is equivalent to 2.54×10^{-8} inches.

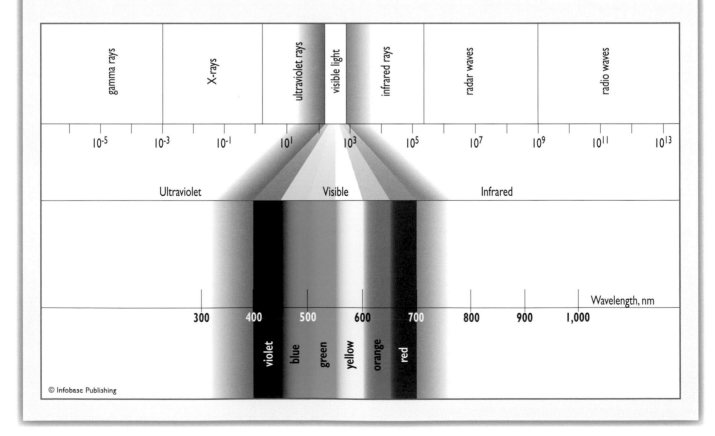

© Infobase Publishing

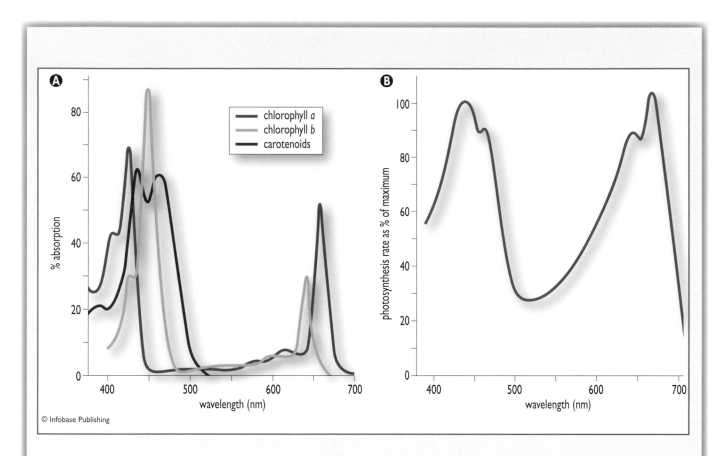

© Infobase Publishing

and plants in all their radiation-dependent activities, from vision to photosynthesis.

Humans have good perception of color and shades throughout the so-called visible spectrum. Three main receptor pigments are involved that have their peak sensitivities at 440 nm (blue), 530 nm (green), and 570 nm (yellow). Birds have similar perceptions, although they are able to pick up shorter-wavelength light in the ultraviolet part of the spectrum. Insects such as the honeybee have efficient sensors within the ultraviolet range. Honeybee receptor compounds have their peak sensitivities at 340 nm (ultraviolet), 463 nm (blue), and 530 nm (green). Their view of the world, and of flowers in particular, is thus rather different from that of people.

Plants require the energy of light for photosynthesis, and they also use a range of pigments for light absorption. Chlorophylls a and b between them absorb light in the range of 400–500 nm (blue) and have a second absorption peak between 650 and 690 nm (red) (see diagram). Additional pigments supplement this range of light absorption, the most common in land plants being carotenoids, which add to the efficiency of absorption in

The absorption and action spectra of a green plant. A. Shown here is the pattern of light absorption by the three main pigments found in green plants, chlorophyll a, chlorophyll b, and the carotenoids. B. The action spectrum of a green plant shows how effectively the plant can photosynthesize when supplied with light of particular wavelengths. Photosynthesis proceeds most effectively when the wavelength of the light received matches the absorption of the main pigments.

the 400- to 500-nm range and also extend the absorption a little way into the green part of the spectrum, at 500–510 nm. Between them, therefore, these pigments trap energy in two major bands, resulting in an action spectrum for photosynthesis shown in the second part of the diagram. This indicates the light wavelengths actually utilized by green plants in photosynthesis. There is a clear zone of lost light energy in the center of the spectrum between wavelengths of 510 and 620 nm. This is the green section of the spectrum, and this light is reflected or transmitted by leaves, giving them a green color when observed by human eyes.

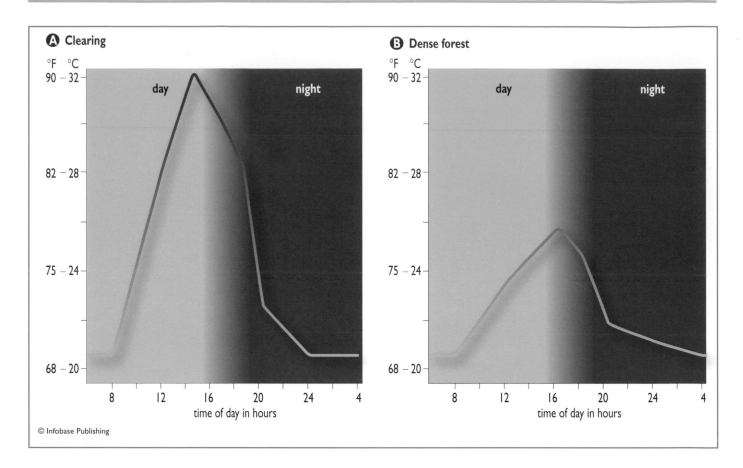

A Clearing

°F °C

90 – 32

82 – 28

75 – 24

68 – 20

day   night

8   12   16   20   24   4

time of day in hours

© Infobase Publishing

B Dense forest

°F °C

90 – 32

82 – 28

75 – 24

68 – 20

day   night

8   12   16   20   24   4

time of day in hours

The temperature variation during the course of a 24-hour period in (A) a clearing in the forest and (B) in dense forest. Diurnal fluctuation in temperature is much greater in a clearing.

the Sun passes across the sky, and these often create a greater range than that of the average monthly temperatures. The diagram above shows the daily range of temperature in a typical equatorial rain forest, comparing values at the top of the canopy (or in a forest clearing) with that of the forest floor. The first part of the diagram illustrates the daily variation in air temperature; at the canopy this can be as great as 22°F (12°C) with a maximum of 90°F (32°C). Beneath the canopy, where very little sunlight penetrates directly, the temperature varies only about 9°F (5°C), with a maximum of 77°F (25°C). Thus, plants and animals living within the forest are not exposed to the same levels of heat in the middle of the day. Some cold-blooded animals may need to supplement their heat intake, hence the sunfleck-seeking and basking activity of reptiles and insects. Nighttime temperatures on the ground and in the canopy are very similar.

The shaded forest soil is a highly insulated environment, its daily range of temperature being less than that of the air temperature. As can be seen in the second part of the diagram, day and night temperatures may vary by only 3.6°F (2°C). Invertebrate animals that live in the forest floor,

therefore, have an extremely uniform microclimate as far as temperature is concerned. Only when a sunfleck strikes are the upper layers of soil likely to be heated, and even then the rise in temperature is unlikely to penetrate as deeply as an inch into the soil. Soils in clearings, however, where the sunlight can warm the surface throughout the day, show greater diurnal variation in temperature. In the daytime the soil temperature often reaches 86°F (30°C), and in very sheltered but well-lit situations it can even reach 122°F (50°C), and this heat has a strong impact on both the invertebrates and the microbes in the soil.

Light penetration into a forest is affected by the *topography,* especially the degree of slope and its *aspect* (see sidebar on page 31). The occurrence of sunflecks, the heating of the ground, and the general humidity are all influenced by the steepness of a slope and the compass direction in which that slope faces, as shown in the diagram on page 31. In areas of varied topography, where hills and valleys are frequent, the microclimate of

(opposite page) A profile of a Northern Hemisphere forested valley, showing the effect of aspect on the incidence of sunlight. The south-facing slope receives direct sunlight for a longer period in the day than the north-facing slope. The daily mean temperature and the diurnal variation in temperature in the canopy layer will therefore be greater on the south-facing slope.

Aspect

Aspect is a term used to describe the direction in which a surface faces. It is most often used for compass orientation of an object or an area of land surface. It is an important consideration in ecology because it can influence the energy balance of a surface, depending on whether it is shaded from or facing the main source of solar radiation. The diagram below shows an east-west-running valley in cross section and illustrates how the steepness and orientation of a slope can affect the quantity of solar energy received. In general, aspect is a less important ecological consideration in equatorial regions than in the higher latitudes because the noonday Sun is overhead on two occasions during the year and swings between north and south in the intervening periods. At the equator, therefore, north- and south-facing slopes would receive, overall, the same amount of energy, but it would occur at different times of year depending on aspect. A site away from the equator displays increasingly large discrepancies in energy input between north- and south-facing aspects. In the Northern Hemisphere a south-facing slope receives a greater supply of solar energy (as shown in the diagram), while in the Southern Hemisphere it is the north-facing slope that receives most of the Sun's rays. A sunny slope will be better lit, with more sunflecks, have a higher daytime temperature, and often experience reduced relative humidity. All of these factors play a part in determining which plants and animals can live there.

Aspect may also be important on smaller scales. The opposite sides of a rock or the trunk of a tree may differ in their microclimates according to their aspect. This difference can affect distribution of small plants and animals at a microhabitat level.

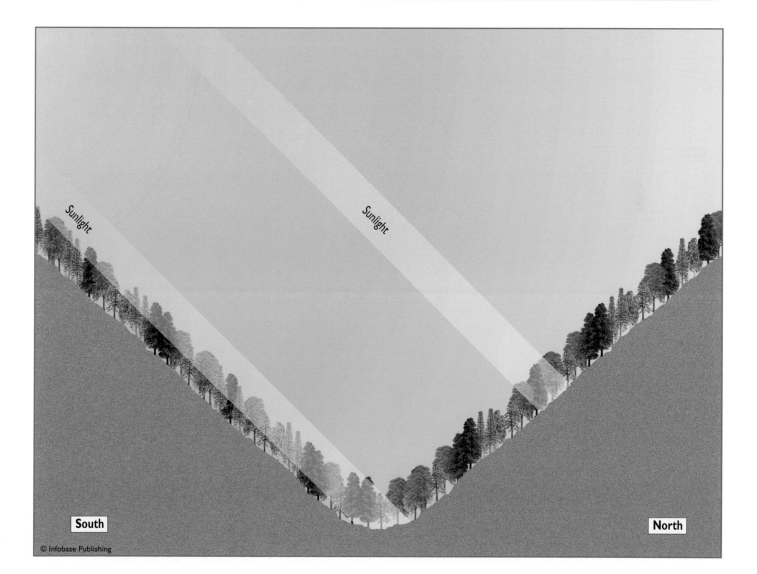

South

North

the forest can be highly diverse. The opposite sides of a valley may have quite different microclimates and may offer a range of opportunities for various plants and animals to make a living. The diversity of microclimates available is often closely linked to the biodiversity of a region.

Wind speeds within the forest are also greatly reduced compared with those above the canopy, for the air within the shelter of the trees may hardly be stirred by the movements up above. As a consequence, air turbulence and mixing is limited, and there is little exchange of air between the deep forest interior and that above the canopy. Moisture evaporating from the soil and the lower canopy leaves tends to remain within the enclosed masses of air, maintaining a high relative humidity, often over 90 percent (see sidebar "Humidity," page 10). Relative humidity falls to its lowest in the middle of the afternoon, as shown in the diagram, and rises to its maximum in the cool of the early morning, when it may approach 100 percent. In clearings in the forest, the humidity may attain similar levels during the night but falls much lower during the daytime, especially the after-

noon. The diagram is based on data from the dry season of a tropical forest. The low relative humidity of about 70 percent is unlikely to occur in the interior of an equatorial rain forest.

Humidity is important to plants because they need to take up the gas carbon dioxide for photosynthesis; they do this through pores called *stomata* in their leaves. When they open these pores to take up the gas from the atmosphere, they are liable to lose water vapor from the saturated atmosphere within the pore spaces (see sidebar "Transpiration, Water Potential, and Wilt," page 114). This is the source of water loss in transpiration. The rate at which water is lost from a leaf depends on the relative humidity of the atmosphere because this determines how much more water the air can absorb. If the relative humidity is high, then the loss of water vapor from a leaf is slow and the plant can afford to keep its stomata open longer, enabling it to conduct more photosynthesis. If the relative humidity of the air is low, by contrast, leaves lose water quickly and may begin to run short. They then have no alternative but to close down their stomata to reduce water loss, but in doing so they cut down their rate of photosynthesis. The high humidity of a tropical forest interior enables plants to keep their stomata open longer and this contributes to the high level of plant productivity (see "Primary Productivity in Tropical Forests," pages 78–82).

The variation in relative humidity in the course of a 24-hour period in (A) a clearing in the forest, and (B) in dense forest. The relative humidity drops strongly during the daytime in the clearing.

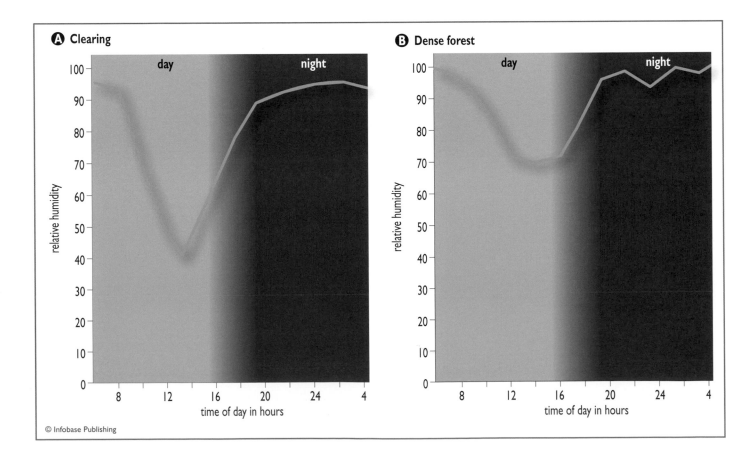

Many animals also benefit from high humidity. Amphibians, such as frogs, lose water through their skins and can quickly desiccate when exposed to dry air. The high humidity of the forest enables them to live away from water and occupy layers of the forest canopy where they could not survive if humidity fell to low levels. Many insects are also sensitive to low humidity and are easily killed by heat and drought in combination. The stable warmth and high humidity of the forest interior provide an ideal microclimate for very large numbers of invertebrates.

Light intensity, light quality, wind speed, temperature, and relative humidity thus all differ considerably between the rain forest canopy and the forest floor. The range of daily variability in all of these factors is also much greater in the canopy than in the sheltered interior. The differences in microclimate within tropical forests provide one of the more important factors determining the composition of communities of living organisms in the various layers of the canopy.

■ CONCLUSIONS

Tropical forests occur in a variety of different forms, mainly depending on the climatic conditions under which they develop. The Tropics display a pattern of climate that varies with latitude. The equatorial regions are the location of high solar energy input, much of which is redistributed to higher latitudes by movements of the atmosphere and the oceans. The heating of the Earth's surface at the equator creates atmospheric turbulence and convection currents. Rising air cools, and this process results in the condensation of water to form rain. As a consequence, the equatorial zone receives abundant rainfall and typically bears a vegetation cover of rain forest. The quantity of precipitation becomes lower away from the equator and is often restricted in its season of occurrence. Converging air masses over the equator produce the trade winds that meet in the intertropical convergence zone (ITCZ). The latitude of the ITCZ varies with season as the Earth's tilt causes shifts in the position of the overhead noonday Sun. Wind patterns thus change in the Tropics according to season, and these changes underlie the seasonal drought patterns away from the immediate vicinity of the equator.

Climate diagrams are a simple way of expressing climate patterns in a visual form. A dry season creates problems of water availability for the forest, and a frequent response is an increasing proportion of deciduous trees. As the dry season becomes longer and subject to increasingly severe water shortage, deciduous forest becomes degraded into savannah woodlands and grasslands and eventually desert scrub. Climate diagrams provide a means of predicting the vegetation that could potentially develop at any location.

Changing wind patterns with season cause intense wet seasons in some locations, especially where a shift in wind direction brings moist air from a warm ocean, as in the northern Indian Ocean. The outcome is a monsoon season that enables areas with a predominantly dry climate to support tropical forest. Rainfall concentrated in one season brings with it the inevitability of flooding, which is an important seasonal factor in some rain forests, such as the Amazon.

Turbulence in the tropical atmosphere creates conditions in which storms are frequent. Cyclones consist of a circulating pattern of high winds, generally commencing over the oceans but often hitting land and bringing devastation to coastal forests. Whereas the monsoon rains are predictable, storms are not—they bring unexpected disturbance to the tropical forest. Drought can also bring unpredictable disruption that can result in high mortality among forest trees and increase the risk of fire, which can be even more destructive. All of these factors affect the structure and composition of the tropical forests.

Mountains create their own distinctive climate patterns. Temperature falls with altitude, but rainfall may increase up to certain limits. On tropical mountains the forests develop distinctive forms, especially the cloud forests that occupy a wet altitudinal zone in some tropical regions.

Overall, climate is by far the most important physical determinant of tropical forests, affecting not only their geographical range but also the form that they take. Temperature is generally quite uniform in the Tropics, except at high altitude. Rainfall, however, varies in space and time, and is the most frequent limiting factor in the latitudinal extent of the tropical forests.

2

Geology of the Tropical Forests

Geology is the study of rocks, their formation, and their breakdown. Rock provides the solid basis on which ecosystems are built, so understanding the way in which a complex ecosystem such as the rain forest works demands a knowledge of the underlying geology. Geologists define rock as any mass of mineral matter found in the Earth's crust that has formed naturally. Most rocks are composed of hard materials, solid stony masses built up by the fusion of different minerals. But the geologist does not limit the word rock to hard materials; rocks can also be soft. Deposits of peat beneath the tropical peat swamps, the layers of clay sediments in a lake bed, and the detritus left behind following a tropical storm or flood are all rocks according to the geological definition. Only those materials constructed by human activity, such as concrete and plastics, lie beyond the confines of the word rock.

Neither the Earth nor the rocks that compose its crust are static; they are constantly changing. Rocks are forming even today as sand, silt, clay, and the bodies of microscopic animals and plants sediment to the floor of the ocean and become compacted. Rocks formed in this way are called *sedimentary rocks.* Some rocks are also modified by the intense pressure and high temperatures they experience within the Earth's crust, and these are called *metamorphic rocks.* Slate is an example of a metamorphic rock that began its existence as a sedimentary deposit, but has been subjected to pressures and temperatures that have compressed and changed its constitution. Metamorphic rocks are often produced in the vicinity of volcanic activity or at great depth within the crust. *Igneous rocks,* the third major rock type, come into being when molten rock, or magma, from the Earth's core breaks up through the crust and solidifies as it cools. All of these rock types are found within the polar tundra regions as well as in the high mountains of the world.

■ THE ROCK CYCLE

The interior of the Earth consists of molten rock, or magma, and convection currents are set up within this fluid core that push upward into the solid crust at the Earth's surface. Sometimes these rising currents in the magma force their way into the crust but fail to reach the surface before the rock cools and solidifies. Occasionally the magma bursts through the crust and erupts in the form of a volcano, and the molten rock solidifies on the surface of the Earth as masses of lava. Some areas of tropical rain forest are located in actively volcanic regions, such as the eastern part of the Congo Basin in Africa and in South and Central America. Solidification of the magma involves many chemical changes in its composition, depending in part on the location (above or below ground) where it finally cools. About 14 percent of magma is dissolved gases, mainly water vapor and carbon dioxide, and these may be discharged into the atmosphere. The chemical constituents of the magma, called *minerals,* have different melting points, so they may separate out during the cooling process, depending on the rate at which the temperature falls. Consequently, there are different kinds of igneous rock that vary chemically according to the conditions under which they were formed.

Igneous rocks that have failed to reach the surface and are buried deep in the crust are protected from the process of rock breakdown, or *weathering,* suffered by surface rocks. But eventually even these buried rocks may be exposed to the atmosphere as the crust above them is worn away. Weathering involves the chemical and physical breakdown of the rocks into their component minerals, or even into smaller chemical units. Living organisms also play a part in this decomposition of the rocks, and the process is the basis of soil formation (see "Soil Formation and Maturation," pages 46–50). Solid rock is converted by weathering into smaller particles, some of which may be soluble in water, and they are transported by water draining over and through the soils in which they lie. This transport of particles is called *erosion,* and the movement of dissolved materials is called *leaching.* As a result of these two processes, the rocks exposed at the Earth's surface are constantly being worn away and their components are transported down to the oceans, where they may remain suspended for a while, but eventually sediment to the bottom.

The sedimentation of particles derived from land-based rocks is accompanied in the oceans by the deposition of tiny remains of planktonic organisms that have spent their lives in the surface layers of the ocean. Creatures such as foraminifera and coccolithophorids, single-cell organisms belonging to the Kingdom Protoctista, accumulate calcium carbonate (lime) from the seawater and build cases for themselves that survive and sink into the ocean sediments after the death of the living creature. The diatoms (also members of the Kingdom Protoctista) build cases out of silicon dioxide, and this relatively inert material, with the same composition as sand, also joins the other particles in the steady rain of fine materials that constantly accumulate in the sediments of the ocean floor. These sediments build up over the course of time and are converted into harder rocks by the pressure from above. The sedimentary rocks produced in this way may eventually be forced beneath the crust to enter the magma once more. Alternatively, they may be crushed, folded, and buckled by movements in the Earth's crust, raised above the surface of the ocean, and exposed to the forces of weathering and erosion once again. In this way, the rocks of the Earth are involved in a constant cycle that takes many millions of years to complete, as shown in the summary diagram below.

The rock cycle. Volcanic activity creates igneous rocks derived from the molten magma of the Earth's interior and modifies the physical and chemical nature of adjacent materials, forming metamorphic rocks. Uplifted landmasses are weathered and eroded by the atmosphere and climate, and the transported materials are carried to the oceans where they are deposited over time to create sedimentary rocks.

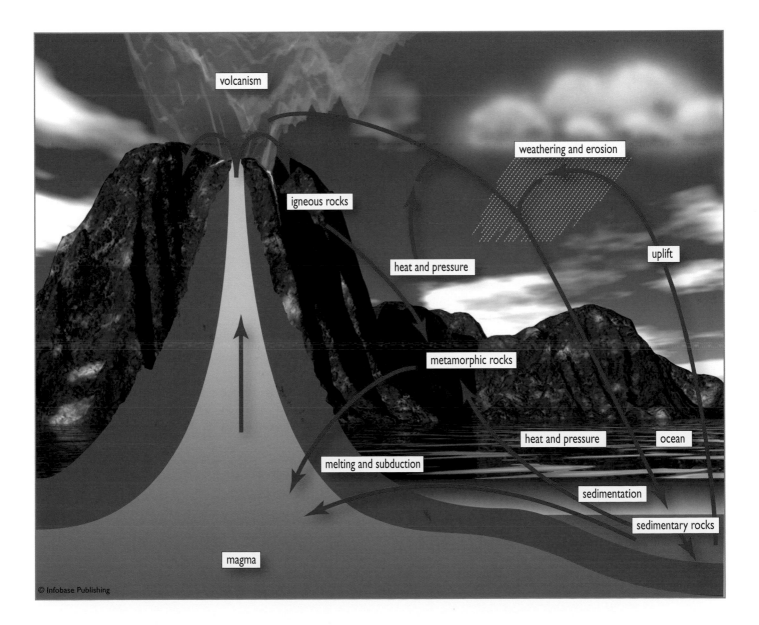

■ PLATE TECTONICS

The rock cycle describes the movement of materials between the Earth's core and the Earth's crust, and this exchange involves other processes that play an important part in the patterning of continental landmasses over the face of the Earth. The location of the landmasses, the continents of the world, is itself a major factor in determining the global distribution of the tropical forests.

Geologists have long known about the workings of the rock cycle, but until the last 50 years they were unaware of the remarkable mobility of the Earth's crust. They pondered over such questions as the origin of the world's oceans and the difficulties of interpreting some of the fossils found in unlikely places, such as tropical animals found in London, England, and fossil forests in the heart of Antarctica. Past climates must have behaved in a very strange and unpredictable way. The answer to these questions came from a very unlikely person, a professor of astronomy and meteorology in Berlin, Germany, Alfred Wegener (1880–1930), who had never studied geology. He noticed that the east coast of South America and the west coast of Africa are mirror images and would fit together very comfortably if the South Atlantic Ocean closed up. The same is true of eastern North America and the west coast of Europe, especially if one considers the edges of the continental shelves rather than the current coastlines. He concluded that there was once a single "supercontinent" that subsequently split into a number of components and gradually drifted apart. His first publication on the subject came in 1912, but the outbreak of World War I and the limited circulation of his German publication meant that it was not noticed by the scientific community. His subsequent publications, including a book on the subject, were received with ridicule. Scientists throughout the world considered his views eccentric and pointed out the lack of any mechanism to explain the wandering continents. Wegener died in an accident in Greenland in 1930, his ideas still universally rejected.

Geological evidence supporting a close association between continents in the past continued to accumulate over the years. The rocks of the Andes Mountains in the west of South America, for example, are also found in the Antarctic Peninsula and in New Zealand. Rocks from eastern Australia also continue in a band across Antarctica. Fossils of a seed fern were found in southern South America, southern Africa, Antarctica, Madagascar, India, and Australia, which added weight to the hypothesis that these landmasses had once been much closer together, or even joined in one supercontinent. Then geophysicists discovered a new technique that changed many opinions and attitudes, called *paleomagnetism,* which is based on the observation that when rocks are first formed they are strongly affected by the Earth's magnetic field. Any particles that can be magnetized, such as iron compounds, become aligned with the magnetic field as the rock solidifies, and they subsequently retain this orientation of magnetization. By studying rocks of different ages in Australia, geologists were able to work out that this continent had migrated around the South Pole and then northward into its present position over the past 500 million years. The only alternative explanation was that the South Pole itself had wandered during this time, but further studies in many parts of the world during the 1950s and '60s confirmed that the continents had indeed been on the move. Wegener was at last proved right.

Proving that continents such as Europe and North America are moving apart is not easy because the rate of movement is very slow. Geologists also needed to find a mechanism that would account for the movement, and they discovered this at the bottom of the sea. The gradual separation of Europe and North America, for example, can only be explained if there is a dividing line out in the Atlantic that is pushing apart the ocean floor as well as the two continents. This is found in the mid-Atlantic ridge, a line of submarine mountains rich in active volcanoes that runs the entire length of the Atlantic Ocean and even emerges above the surface in the island of Iceland. Along this line, magma is rising to the surface and solidifying into new crustal rocks that are pushed either east or west, forcing the continents apart. Lines of separation of this kind are now known to occur around the world, dividing the Earth's crust into a series of plates and forcing them apart, as shown in the diagram on page 37. As two plates are pushed apart, an ocean basin may be formed between them, as in the case of the Atlantic Ocean, which is continuing to enlarge by between two and four inches per year (5–10 cm per year).

If new crustal material is constantly being created, then there are two possible consequences. Either the new material will force older parts of the crust to buckle and fold, forming new mountain chains, or there must be places where the crust is being destroyed or is slipping down into the Earth to become magma once again. In fact, both of these possibilities occur. Moving plates may collide and the force of the collision can cause crumpling and up-thrusting of the crust, creating high mountains. This type of mountain-building collision occurs when both of the plates involved bear continental land masses. India, for example, is part of the same plate as Australia (see diagram), and when Australia drifted northward, India also moved and eventually collided with the Eurasian plate. The result was the creation of the Himalaya Mountains along the collision boundary. A similar collision between the African plate and the Eurasian plate resulted in the building of the Alps in central Europe, the Caucasus Mountains of eastern Europe, and the Alborz Mountains of Iran; the ripples caused by that collision created hills in southern England.

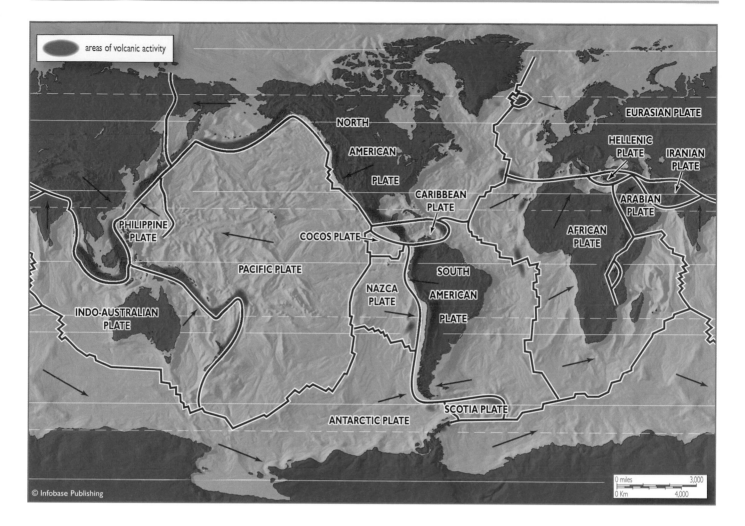

areas of volcanic activity

EURASIAN PLATE

NORTH AMERICAN PLATE

HELLENIC PLATE

IRANIAN PLATE

CARIBBEAN PLATE

ARABIAN PLATE

PHILIPPINE PLATE

COCOS PLATE

AFRICAN PLATE

PACIFIC PLATE

SOUTH AMERICAN PLATE

NAZCA PLATE

INDO-AUSTRALIAN PLATE

SCOTIA PLATE

ANTARCTIC PLATE

© Infobase Publishing

0 miles 3,000
0 Km 4,000

A map showing the locations of the crustal plates of the Earth. Plate boundaries are the locations of intense tectonic activity. Some boundaries are constructive, the lines along which new crustal material is being formed; other boundaries are destructive, the places where crustal material plunges back into the Earth's interior, called subduction zones. Volcanic activity is particularly strong along the boundaries of subduction zones, shown here by red shading.

The crust beneath the oceans is often only about four miles (6.5 km) thick, much thinner than the continental crust, which may be up to 20 miles (32 km) thick and is generally composed of older rocks. When an oceanic plate collides with a continental plate, it usually results in the oceanic crust sliding underneath the continental one, where it is pushed back down to rejoin the magma. This is called a *subduction zone* and is illustrated in the diagram on page 38. The existence of subduction zones ensures that there is a balance between the production of new crust at the ocean ridges and its destruction by burial. The oceanic crust is heavy, so on collision it tilts downward and begins its descent back into the magma below. As it does so, it drags some of the lighter

continental plate materials with it and creates a trench in the zone of subduction. As this mix of rock, often saturated with water, descends into the Earth, it begins to melt in the increasing heat, and the lighter, more volatile constituents form an unstable mass of material that forces its way back up to the surface. The consequence is that subduction zones are often accompanied by intense volcanic activity. One of the most active volcanic regions of the world lies along the junction of the Indo-Australian plate with the southeastern edge of the Eurasian plate in the vicinity of Java. The famous island of Krakatoa, which erupted and exploded in 1883, is just one of the volcanoes that have been created along this subduction zone (see "Tropical Volcanoes and Islands," pages 41–43).

Plate boundaries are often zones of stress for the Earth's crust, but their movements are not always in direct opposition to one another, as in the case of plate collisions. Sometimes plates are moving alongside but in opposite directions and they slide past one another. This is called a conservative plate boundary. A good example of this type of boundary lies immediately west of the Sierra Nevada in California, where the Pacific plate is moving northward with respect to the North American plate, having a relative

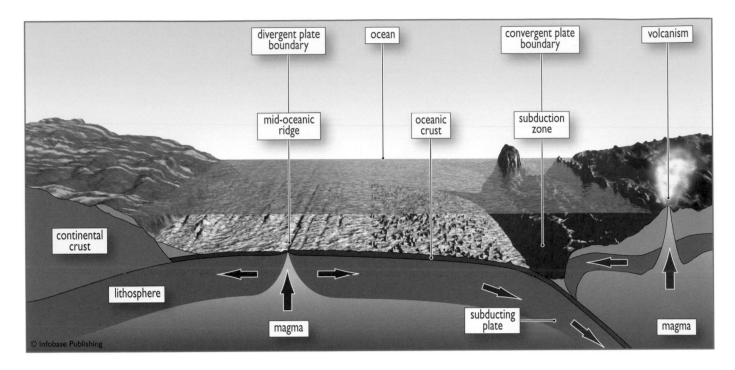

Convection currents in the Earth's molten interior cause a constant movement of the surface crust. New oceanic crust is created at a mid-oceanic ridge, forming a divergent plate boundary. The outward spreading of the oceanic crust results in a collision with continental crust at a convergent plate boundary, and the thinner oceanic crust is forced beneath the continental crust, creating a subduction zone. As the submerging crust melts, it results in a high incidence of volcanism along the continental edge of the plate boundary.

motion of about half an inch (1.2 cm) per year, although it has been recently accelerating. The friction between the two plates causes them to stick at times and the pressure is then released by a sudden and catastrophic movement that causes a major earthquake. The San Francisco earthquake of 1906 was due to one of these sudden releases of pressure between the sliding plates, and it resulted in a sudden movement of 21 feet (6.4 m). A similar sudden shifting of plates took place in late 2004 beneath the sea in Southeast Asia, the outcome being a devastating tidal wave, or tsunami, which resulted in hundreds of thousands of deaths in the coastal regions of lands surrounding the Indian Ocean. Plate boundaries are the places where the Earth is at its most dynamic, and some junctions lie within the tropical forest zone, such as the very active Indian–Eurasian plant boundary in Southeast Asia.

The world map shown in the diagram on page 37 illustrates the major plate boundaries, although there are smaller plates and subplates within the system and some of these are important in the geological processes within the tropical forest zone. In East Africa, a section of the east coast occupies an East African subplate, as shown in the diagram on page 39, and Arabia and Madagascar occupy relatively small plates that are separate from both the African plate and the Indo-Australian plate. The junction between the main African plate and the East African subplate demonstrates that splitting and separation of plates can take place within a continent as well as in the ocean. This is called continental rifting, and the initial outcome is the formation of a rift valley. The East African rift valley is a major geological feature of the regions and continues northward through the Red Sea and up into the Jordan rift valley of southwest Asia.

■ TECTONIC HISTORY OF THE TROPICS

By studying the patterns of rock arrangements, their fossils in different continents, and their residue of magnetic alignment, geologists can trace the ways in which landmasses have moved in the past. The complicated story of past continental wandering is now more or less completely unraveled.

In Devonian times, around 380 million years ago, the continents then in existence bore no resemblance to those found today. There were three major landmasses: one consisted of the lands that would one day become North America and Europe, a second was a block of land destined to become Siberia, and the third was a massive supercontinent called Gondwana, consisting of most of the remaining land areas of the world. By the late Permian period, some 260 million years ago, all of the world's landmasses had fused into one supercontinent, called Pangaea. This arrangement is shown in the accompanying diagram on page 40, part A. At that time the region now called Asia was fragmented and its component parts well separated around the edge of the

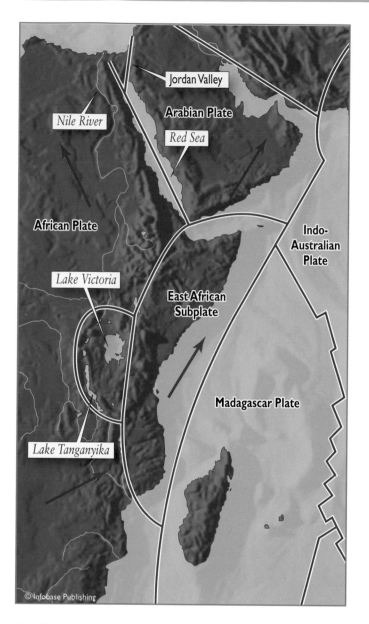

Small plates, or subplates, are created within the main crustal plates of the Earth, as in the case of the East African subplate, separated from the main African plate by the African rift valley. This rift runs northward up the Red Sea and continues as the Jordan rift valley.

supercontinent, but the other major continental masses were intact and joined together. But this single, unified world soon began to break up, forming two major land areas, one called Laurasia, which consisted of North America, Europe, and Asia, and the other, the original Gondwana, consisting of South America, Africa, India, Australia, and Antarctica. By the Early Cretaceous period (B), some 140 million years ago, even these continental masses were fragmenting as epicontinental seas began to form on top of the supercontinental crusts, and the major continents that make up the modern world started to emerge.

In Late Cretaceous times, 90 million years ago (C), Gondwana had split into its component parts, with Africa, Madagascar, India, and New Zealand completely separated off, and South America and Australia were linked to Antarctica only by tenuous land bridges. Even these links were broken by the Eocene epoch, about 40 million years ago (D), and the world had begun to take on its familiar form. India had collided with southern Asia, creating the Himalaya Mountains in the process, and northeast Africa was about to make contact with western Asia to create the Arabian Peninsula. The bridge between North and South America was not yet in existence.

The wandering of the continents over the face of the Earth has had important climatic consequences. Eastern North America and western Europe, for example, were far south of their current geographical positions 60 million years ago and experienced subtropical climatic conditions. At that time, what is now Greenland and the Hudson Bay region of Canada would have been considerably warmer. In northern Greenland fossil leaves of the breadfruit tree (*Artocarpus dicksoni*) indicate that conditions must have verged on the tropical. Geologists have taken corings into the sediments of the Arctic Ocean and discovered fossil materials from 70 million years ago demonstrating that much warmer conditions existed even at very high latitudes. They speculate that the average annual temperature in the Arctic Ocean at that time was 60°F (15°C) and that the region was ice-free throughout the year.

When all continents were fused into one massive landmass, oceanic waters had free access to the polar regions, so the redistribution of energy over the face of the Earth was uninterrupted and the poles were considerably warmer than today, so there were no polar icecaps. The southern supercontinent of Gondwana experienced mild conditions with high precipitation as a result of the evaporation of water from warm oceans and its condensation over the land. During mid-Carboniferous and Permian times, from about 330 million years ago, Gondwana was drifting toward the South Pole and this prevented any warm tropical water from reaching this part of the world where the Sun's energy was in low supply. As a consequence, an ice sheet began to form on the Gondwana continent. Meanwhile, the northern landmasses, including what is now North America and Europe, were close to the equator, so tropical swamp forests prevailed there, laying down deep deposits of peat that would later turn to coal. Thus tropical forest, although it contained a very different assemblage of plants and animals, has existed on Earth for hundreds of millions of years. Despite the formation of a southern polar ice cap, Gondwana still retained a relatively mild summer climate around its edges, as the fossil floras from Carboniferous and Permian rocks record. One particular plant fossil dominates, namely a shrub or small tree with relatively thick woody stems and broad

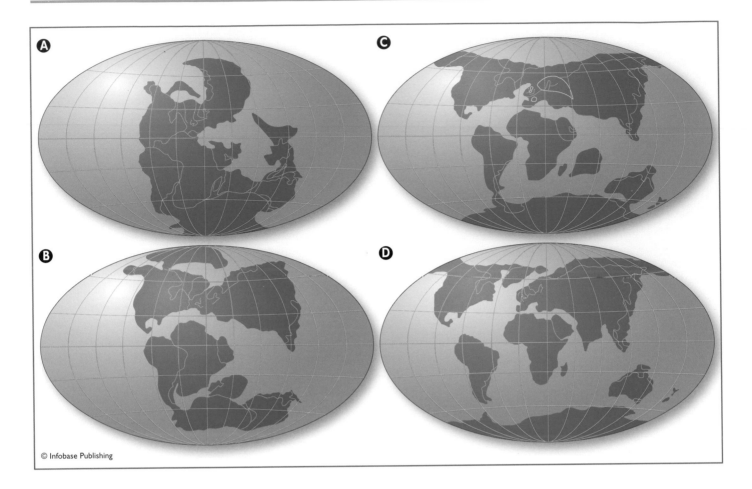

© Infobase Publishing

Tectonic plate movements have been taking place throughout geological history. In Permian times (A), about 260 million years ago, all the Earth's land masses were linked in one supercontinent. This split into two by Cretaceous times (B), about 140 million years ago, and the breakup continued through the Cretaceous period until the continents had begun to assume their familiar shapes by the late Cretaceous (C), 90 million years ago. By the Eocene epoch (D), 40 million years ago, India had collided with Eurasia, and South America had broken its links with Antarctica, but had not yet formed a bridge to link it with North America. The North Atlantic Ocean had formed, separating the Old World from the New World.

leaves. It has been given the name *Glossopteris*; together with the associated plant species, the community of that time is called the *Glossopteris* flora. The fossil remains of woody parts of *Glossopteris* have very distinct annual growth rings, indicating that the plant exhibited summer growth and winter dormancy. It is remarkable that this lush vegetation survived even within 5° latitude of the contemporaneous South Pole, indicating that the temperature of the entire globe must have been warmer at that time. The *Glossopteris* flora has become important in the reconstruction of tectonic processes because its fossil remains have been found in the rocks of Australia, Antarctica, South America, and southern

Africa. The fossils thus provide important evidence for the former close contact between these continents.

The breakup of Gondwana during Cretaceous times was accompanied by the development of extensive forests in what is now Antarctica. Geologists have found the remains of tree trunks fossilized in their original positions, and this finding has enabled them to reconstruct the ancient forests. The trees were widely spaced, about 10–16 feet (3 to 5m) apart, and held their foliage in a vertical rather than horizontal plane. This would have allowed them to catch the maximum sunlight that arrived at a very low angle. Among the trees present was a species that closely resembles the modern maidenhair tree (*Ginkgo biloba*). Late Cretaceous times (99 to 65 million years ago) saw a great diversification of flowering plants, and these made their way through the fragmenting supercontinent on migration routes that passed from South America, through Antarctica and South Africa, to Australia. Among these plants, which still persist in all of these continents apart from Antarctica, was the southern beech (*Nothofagus* species) and members of the family Proteaceae. The migratory routes became disrupted as the supercontinent broke up and Antarctica no longer provided a migration route for plants and animals to move between South America, Africa, Madagascar, India, and Australia. By this time many major groups of plants and animals had found their way over the land bridges and were present on all the major continents. The separation of Madagascar from the mainland of Africa occurred quite early in this process of continental fragmentation, perhaps around 160 million years ago in the Jurassic Period. This long isolation of Madagascar has strongly influenced the composition of its flora and fauna.

■ TROPICAL VOLCANOES AND ISLANDS

Tectonic activity in the course of geological history has thus influenced the displacement of the continents and has brought the world to its current pattern of oceans and landmasses. The activity continues, however, both in the creation and subduction of crustal material and, more dramatically, in the eruption of volcanoes. Some tropical areas are rich in volcanoes, as shown in the diagram on page 37. These volcanoes can have a profound influence on the development of tropical ecosystems.

As can be seen on the map, the global distribution of volcanoes is closely associated with plate boundaries. The islands of Melanesia, to the north of Australia, lie along the most active of plate boundaries and are particularly rich in volcanoes. Here the Indo-Australian, Eurasian, Pacific, and Philippine plates come together and the instability results in frequent earthquakes and volcanic eruptions. Among the most famous, because of its intensity and its relatively recent occurrence, is the explosion of the island of Krakatoa (see sidebar below). Other tropical regions disturbed by volcanic activity include the Andes of South America, the spine of Central America, islands in the Caribbean, and East Africa in the region of the Rift Valley.

Volcanoes are major sources of disturbance in tropical vegetation, along with hurricanes, floods, and fire; regions where volcanic activity is abundant must experience regular, if infrequent, catastrophe. This raises the question of whether such a disturbance has a negative impact on the biodiver-

Krakatoa

The volcanic island of Krakatoa lay between the much larger islands of Java and Sumatra in Southeast Asia. It was one of a series of volcanoes lying along the boundary between the Indo-Australian plate and the Eurasian plate. As the Indo-Australian plate moves northward its oceanic crust is forced below the continental crust of the island ridge, creating a subduction zone. The rate of movement is slow but inexorable. Each year it shifts about four inches (10 cm), or 100 yards (90 m) every thousand years. As is often the case, the adjacent ridge of continental crust is particularly unstable as magma forces its way back to the surface. Most volcanoes are not permanently active but may lie dormant for hundreds of years as pressure builds up below, eventually erupting with explosive force. This was the case with Krakatoa, which had evidently been dormant for generations before its massive eruption in 1883. The lower slopes of the mountain were covered with tropical rain forest at this time. Captain Cook visited the island on at least two occasions during his voyages of discovery and recorded that people lived there and cultivated parts of the island, growing peppers and raising chickens and goats. During his visit in 1780 an artist on board recorded the lush tropical vegetation of the coastal zone of the island. It is difficult to say how long a period had elapsed since any previous eruption, but it is likely that Krakatoa had

erupted in 1680, so the jungle recorded by Cook must have been about 100 years old. In 1883, following some preliminary tremors and eruptions, Krakatoa exploded at 10:02 AM on Monday, August 27. The sound of the explosion was heard 3,000 miles (4,800 km) away in the western Indian Ocean and in Perth, Australia. It may well be the loudest noise ever recorded on Earth. The tropical forest was destroyed along with the entire island, and tsunami waves up to 100 feet (30 m) in height were generated that flooded the coastal forests of Java and Sumatra. Surrounding islands within a radius of 12 miles (19 km) also lost their vegetation, buried in volcanic ash and burned by the fall of hot rocks.

In 1927 a new volcanic island rose out of the sea in the center of a ring of fragmented islands that formed the remains of Krakatoa. Scientists have used these devastated islands as a living laboratory in which to study the recolonization of vegetation and animal life. In the tropical climate, this process is surprisingly rapid; some plants returned within a year of the eruption and by 1906 a fringe of woodland had developed along the coastline of the islands. By 1928 rain forest extended to an altitude of 1,300 feet (400 m) and some trees were (30 m) tall. A true soil had begun to develop from the volcanic ash, and the first stages of ferricrete (laterite) formation had begun (see "Soil Formation and Maturation," pages 46–50).

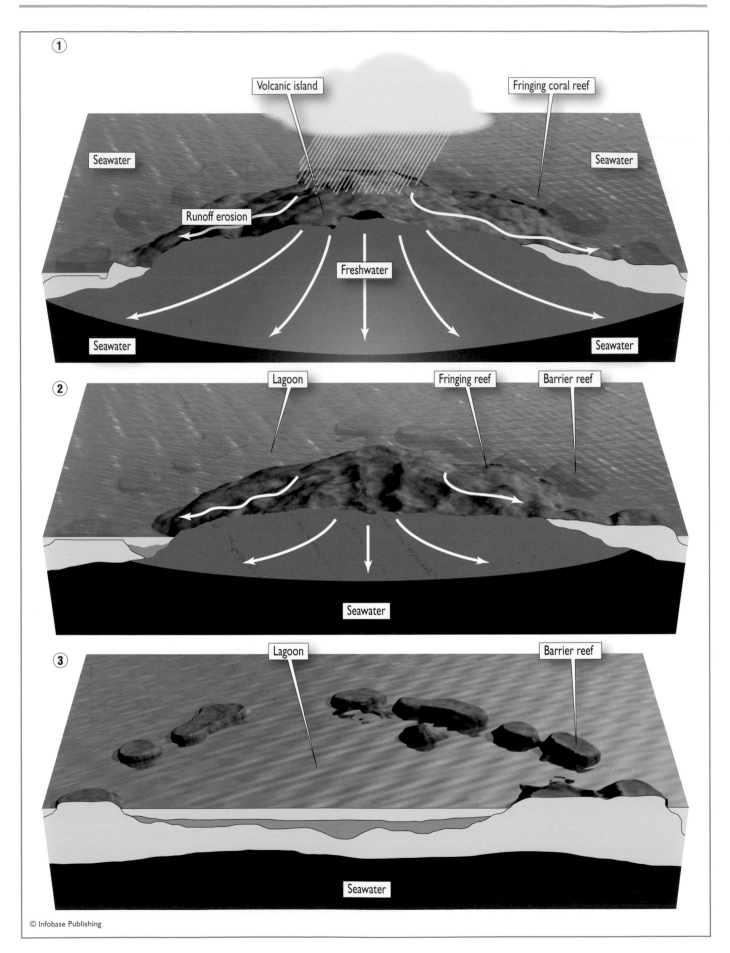

sity of a region. One might expect the cataclysmic eruption of volcanoes, accompanied by fire, to be a cause of extinction among plant and animal communities. The diversity of plants and animals found in the forests of Southeast Asia, an area which is highly geologically active, is actually much greater than that of the Congo Basin, which is geologically stable. It appears, therefore, that geological instability does not have an inhibitory effect on the evolution of diversity; indeed, it may have precisely the reverse impact. Perhaps disturbance creates more opportunities for different plants and animals to make a living. This possibility will be considered in greater detail in chapter 5 (pages 98–108).

Some tropical forests occupy remote islands, far from any continental mainland. The Pacific Ocean is particularly rich in tropical islands and early explorers and geologists found it difficult to explain their presence so far from coastlines. Charles Darwin (1809–82) proposed a theory that has since proved to be substantially correct. He claimed that these isolated tropical islands were formed from volcanoes that forced their way up to the surface of the ocean. He had no knowledge of tectonic plates and their movements, of course, so he could not account for the origins of submarine volcanoes, but now their presence is explained in terms of plate boundaries and subduction zones. Volcanic islands may still be active, but many of them are peaceful. Some have peaks so high that despite their tropical position they are snow-capped. They vary climatically from very dry to very wet, depending on their position in relation to the trade winds, the wet sites supporting rain forest. Their soils are derived from lava and ash, so their acidity is extremely variable.

The volcanic islands are gradually eroded by the action of rain and wind, so their topography is often rugged and dissected by erosion channels. The rainfall produces streams that run down the sides of the islands and soak into the porous rock, as shown in the diagram on page 42. In wet climates the freshwater forms a deep lens that extends below sea level, forcing out the percolating seawater and maintaining conditions suitable for forest growth. Around the edges of the island a fringing reef of coral often forms.

(opposite page) The formation and development of a tropical island. 1. Volcanic activity brings the island into being, often resulting in a rapid rise of land from the ocean floor. A fringing coral reef forms around the edges of the island. Rainwater erodes the surface rocks, but also soaks into them, creating a lens of freshwater even below the level of the sea. 2. Continued erosion over time reduces the size of the volcanic island, while the extension of the fringing reef may give rise to a circular barrier reef and enclosed lagoon. The freshwater lens is now smaller. 3. Erosion causes the original rocks of the island to disappear beneath the waters of the lagoon, and all that remains is the circle of the coral barrier reef.

Over the course of time, tectonic movements may cause the volcanic core of the island to sink, or sea level may change, resulting in the original island dropping below sea level. Coral may continue to build up in the shallow water above the sunken volcano and produce a coral atoll with a circular reef enclosing a lagoon. Tectonic uplift can sometimes result in coral being lifted above the sea level once more, creating an island formed of coral limestone.

■ TROPICAL ROCK WEATHERING

The rocks of the tropical regions are as varied as their individual geological histories. The formation of continental masses began with the cooling and development of the first crust upon the surface of the Earth, in Archaean times, early in the Precambrian. These ancient crusts form shields that underlie the continents and in places come to the surface. Changing patterns of oceans and epicontinental seas led to the formation of sedimentary rocks, sometimes overlying the shields. Mountain-building events, orogenies, led to uplift of both the ancient igneous and the more modern sedimentary rocks, exposing them to the atmosphere and to the erosive force of wind, frost, and rain, leading to commencement of the rock cycle (see "The Rock Cycle," pages 34–35). The degradation and decay of rocks exposed to the atmosphere, a process called weathering, set into motion the formation of soil.

The mineral component of the soil consists of the various rock fragments and forms the main soil skeleton. The chemical constitution of these particles depends upon the rocks from which they were derived, and this has a lasting effect on the nature of the soil that develops. Weathering takes place as a result of the activity of physical, chemical, and biological factors.

Physical weathering in tropical soils results from heat. The heating of rocks in the sunlight can lead to their physical breakdown. As the dark rocks absorb the light, their temperatures rise, causing cracking and breakage along planes of weakness. Rocks consist of mixtures of crystalline minerals, and these become separated from one another during this weathering process. Granite, for example, is broken down into three main minerals—feldspar, mica, and quartz. During its formation these three minerals crystallize separately from the molten magma in the cooling process, and the physical weathering of granite breaks them apart and liberates them into the developing soil. Heating splits layers from rocks, rather like the peeling of an onion, in a process called exfoliation. Physical weathering thus leads to the fragmentation of rocks into ever smaller particles. The proportions of different-sized particles found in a soil, termed its *texture,* influences many soil properties, such as drainage. Soil scientists give different-sized particles specific names, defined in the sidebar on page 44.

Particle Size

Soil scientists and geologists have devised a scheme for classifying the different particles of rock according to their size. In this way they can provide a precise definition for terms that could otherwise be used very loosely, such as sand and clay. Unfortunately, scientists in different parts of the world have failed to agree on a universally accepted set of definitions, but within the United States there is some measure of agreement. A rock fragment with a diameter greater than 10 inches (25.4 cm) is called a boulder, while smaller particles that exceed 2.5 inches (6.4 cm) are cobbles. Any particle greater than 0.08 inches (0.2 cm) is called gravel. Sand consists of particles between 0.0008 and 0.08 inches (0.002–0.2 cm) in diameter, while silt is still smaller, having a diameter of 0.00008–0.0008 inches (0.0002–0.002 cm). Clay is the smallest of all particles, having a diameter of less than 0.00008 inches (0.0002 cm).

Clay is distinctive not only as a consequence of its small size but also because of its chemical structure. The tiny particles of clay are composed of chemical sandwiches in which layers of silicon oxides alternate with layers of aluminum and iron oxides. The outcome of this complex crystalline structure is that the surface of the clay particle is negatively charged, and when clay particles are suspended in water the effect of these negative charges is to repel any neighboring clay particle. Clay is thus suspended in water both by its small size, which makes sedimentation slower, and by the repulsion between the fine particles. Clay remains in suspension in water for considerably longer than all other particles and eventually settles only under very calm conditions. The negative charges on clay particles are also important in the chemical reactions of a soil. They attract positively charged particles (*cations*) and thus act as a store for certain important elements, such as calcium and potassium. Clays are said to have a high *cation exchange capacity*. Negatively charged groups of elements (*anions*), such as nitrates, NO_3^-, are not held by clays so are very easily leached from soils.

Following physical weathering, chemical weathering acts upon the fragmented rock particles. Water is an important medium for chemical weathering because it is an excellent solvent and contains many dissolved substances. Rainwater contains a range of dissolved materials, most importantly dissolved carbon dioxide. When dissolved in water, this gas forms a weak acid, carbonic acid, and this operates on the minerals in the soil, reacting with them to form new compounds. The atmosphere also contains small amounts of other gases, such as oxides of sulfur and nitrogen, which are derived from both natural sources and from pollution. The oceans are a source of dimethyl sulfide and volcanic activity produces hydrogen sulfide, as do some wetlands. In the atmosphere these sulfur compounds become oxidized and then dissolve in water to produce sulfuric acid, a very much stronger acid than carbonic acid. The burning of fossil fuels in industrial processes and automobiles has added to the atmospheric load of sulfur, and this form of pollution affects even the most remote of tropical forest habitats. Nitrogen oxides are also produced by fossil fuel combustion and by burning vegetation in forests and grasslands. In the atmosphere these oxides dissolve in water to form nitric acid, another strong acid. When this collection of acids reacts with the mineral components of soils the result is chemical breakdown. The feldspar mineral from granite, for example, is broken down by carbonic acid to form potassium carbonate and a relatively inert material, kaolinite. Kaolinite, or china clay, is a clay mineral (see sidebar [left]), and is not broken down further. It plays an important part in controlling the behavior of certain elements in the soil. Carbonic acid also acts upon the mica mineral derived from granite, breaking it into potassium and magnesium carbonates, along with oxides of iron and aluminum. The third mineral derived from granite, quartz, however, is relatively inert, being composed of silicon dioxide (silica), better known in the form of sand.

Granite thus weathers through physical and chemical processes liberating potassium, magnesium, iron, and aluminum compounds, together with a clay mineral and silica. Although several of these elements are needed for plant growth, aluminum can be toxic, and many of the other important elements plants need are absent, such as phosphorus and calcium. The absence of calcium in particular is important because the acids that accumulate in the soil are not neutralized and the soil becomes acidic in reaction. In other words, it has a low *pH* (see sidebar on page 45).

The degree of acidity of a soil is important because it affects the growth of plants. Not only does acidity have a direct effect on plant roots but it also influences the availability of many other elements in the soil. For most plants a pH of less than 5 begins to cause problems. Many elements essential for plant growth, such as potassium, phosphorus, calcium, and magnesium, become increasingly difficult for a plant to obtain at such high levels of acidity. Iron and aluminum, by contrast, are more soluble in very acidic solutions, but both can be toxic to plants, so this is not a great advantage. In some areas, however, less acidic rock, includ-

Acidity and pH

An acid can be defined as a source of hydrogen ions, or protons, H^+. The more hydrogen ions present in a solution, the greater the acidity. Chemists therefore measure the acidity of a solution in terms of its hydrogen ion concentration. But this varies greatly and the numbers involved would be extremely great, so a scale, called the pH scale, has been devised to express this concentration in terms that can be more easily handled. As in the case of any very long scale of numbers, the simplest way of reducing it is to express the numbers as logarithms. Using a base 10 logarithm, the difference between one unit and the next is a factor of 10 times. In the case of pH, the scale is also expressed in a negative fashion, so that high numbers represent low hydrogen ion concentrations and low numbers indicate high concentrations, in other words, greater acidity.

The pH scale is shown in the diagram below. The scale runs from 0 to 14 and the figure of pH 7 is taken as an indication of neutrality, being neither acidic nor basic. At this pH, hydrogen ions have a concentration of 10^{-7} molar, or 0.0000001 g/l. Above the figure of 7 the pH indicates basic conditions and below 7 the pH denotes acidity. The diagram shows the pH of certain common solutions to demonstrate how the scale works out in practice. It is important to remember that the scale is logarithmic, so pH 4, for example, indicates 10 times the acidity of pH 5 and 100 times the acidity of pH 6.

The pH scale of acidity and alkalinity. Neutrality is marked by pH 7. The scale is logarithmic and negative, so pH 4 is 10 times as acid as pH 5. Various familiar substances have their pH recorded along the scale.

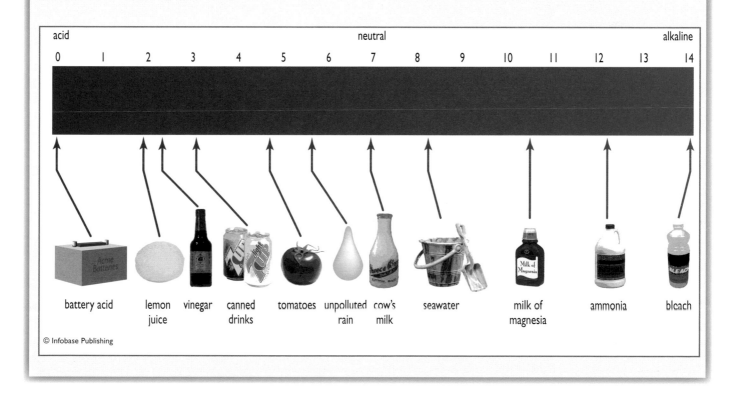

| acid | | | | | | | neutral | | | | | | alkaline |

0 1 2 3 4 5 6 7 8 9 10 11 12 13 14

battery acid — lemon juice — vinegar — canned drinks — tomatoes — unpolluted rain — cow's milk — seawater — milk of magnesia — ammonia — bleach

© Infobase Publishing

ing calcareous limestone, forms the bedrock, and here very different soils develop with neutral or slightly alkaline pH. In general, more of the essential elements for plant growth are easily available at neutral pH, but very high pH can cause problems. Above pH 8.5, phosphorus, manganese, and iron can be difficult for the plant to obtain.

The consistently high temperature of tropical regions has its impact on soil chemical weathering. The breakdown of clay minerals in the soil follows a different course from

that found in temperate regions. In the temperate zone, the decay of clay minerals usually results in the mobilization of iron and aluminium oxides, leaving bleached silica behind. In tropical soils, however, it is often the silica that becomes more mobile, leaving soils that contain high iron oxide concentrations and are consequently colored red.

Living organisms rapidly colonize any site where they can make a living. Only the most severe environments on the planet, such as the vent of an active volcano, are totally devoid of life. In the Tropics there is no shortage of living organisms to invade and exploit the developing soil environment, and these also have an impact on the physical and chemical properties of the soil. The roots of trees may penetrate into bedrock, especially where the soil is shallow, and the force exerted by an expanding root is often sufficient to split open rocks and hasten their breakdown. At a much smaller scale, the respiration of fungi, bacteria, and the many other tiny organisms that occupy a soil produces carbon dioxide. This element dissolves in soil water and acidifies it, enhancing the chemical weathering of soil particles. Some of the breakdown products of plant detritus also have a chemical impact on the soil. Lignin, the main component of wood, decays to produce polymeric acids that interact with clay minerals and other soil components. These processes associated with living organisms constitute biological weathering.

The combination of physical, chemical, and biological weathering gradually wears the rock particles into smaller parts and releases increasing quantities of chemical elements into the local environment. As the rocks are degraded, a soil is gradually born.

■ SOIL FORMATION AND MATURATION

Soil consists of more than just rock fragments randomly thrown together. Soil contains spaces between these fragments, which may be filled with air or with water, depending on the rate of water input and the nature of site drainage. Soil also contains dead organic matter derived from the plants and animals living in the soil or on the surface. These living creatures can also be considered components of the soil because their activities have a considerable effect on soil development. Soil, therefore, develops over time. Its character changes, both physically and chemically, as it is subjected to the continued effects of vegetation and climate.

Water is abundant throughout the year in some tropical soils. In other soils, developed where there is a pronounced dry season, water may become scarce for part of the year. Tropical regions with more extensive droughts may be deficient in water for much of the year, but if water is constantly scarce, forest is unlikely to develop. The soils of the tropical forest regions are generally well supplied with water for at least part of the year (see "Tropical Climates," pages 9–14). When a soil receives an abundance of water, the overall direction of water through the soil is usually downward. This type of soil is called a pedalfer. During periods of drought accompanied by high temperature, water may move in the opposite direction. Evaporation of water from the soil surface can lead to a general upward movement of water through the soil; this type of soil is called a pedocal. This schema is a very simple way of classifying soils, but the difference between them has great significance for the way in which chemical elements are carried within the soil. In the pedalfer, draining water carries chemical elements down the soil profile, whereas in pedocals the movement of chemicals is up toward the surface, where they may accumulate. The behavior of water in a soil is therefore important in understanding its development and properties.

Water moves through the soil, upward or downward, displacing air from its abundant small pockets and air channels (*capillaries*), clinging tightly to small soil particles, and held by the forces of *capillarity*. Some water becomes chemically bound to crystals in the soil's mineral particles and is very difficult to displace. A very simple but very expressive classification of soil water is into three components: gravitational water, capillary water, and chemically bound water. Gravitational water does not remain in the soil for very long. Following a soaking by snowmelt or rainfall, the soil is saturated with water, but within 24 hours of the water supply ceasing, some of the water drains away under the influence of gravity. From the point of view of plant life, this component of the soil water is of little value because it is present for such a short period of time. Gravitational water, however, is mobile and does influence the nature of the soil by taking with it some of the smaller particles and the dissolved chemicals in the soil solution. The removal of elements from the soil by the water moving through it is called *leaching*.

In pedocals water is scarce, and any supply from precipitation is unlikely to do more than soak into the soil without providing the excess needed for gravitational drainage. It is then gradually lost by evaporation, moving back up the soil profile as it is drawn by the evaporative pull, similar to that of plant transpiration (see sidebar "Trees, Water, and Climate," page 18).

Capillary water is held longer in the soil than gravitational water because of the forces of surface tension. Thin layers of water on small soil particles and small columns of water penetrating into the complex architecture of the rock fragments are held in place by these surface forces, and the pull of gravity is not sufficient to remove them. Plant roots exert greater pressures than gravity. Water is constantly being lost from the leaves of plants as vapor moves out of the leaves and into the atmosphere, and this creates a strong pull on the columns of water that extend unbroken from the leaf down the stem and into the root. As water is withdrawn up the stem, the root tissues are placed under a negative water pressure and this has the effect of drawing in water from the soil outside the cells. It is the capillary water in soils that forms the main resource for the plant's water requirements, because this water is available over long periods of time and yet is not held so strongly in the soil that it cannot be accessed by the suction of the plant root. The evaporative pull produced by the hot, dry air is capable of removing this capillary water in pedocal soils. The third kind of water in the soil, the chemically bound water, is too strongly held for plant roots to withdraw it, so it is effectively useless to them.

The capillary water present varies with the architecture of the mineral skeleton of the soil and the arrangement of organic matter within the soil. If a soil consists of large particles, such as sand or gravel (see sidebar "Particle Size," page 44), then there are relatively few capillaries in which water can be held. Most of the water present is of the gravitational type and is quickly lost in times of drought. Soils with large particles are thus free-draining and plants living on them may experience water shortages under dry conditions. Soils with an abundance of fine particles, such as silt and clay, however, are rich in fine capillaries and have a high capacity for water retention. They are less prone to water shortage and plants generally find them more suitable for growth. They may, however, suffer from the opposite problem, namely long-term water saturation or waterlogging. Excess water over long periods creates problems for plant roots because the water occupies all the spaces where air would normally penetrate the soil, so that roots can effectively be drowned.

The movement of water through soils has an important impact on its maturation. The removal of chemical elements from a soil by downward leaching in pedalfer soils leaves the soil progressively more deficient in some compounds. Materials may be removed completely from the site and washed out into streams, rivers, lakes, and ultimately the oceans, or they may simply be redeposited deeper in the soil. If this happens, then the soil takes on a layered form that can best be seen in cross section by digging a soil pit. Such a cross section of a soil is called a soil profile. A soil type that illustrates the development of a soil profile most clearly is the *podzol* or spodosol (see sidebar on page 48). The process by which a podzol is formed is called podzolization.

Podzols, or spodosols, occur in tropical environments where soil drainage is good and where precipitation is high. They are most common in sites where soils are deep and sandy, as in parts of Australia, South America, and Sarawak, but are scarce in Africa. Unlike the podzols of the temperate zone, tropical podzols are usually many feet deep with an extensive eluviated A (E) horizon. The B horizon can be up to three feet (1 m) in thickness and may become concreted to rock hardness by the deposition of organic and iron compounds. These leached soils can be extremely acidic, with a pH of 3 or below, and are very poor in plant nutrients. The acidic, organic materials, especially polyphenols derived from decaying vegetation, may be leached through the soil and end up in the drainage water in streams and rivers. The "blackwater" tributaries of the Amazon River owe their dark color to these compounds (see illustration [right]).

Among the commonest of tropical forest soils are the kaolisols, which are characterized by the deep red color of their profile, caused by the presence of oxides of iron and aluminum. They have been called by many different names in the past, such as *laterites* and tropical red soils, and their description and classification are still confused. They form by a process similar to podzolization. Clay minerals break down to produce silica, together with iron and aluminum oxides, but instead of the iron and aluminum being leached down the profile, silica is removed and the iron and aluminum is left behind. As in the case of podzols, organic matter accumulates on the surface, but in the case of kaolisols it is rapidly decomposed and incorporated into the upper layer of the soil. Litter is decomposed within a few weeks of falling on the soil and enters an organic-rich upper A horizon, where most of the tree roots are concentrated, as shown in the diagram on page 49. Like podzols, kaolisols are pedalfer soils, so the leaching process takes place in a downward direction, rapidly removing nutrients from the reach of roots. Tree roots therefore need to take up elements released by decomposition before they begin to move down the soil profile, hence they are most active in the surface layers. In some kaolisols the iron, manganese, and aluminum combine to form a concreted soil that restricts any root growth into the deeper soil.

The Negro River in the Amazon Basin as seen from Landsat. The forest is permeated by numerous, branching streams and tributaries of the river.   *(Large-Scale Biosphere-Atmosphere Scientific Conference)*

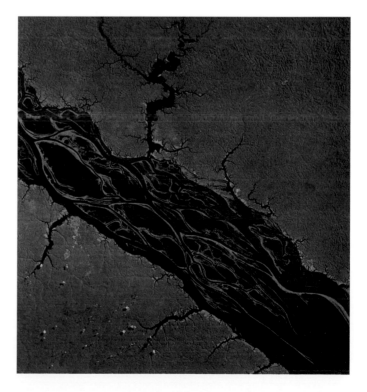

The Soil Profile of a Podzol

Soil scientists, or pedologists, who seek to understand the development of a soil over time usually dig a pit so that they can observe a cross section, or profile, of the soil. A soil profile reveals layers that have resulted from the operation of soil processes as the soil has matured, and one of the most important processes in temperate and tundra soils is leaching. Leaching occurs where water moves downward through the soil profile, dissolving some of the components in the upper soil layers and sometimes depositing these lower down. Downward leaching of this sort can only occur if the input of water to the site from the surface is greater than the evaporative forces that pull water back into the atmosphere, so it is a process found in both cool and hot parts of the world, as long as they have adequate precipitation. Leaching is assisted by acids dissolved in the descending water because these help dissolve elements held in the soil. Some of the acidity is derived from carbon dioxide in the atmosphere and in the soil (produced by respiring bacteria, plant roots and small animals). Other acids come from atmospheric pollution, including sulfuric and nitric acids. The soil organic matter also adds to the acidity of the soil water because dead and decomposing material liberates many acidic compounds, such as polyuronic acids and polyphenols. All of these chemical components contribute to the acidic nature of the water as it leaches the surface layers of soil.

In the case of the podzol or spodosol profile, which is found in many parts of the world, from the humid Tropics to the boreal regions of coniferous forests, the outcome of the leaching is a very distinctly layered soil profile, as shown in the diagram. The top layer of the soil profile consists of a layer of organic matter produced by the plants that occupy the site and is usually called the O horizon (denoting organic). Often this can be subdivided into L, F, and H layers, standing for *litter* layer (relatively intact plant fragments), fragmentation layer (broken and comminuted parts of plants), and humic layer (dark, compacted material in which few fragments are recognizable). In a mature podzol there is very little mixing between these layers and at their base is a distinct line of demarcation, separating the O horizon from the pale gray zone below in which organic matter is scarce. If a transition layer of mixed organic and mineral matter is present, it suggests that the podzol is still in the process of maturing. A lack of mixing indicates that few soil invertebrates, such as earthworms, are active in the

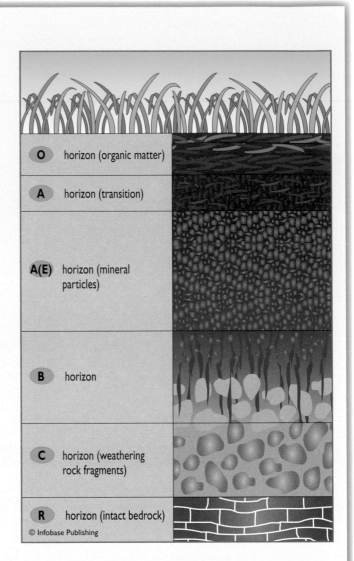

The soil profile of a podzol. A cross section of a soil can be divided into a series of layers, or horizons. The uppermost O horizon consists of decomposing organic matter derived mainly from the litter of plants on the surface. The A horizon may have some mixing between the organic matter and the mineral soil, but much of this layer (sometimes called the E horizon) has been stripped of organic matter, together with the iron and aluminum from degraded clay minerals. This process is called eluviation (hence E horizon) and leaves the soil in this layer bleached and pale. The B horizon is where much of the leached material is deposited, including an organic layer and a red-stained layer of iron oxides. The C horizon consists of rock fragments in the course of weathering and breakdown, and the R horizon is the intact bedrock.

soil profile, and the numbers of these organisms decline as the soil profile matures. Some detritus-eating insects, such as springtails and flies, are present in the O horizon, but they do not take the organic matter farther down the profile. A clear layer of organic matter that decays slowly and is not incorporated into the deeper soil is called mor humus.

Organic acids are produced in the O horizon, and these contribute to the low pH of the descending water, which dissolves many soluble materials out of the A horizon below. Clay minerals are broken down into oxides of iron, aluminum, and silica, and these breakdown products are available for leaching. The A horizon becomes bleached and relatively colorless as iron is leached from it. Many other elements, including calcium, potassium, and phosphorus, are also leached, as is aluminum. The acid waters also carry down any organic compounds present, leaving the A horizon devoid of all

but bleached silica particles. It is said to be an eluviated horizon, labeled by some soil scientists as the E horizon. In the B horizon, immediately below the A horizon and usually separated by a sharp transition, many of these leached materials are redeposited. Organic compounds may emerge from solution and create a thin dark line. Iron and aluminum are then deposited, forming a reddish-orange layer that can become hard and concreted, forming an iron pan. This iron-rich layer may be so tough that it prevents water moving downward and blocks drainage. Below this B horizon is the subsoil, consisting of fragmented and weathering bedrock in the C horizon. The nature of the C horizon depends on the original geology of the area, which is determined by the bedrock exposed in the lowermost R horizon.

This happens particularly in soils where there is fluctuation in seasonal water availability. Such soils are now termed *ferricretes,* but they have in the past been called laterites.

Kaolisols are widespread through the moist tropics. They are found in South America, Africa, and India, being most common in areas that are tectonically stable, away from plate boundaries where conditions are less stable, such as Central America, the Caribbean, and Southeast Asia. They are among the most intensively weathered and matured of

tropical soils. Waters draining from kaolisols may be turbid, but they do not have the tea-colored staining of the blackwaters draining from podzols.

Kaolisols that develop from basalt rocks produce soils that are relatively rich in mineral elements and these are the most suitable for agricultural use, but most of the kaolisols of the Tropics are very poor in nutrients, having developed from rocks rich in quartz. One of the most surprising features of the tropical forests that such a massive bulk of vegetation with such a high productivity can be sustained by such impoverished soils. Even soils developed in limestone regions, such as in Belize, are poor in most elements apart from calcium. They are usually rich in clays, but the underlying limestone is so porous that they are not as water retentive as one might expect, and these soils are often subject to drought.

In situations where drainage is poor and the soil becomes waterlogged, such as in valley bottoms and alluvial floodplains, the downward leaching process no longer takes place. Here the soil profile takes on a relatively uniform gray appearance, with only the surface accumulation of organic matter being apparent. The grayness is due to the lack of oxygen in the waterlogged conditions. As a consequence, any iron present in the soil is reduced from its Fe^{3+} form

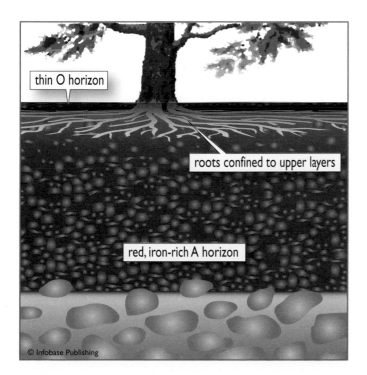

thin O horizon

roots confined to upper layers

red, iron-rich A horizon

© Infobase Publishing

The soil profile of a tropical red soil, or kaolisol. The O horizon consists of a thin layer of plant litter. Tree roots are confined to the uppermost layer of the A horizon of the soil. The red color of the soil is caused by the leaching of silica, leaving red iron oxides behind. These iron compounds can become hardened and concreted to form a ferricrete or laterite soil.

(ferric iron, which is a red–orange color) to its Fe²⁺ form (ferrous iron, a blue–gray color). Soils of this type are called *gleysols* and are frequent in flooded tropical habitats where water accumulates or moves very slowly through the soil profile. A close examination of the profile of a gleysol may reveal thin orange tubes leading through the uniform gray matrix. These are old root channels in which the original plant root tissue has died and decayed, leaving an open tube connecting to the air above. Air permeates these tubes and oxidizes the ferrous iron present, turning it from blue to the orange color of ferric iron.

Permanent waterlogging and the maintenance of a high water table in the soil can lead to the formation of increasingly thick layers of undecayed organic matter on the surface of a soil. If the organic O horizon is constantly saturated with water, the lack of oxygen affects the process of decomposition. Oxygen diffuses about 10,000 times more slowly when dissolved in water than it does when moving through air, so organisms that require oxygen for their respiration find it much more difficult to obtain adequate oxygen resources for their needs. Fungi and aerobic bacteria that feed on organic matter in the soil and assist in its decomposition are suppressed by the lack of available oxygen and organic materials decay much more slowly. As a consequence, the deposition of plant litter and detritus on the soil surface may occur more rapidly than the decomposer organisms can consume them, and *peat* accumulates. Once the peat layer is 16 inches (40 cm) deep, the landscape can be termed peat land, and the soil type is called a *histosol*. Soils of this type occur in some coastal regions of Southeast Asia, where the peat can attain a depth of 65 feet (20 m), forming elevated peat domes (see "Lowland Swamp and Bog Forests," pages 68–71).

At the other end of the moisture spectrum, deciduous tropical forests and monsoon forests merge into the savannah woodlands of climatically drier regions. The soils of the savannah woodlands are generally porous and have a poor water-retaining capacity. One consequence of this is that plant litter and organic matter are rapidly decomposed in the well-aerated conditions, further exacerbating the water-holding problem because organic matter in a soil is able to retain moisture. Fire occurs regularly, removing much of the dry organic matter present on the soil surface. Many savannah soils are ancient and have lost most of the nutrient elements originally released by rock weathering, so they are poor in phosphates. On the other hand, they are often rich in toxic aluminum compounds. The summer drought can result in upward movement of water in the profile, but any salts carried up in this way are usually washed away by the torrential rains of the wet season.

Biogeographers argue about the conditions that control the boundary between true forest and the savannah woodlands. Some support the idea that human use of fire limits the development of forest, because only small, fire-resistant trees can survive if an area is regularly burnt. Supporters of this theory point to the fact that the boundary between the two types of vegetation is often sharp; fire does not penetrate easily into the more humid forest, so once established the division persists through subsequent fires. Others, however, claim that the poverty of the soil is to blame and that the lack of plant nutrients, especially phosphorus, in the savannah soils prevents the growth of luxuriant vegetation. It is likely that each explanation is true for particular situations. In East Africa fire is used as a regular management tool in the savannah, and this practice may well create the sharp boundaries found there, but in the west of Africa, in the Congo Basin, fire is not frequent and the savannah development may simply be a consequence of the poor quality of soils. This latter type of savannah can be termed an edaphic savannah. The word edaphic means influenced by the soil and comes from the Greek word *edaphos,* the floor or ground.

An unusual type of savannah exists in South America, around the Orinoco Basin. These areas are called the *llanos,* and are subjected to seasonal flooding followed by intense drought. Heavy soils, rich in clays, are present, and the predominant vegetation is open savannah woodland. It is possible that the extremes of wetness and drought prevent tropical forests from spreading into this area, but the dry season is also accompanied by extensive fires, so the actual limiting factor is unclear.

■ SOIL CATENAS

There are thus many different types of soils present in the regions occupied by tropical forests. The variety of soils depends in part on the nature of the bedrock and the way in which it weathers, but the soil profile is also influenced by water movements and drainage patterns. Climate is important in determining how much water enters the soil, but the topography of the land is also significant as it affects the way in which water drains through the soil. Soils on ridges and slopes drain more effectively than soils in valleys, so even within a relatively small geographical region a range of different soil types can be found. The sequence of soils present along a topographical gradient is called a soil *catena.*

An example of a soil catena is shown in the diagram on page 51. Soils on ridges, especially under drier forest types, become highly concreted ferricretes, or laterites. If the soil is shallow at the head of the ridge, the bedrock may even project above the soil surface in outcrops. The shallow soils that develop in unstable areas of outcrop are immature and soil scientists give them the name skeletal soils or leptosols. More gentle slopes that are freely drained but moist are covered with the tropical red soils, kaolisols. Water draining through the soils of the slopes accumulates in the lower regions, and

Dry forest   Moist forest   Wet swamp forest

Ferricrete or laterite kaolisols

Kaolisols

Saturated gleysols

Peaty histosols

© Infobase Publishing

Soils vary along topographical gradients, forming a soil catena. Here soils on elevated areas and ridges are ferricretes or laterites, while soils on the gentle slopes are typical tropical red soils, or kaolisols. In the valley, where water accumulates and drainage is poor, gleysols are found. If the water supply is sufficient, surface peat deposits may form to create histosols.

the supply of water may be sufficient to keep the soil profile of the lower slopes permanently waterlogged. Under these conditions a groundwater gleysol forms. The gleysols have a water table within them that fluctuates during wet and dry periods, resulting in occasional aeration of the upper layers, so the activity of decomposers continues and organic matter decays. These soils are increasingly waterlogged in the lower part of the catena around the river floodplain, and as decomposition is increasingly impaired, peat formation on the soil surface increases. Peaty histosols develop in such valley sites.

A cross section of a valley in a rain forest region thus reveals a series of soil types in close proximity to one another. Just as slope, and hence drainage, varies gradually along the

profile, so is the change from one soil type to another. There are no sharp boundaries between different soil types unless there is a sudden change in the topography, such as a break in slope. This is one reason why classifying soil types is so difficult; they blend into one another and every possible intermediate is present along an environmental gradient.

Ecologists debate whether soil type determines vegetation, or whether the vegetation type determines the soil. There is no solution to this question because both statements contain an element of truth. Vegetation does have a strong influence on soil development through the penetration of roots and the supply of organic matter that supports the animal life of the soil. But soils also affect vegetation as they are the source of chemical materials for plant growth and their drainage characteristics affect root function and determine what types of plant can survive. Vegetation thus varies along topographic gradients in the same way as soils. It is often assumed that the productivity and diversity of forests are greatest where conditions are wet and decline as drought becomes more common. But very wet soils are only able to support trees with roots that are adapted to waterlogging, so it is possible to have too much of a good thing. Very wet conditions are often associated with lower productivity, simpler forest structure, and lower diversity.

■ CONCLUSIONS

Rocks and their decay products play an important part in the tropical forest ecosystem. They supply the mineral elements that plants need to grow and they provide a substrate where vegetation can establish. Igneous rocks are produced by cooling magma and be brought to the surface by volcanic activity, while sedimentary rocks result from the accumulation of sediments in the oceans, followed by long periods of compression. If these sedimentary rocks are heated by rising magma they become physically and chemically altered to form metamorphic rocks. All of these rock types are linked in a rock cycle in which formation is followed by uplift, then degradation and erosion leads to the transport of rock components back to the ocean, where the sedimentary cycle begins again.

The Earth's crust is not static, but is constantly on the move. Mid-ocean ridges of rising magma create new oceanic crust that is constantly being forced apart, pushing the continental plates away from one another. Where plates collide they may buckle to form mountain chains, or one plate may be pushed beneath the other to form a subduction zone. Subduction zones are particularly active and are often the source of earthquakes and volcano development. The separation and drifting of the continents over hundreds of millions of years has created the present pattern of distribution, and past links account for some of the similarities between the animal and plant life of the present-day tropical zone.

Tropical areas close to subduction zones, such as the South American Andes and the islands of Southeast Asia, are tectonically active, and the disturbance caused by volcanic eruptions, earthquakes, and tsunamis may have contributed to the diversity of plants and animals they now contain. Remote tropical islands have often been formed by oceanic volcanoes that have become fringed by coral reefs. They evolve through a process of uplift, degradation, and coral formation.

Rocks are gradually broken down when exposed to air and rainfall, a process called weathering. Physical factors, such as heat, chemical factors, such as acidic rainfall, and biological factors, such as plant roots, all play a part in rock weathering. Splitting and exfoliation of rock lead to the formation of increasingly small particles of mineral material, and the proportions of different particle sizes in the resultant soil influences the aeration and drainage properties of the soil. Tiny clay particles are particularly important in soils because the negative charges on their surfaces attract cations and hold them against the process of leaching, retaining them in a state where they are available for plant absorption.

Over the course of time, soils develop a characteristic pattern in their cross section, or profile. In predominantly wet climates the movement of water through the soil is downward, so leaching takes chemical elements through the soil and into drainage water. This type of soil is a pedalfer. When conditions are drier, water may move upward because of evaporation, leading to the development of a pedocal soil. Tropical forests develop mainly on pedalfers. In coarse, sandy soils iron and aluminum oxides are leached down the soil profile and deposited in lower horizons, forming a podzol. Even more frequently, especially in tropical areas, it is silica that moves down the profile, leaving iron oxides behind and giving the soil a red color. This soil profile is called a kaolisol. The iron may accumulate to produce a concreted mass, in which case the soil is called a ferricrete or laterite. In sites that receive excessive drainage water, the soil may become waterlogged and anaerobic, when a gleysol develops. This has a gray color because of the reduced state of the iron compounds. In even wetter conditions, the breakdown of surface litter may be inhibited and peat develops, forming a histosol.

Soils in the Tropics may vary considerably even over short distances, especially if the topography is varied. Soils on ridges will differ from those on slopes or in the valley floodplains. The gradual variation of soils along environmental gradients is called a soil catena. Soil catenas increase the habitat diversity in tropical forests.

Geology thus plays an important part in the ecology of the forest. The range of soils developed from different rock types in varied topography creates a physical background in which the living components of the ecosystem are found.

3

Types of Tropical Forest

Tropical forests are found in the equatorial regions of all the tropical land areas of the globe. They extend north and south of the equator, becoming limited in their extent only as a result of increasing drought as the tropics of Cancer and Capricorn are approached. Given their very wide geographical distribution, it is not surprising that tropical forests occur in a considerable variety of forms. They vary in the height of their canopies, their patterns of leaf fall, the architecture of the trees, and the species of animals and plants that occupy them. Lowland tropical forests occur in extensive floodplains of major rivers, such as the Amazon (see illustration below), and in coastal wetlands, while the upland types of forest grow on the slopes of tropical mountains. Tropical forests grade north and south into the woodlands and grasslands of the tropical savannas. In all of these situations, however, tropical forests have certain general characteristics.

■ CHARACTERISTICS OF TROPICAL FORESTS

Forests are dominated by trees that grow in such density that their crowns form an almost uninterrupted cover of leafy canopy. The closed nature of the forest canopy distinguishes this type of vegetation from woodland, in which tree density is lower and there are many gaps and openings between individual trees. Where the two vegetation types come into contact, as at the borders of the tropical savannas, they blend into one another, so the distinction is not always a clear and simple one. Even in dense forest some openings and gaps occur as a result of the death of old trees or because of damage from lightning strikes or wind damage. So the concept of a totally uninterrupted canopy cover is an ideal that is rarely fully achieved. To provide an arbitrary distinction between forest and the more open conditions of woodland, therefore, ecologists have settled upon the value of a 40 percent canopy cover as the maximum for woodland. If the tree canopies cover more than 40 percent of the ground, then the vegetation can be defined as forest rather than woodland.

The height of the forest canopy varies greatly between different regions and sites. All tree canopies lose water to the atmosphere by transpiration, and tall trees are faced with the problem of conducting water from the soil to their high leaf canopies (see sidebar "Trees, Water, and Climate," page 18). As a result, tall trees can only survive where water supplies are abundant. In areas subjected to seasonal drought or where soils are particularly well drained the canopy is lower. The form of the canopy is also very variable. Some tropical

This image of South America shows the extent of the tropical forests of the Amazon Basin and the Andes Mountains forming a western ridge, where many of the tributaries of the Amazon arise. *(Reto Stöckli and Robert Simmon, NASA's Earth Observatory)*

forest have a fairly uniform upper canopy, such as the forests of Trinidad, whose upper canopy is dominated by the tree *Mora excelsa* and forms a relatively flat surface at a height of about 140 feet (43 m). But more frequently the canopy is composed of a diverse range of tree species and is very uneven in nature. In Nigeria, for example, the rain forest has some trees that exceed a height of 150 feet (46 m), while the general forest canopy lies at a height of about 100 feet (31 m). Trees that penetrate to heights well above the general canopy of a forest (see illustration on page 55) are called *emergents*. These isolated tall trees rarely exceed 200 feet (60 m) in height, although there are records of emergents in the rain forests of Sarawak reaching a height of 275 feet (84 m).

Beneath the emergent trees of the equatorial rain forests there are usually at least two other layers of canopy, so that the tropical forest has a distinctive *stratified*, or layered architecture that ecologists call its structure (see sidebar [right] and illustration [below]).

Lowland rain forest in Uganda, showing the forest structure. *(Peter D. Moore)*

Forest Structure

Structure is the term used by ecologists and vegetation scientists to describe the architecture of vegetation. Just like a building, the plants that cover the ground are constructed in a distinctive manner. Plants compete with one another to gain access to the commodity that is vital for their growth—sunlight. By growing tall, therefore, an individual plant is at a distinct competitive advantage over its neighbors. But in the high latitudes (approaching the polar regions), extending above the general canopy brings its own hazards, particularly the prospect of frost damage. In tropical forests, however, there is no such limitation because frost does not occur, so tall growth is generally the most competitive option in the struggle for light. It may take a tree sapling many years to grow high enough to reach the canopy, and some may remain in the shade of their neighbors until a canopy tree eventually dies and falls, leaving a gap into which it can expand. Some trees do not have the genetic capacity to grow to great heights and may have adapted to cope with a life spent entirely in the shade. The tropical forest, therefore, has a complex architecture, with tree canopies layered at different heights above the ground, a feature first described in detail by the British tropical botanist Paul Richards (1908–95). Ecologists debate the reasons why tree canopies have become structured in this way, but it seems likely that light capture by the different tree species is a factor. Evidently the layered arrangement results in the most efficient system for catching all the available light.

Canopy structure is of great significance in ecology because it has a strong influence on microclimate. Pockets of shaded space exist between the canopy layers in which humidity is high and temperatures are relatively cool. Many animals have specific preferences for these conditions and spend their lives at particular levels within the canopy. The complex structure of the tropical forest is thus a major contributor to its diversity.

The tropical ecologist Paul Richards proposed that the three tree layers of the typical tropical rain should be given labels. The emergent trees of the A stratum form an irregular and discontinuous layer, often attaining a height of between 100 and 140 feet (30 to 42 m). The B stratum

© Infobase Publishing

Profile of an evergreen tropical rain forest. Occasional trees may reach heights of 150 feet (50 m) and emerge above the general canopy. Below them, trees often occupy distinct canopy layers, while the forest floor is relatively bare of vegetation.

generally forms a continuous canopy, its component trees being between 60 and 90 feet (18 to 27 m). The C stratum lies in the shade of the upper canopy layers at a height of between 26 and 46 feet (8 to 14 m). This scheme of layering in tropical forests is variable, and some types may have additional or fewer layers. Ecologists also argue about the validity of this layered model and suggest that the layers can merge into one another and cannot always be distinguished. Undoubtedly this can be the case, but many analyses of tropical forest profiles reveal the distinctive structure, as shown in the diagram above.

When ecologists wish to survey a forest, one of the first requirements is the construction of a profile diagram. Usually this involves running a sampling line, or *transect,* about 200 feet (61 m) long and 25 feet (8 m) wide through a typical section of the forest. The surveyor then records the positions and measures the heights of the upper and lower limits of the canopies of all of the trees within this belt, and displays them as a profile diagram similar to that shown in the figure above. This is not as easy to achieve as it sounds because measuring tree height in a dense forest is extremely difficult. Some studies have required the felling of strip of forest to achieve the required data, but this is not ideal because of the destruction involved. An alternative is to construct a tower within the forest that allows measurements to be taken and enables the researcher to observe other aspects of forest biodiversity in the different canopy layers. The profile diagram that emerges from such studies is thus a simple representation of the structure of the forest and provides a useful basis for the comparison of different types of tropical forest.

After the canopy structure, the next most noticeable characteristic of the forest is the nature of the leaves in the canopy. Some trees may bear photosynthetically active leaves throughout the year and are said to be *evergreen.* Others lose all their leaves at one particular time of year so that they are totally bare of leaves for awhile. Such trees are said to be *deciduous.* Tropical forests may consist of entirely evergreen or entirely deciduous trees, or they may contain a mixture of the two. The proportion of evergreen and deciduous trees in the canopy, together with the overall structure, is one of the most important characteristics used in the classification of different types of tropical forest.

■ TROPICAL EVERGREEN RAIN FOREST

The equatorial regions of the world experience very high levels of precipitation (see "Tropical Climates," pages 9–14). Land lying close to the equator is largely covered by evergreen rain forest, with the exception of high mountain ranges and regions cleared for agriculture. This forest type contains the tallest trees and has the most complex structure of all tropical forests. Many trees exceed 100 feet (30 m) in height and undisturbed regions exhibit a three-layered structure. Beneath the three canopy layers there may be additional layers of vegetation close to the ground, including shrubs and saplings awaiting their opportunity to join the upper layers, and a weakly developed herbaceous layer at ground level. Sometimes these layers are referred to as the *D stratum* and *E stratum*, respectively. Relatively little light penetrates to the ground, however, so the growth of vegetation at this level is rarely dense. The idea that the rain forest consists of an impenetrable jungle is quite false because so little herbage is present beneath the tall trees. It is probable that initial descriptions of the jungle were based on the tales of travelers making their way into the forest interior by river, and the forest edge along a riverbank often carries a dense growth of shrubs and climbers, giving the forest an impenetrable appearance. Or the early explorers may have come across forests that were recovering from earlier disturbance (see "Secondary Forests," pages 62–65), which are often denser in their undergrowth than undisturbed forest.

The microclimate (see "Microclimate," pages 25–33) of the various forest strata changes from the upper to the lower layers. The A stratum canopy of the emergent trees is exposed to high light intensity throughout the daylight hours. Skies in the equatorial zone are often cloudy and rainfall is frequent and heavy, but light in the daytime, even on a cloudy day, is never limiting to plant growth. Emergent trees experience the force of high winds and may be subjected to lightning strikes. The leafy branches of the emergent trees therefore undergo more environmental stress than other parts of the forest. They experience a degree of drought that does not occur in the lower layers, thus these emergents are the rain forest members most likely to become short of water. Animals living within their canopies and epiphytes growing on their branches are periodically exposed to high temperature and low humidity, which influences the composition of the flora and fauna supported by the emergents. The upper roof of the B stratum also receives high light because the emergent trees, with their interrupted canopies, offer limited shade, but there is a gradient of cooler and more humid conditions as one descends through the canopy of the B layer. In the C stratum direct light from Sun, sky, or cloud is increasingly scarce (see sidebar "Light, Vision, and Photosynthesis," pages 28–29). The temperature is cooler and less variable throughout the course of the day and night, which makes conditions more tolerable, especially for invertebrate animals. The maintained high humidity is also favorable to many invertebrates. Finally, on the forest floor, shade is almost permanent, temperature is relatively low and very uniform, and humidity is constantly high. Only the occasional high light intensity and warmth of a direct beam of sunlight, or *sunfleck,* interrupts the uniform conditions of the forest floor.

Some rain forest ecologists believe that the most significant boundary in the profile of the forest lies where the open upper layers of the canopy are replaced by the dense tangle of vegetation in the lower part of the canopy, generally within the B stratum. Light penetrates relatively freely through the upper zone, which they term the *euphotic zone,* while the density of leaves makes the lower layers far more poorly lit, called the *oligophotic zone.* These terms are borrowed from aquatic studies where they are used to describe the degree of light penetration through increasing depths of water; similarly, they convey a drop in light values as one descends through the forest canopy to the deep shade of the forest floor.

The density of trees in the forest is extremely variable, depending on such factors as past disturbance (including natural disturbance by wind, fire, and flood). Usually there are between 740 and 1,700 trees with a trunk diameter of greater than four inches (10 cm) per acre (300–700 trees h^{-1}). This means that each tree occupies a space of between 10 and 20 square yards (8–16 m^2), so there is plenty of walking space between the trees. The defining of trees, saplings, and shrubs is itself a problem, which foresters address by measuring the diameter of a trunk at a height of 4.25 feet (1.3 m) above the ground; this is termed the *diameter at breast height* (DBH). The circumference at that height is known as the *girth.* This value is easily measured and does not involve the felling of the tree. When trees are felled for forestry, however, their height can be measured accurately and it usually relates closely to the DBH. When foresters have established the mathematics of this relationship for each tree species, it enables them to calculate the height of trees and hence their timber volume in a nondestructive way.

Trees in the rain forest are characteristically tall but with relatively narrow trunks. Even the taller trees may have a DBH of only one foot (0.3 m), although very large trees may attain a DBH of 3.3 feet (1 m) or more. Emergent trees often have straight, unbranched stems and an umbrella-shaped canopy, while trees of the lower layers are more cylindrical in their canopy structure. Relatively little is known about the age of rain forest trees because their trunk wood does not contain the annual growth rings found in temperate forest trees. There is no significant change in the growth of the tree throughout the year, so no record is left in the wood to indicate the age. Some measurements have been taken using such techniques as radiocarbon dating, but these have usually been conducted on larger trees considered ancient and thus only provide an indication of maximum ages for forest trees. Ages of 800 to 1,000 years are rare, while many of the larger trees reach ages of 200 to 600 years. One rain forest tree in Malaysia, *Shorea curtisii*, has been recorded with an age of 1,400 years, but this is exceptional. Trees of the rain forest cannot compete with those of the arid lands in terms of longevity, for the bristlecone pine (*Pinus aristata*) of the southwestern deserts of the United States is known to live for over 4,000 years. It seems likely that pioneer trees that first colonize gaps in the forest are relatively short-lived and are eventually replaced by slower-growing trees that live much longer.

As the name of this forest type implies, most of the trees are evergreen. This does not mean that the leaves last forever; they usually survive for only two or three years. A botanist in Costa Rica marked leaves so that their individual history could be recorded and he found that after two years just 40 percent of the original leaves were still present. But this probably varies from one tree species to another. Leaves are shed individually, however, so that the tree always has a green cover of foliage, whatever the time of year. It may seem surprising that a growth habit associated with the temperate coniferous forests of the higher latitudes should also prove successful in the warmth and wetness of the Tropics, but evidently the evergreen leaf has proved the most efficient organ for photosynthesis in the equatorial regions (see sidebar [right]).

The leaves of the many species of rain forest trees are remarkably similar in shape, making tree identification extremely difficult for tropical botanists. Most tree leaves in the Tropics are simple, undivided, untoothed, elliptical structures, often extending into a fine, bent tip, or drip-tip (see "The Tropical Leaf," pages 118–119). The leaves of trees in the lower canopy layers are often larger and thinner than their counterparts in the upper layers, which is an adaptation for light trapping in an increasingly shady environment. Unlike trees of the temperate zone, tropical rain forest trees have no dormant buds. Their growth is continuous throughout the year, so there is no need for protective bud scales.

Palms are often conspicuous members of the lower layers of the rain forest (see illustration on page 58). Palms have large leaves that are particularly long-lived; individual leaves often last five or more years. Herbs are infrequent, providing only a very small proportion of the total flora, and are found as scattered individuals on the forest floor. Some grasses and sedges are found in the rain forest, but these often have much broader leaves than their equivalents in the temperate zone, a response to the low light levels. Ferns occur mainly in particularly shady and damp loca-

Why Be an Evergreen?

The obvious advantage of being evergreen in the equatorial Tropics is the fact that warmth and wetness are available all year round, so a tree can continue to photosynthesize at all times of the year if it is clothed with leaves. Although leaves are the main means by which the tree obtains its energy, they are also the main source of water loss, so maintaining a leafy canopy is costly in terms of water use. If water is abundant, however, as is the case in most equatorial regions, a tree can afford to be profligate in its use of water, as it is assured of new supplies from the soil to maintain its cell turgor. But this does not explain why evergreen leaves usually last longer than deciduous ones. A tree could be an evergreen and yet have short-lived leaves. It is more economical for the plant to maintain a leaf for a longer period of time if other factors (such as drought or cold) do not place a strain on the plant and result in leaf loss. Constructing a leaf takes energy, but this is probably not a problem for tropical trees because solar energy is in rich supply. Leaf building also requires certain elements, such as nitrogen (for proteins) and phosphorus (for membranes and energy transfer), and these are in short supply in the rain forests (see "Soil Formation and Maturation," pages 46–50). Because soils in the Tropics are often poor in these elements, the plant cannot afford to shed the leaves that contain them. The construction of a leaf is therefore expensive in terms of nutrients as well as energy, and once a leaf has been constructed it is strategically sensible to retain it as long as possible.

An additional reason for long-lived evergreen leaves being advantageous to a plant is related to defense. In an environment where a rich array of animals is assembled to eat all things palatable, the plant can best defend itself by making its leaves toxic. Many distasteful compounds are constructed in the leaves of rain forest trees, which serve the purpose of deterring grazers, but these compounds take time to build up. Young leaves are relatively poorly defended, but the longer a leaf lasts, the better its chance of developing an assortment of unpleasant chemicals to escape the attentions of herbivores. Nutrient economy and defense thus encourage a plant to adopt an evergreen strategy despite the heavy demand for water that results from being an evergreen.

Palms are almost entirely tropical and subtropical in their global distribution. They are particularly well adapted to relatively well-lit, open conditions in forest clearings, but are also found in shaded locations beneath the canopy. *(Amanda Rohde)*

tions, such as along the edges of streams in the forest. Life for a photosynthetic plant on the forest floor, where light levels may be only 2 or 3 percent of that above the canopy, is extremely difficult. Some of these plants have developed effective mechanisms for overcoming the problem. *Lianes,* or climbers, germinate on the ground, but rapidly make their way into the upper layers of the canopy, using the trunks and branches of trees for support (see "The Fight for Life: Upwardly Mobile Plants," pages 125–127). Other plants germinate and establish themselves in crevices and along the branches of tall trees, using their structure for support but causing the host tree no damage beyond the additional weight load. These plants are called *epiphytes* (see sidebar "Epiphytes," page 121). Both of these life forms are found within the equatorial lowland rain forests but become even more abundant in the montane and cloud forests of the highland regions of the Tropics.

The layered structure of the rain forest means that the light-trapping leaves of the canopy must overlap one another extensively. Ecologists express the degree of overlap by measuring the *leaf area index* (see sidebar on page 59).

Measuring the leaf area index of a rain forest is a very difficult process. In one experiment in the rain forest of southwestern Cambodia, ecologists felled a sample of 100 trees and measured their leaf area. Dividing the total leaf area by the area of ground occupied by the sample of trees, they came to the conclusion that the leaf area index was 7.4. They were able to divide this figure into layers. The A stratum accounted for 2.0, B and C strata for 3.8, and the D and E strata for 1.6. Obviously, this is just one set of measurements, but it shows that successive leaf overlapping takes place within the forest, taking advantage of almost all the light that penetrates to the ground.

The rain forest is not uniform in its structure and composition. Variations in topography, such as wet hollows, stream courses, hills, and ridges, lead to differences in tree composition. There are also small gaps where old trees have fallen and the trees of the lower layers are presented with new opportunities to grow up into the higher layers. Light penetrates more effectively into these gaps. Consequently the vegetation close to the ground is often denser and a new

Leaf Area Index

The leaf area index (LAI) is a measure of how many times a given area of ground is covered by plant leaf tissue. It is obtained from the simple expression:

Total leaf area of plants / Ground area

Thus, if every square inch of ground supports plant life with a total leaf area of one square inch, the LAI is 1.0. In a habitat where there is much open ground visible, such as in a desert or on a sand dune, the LAI is less than unity, but in most natural habitats it is greater than this. Even in a closely mown lawn or in grazed grassland the LAI is often about 3.0. The LAI can be measured by passing a thin needle vertically down through the grass canopy and measuring how often it strikes a leaf. The measurement needs to be repeated many times to obtain an average LAI value for the vegetation. Woodland may have an LAI of 6 or more, and rain forests may reach about 10, although it is very difficult to measure leaf area indices in complex ecosystems like forests. This is best achieved by felling trees and taking detailed measurements of total leaf areas. Photographic techniques have been used; although they give a good indication of overall leaf cover, they do not adequately display leaf overlap.

The LAI gives some indication of the productive potential of an ecosystem as well as supplying a very simple index of the complexity of its structure. An ecosystem with a high LAI is likely to have a greater capacity for light capture and photosynthesis.

growth of climbers often begins at this stage. In a relatively pristine forest, undisturbed by human clearance, gaps of this sort usually cover about 5 percent of the forest area. Undisturbed forest, however, is becoming rarer and patches of forest that have been cleared for human agriculture or forestry and then abandoned are increasingly frequent. The vegetation of this type of clearing is called *secondary forest,* in contrast to the *primary forest* of the pristine regions (see "Secondary Forests," pages 62–65). Natural gaps in the primary forest are normally much smaller than areas of secondary forest regeneration.

Rivers passing through the rain forest create larger, linear openings that allow the penetration of light. As a consequence the layered structure of the forest is less apparent at the river's edge because climbers and smaller trees flourish in the additional light and fill all gaps with their foliage.

Various types of palms, including the rattans, or climbing palms, are characteristic of this type of forest, as is the Swiss cheese plant (*Monstera deliciosa*), well known as a popular house plant, that clings to the trunks of trees with its adhesive roots and uses them to support its growth into the light of the upper layers. The river's edge thus presents a solid wall of vegetation to any one approaching the forest from the water.

Evergreen tropical rain forests thus have some general features in common throughout the world, in South America, Africa, and Southeast Asia. They are dominated by evergreen trees arranged in a series of layers, but they also exhibit a wide diversity of forms depending on topography and the formation of natural gaps of regenerating vegetation.

■ TROPICAL SEASONAL FORESTS

The equatorial regions experience little seasonal variation in climate. With increasing distance from the equator, however, both north and south, annual changes in seasonal conditions become more pronounced. The temperature still varies only a little seasonally as a consequence of changes in the angle of the midday Sun, but a shift in the position of the intertropical convergence zone (ITCZ) (see sidebar "The Intertropical Convergence Zone," page 8) results in changing patterns of precipitation. Rainfall is highest at the time of year when the region falls within the ITCZ (the "summer") and diminishes when the ITCZ has moved into the other hemisphere (the "winter"). This can be seen in the climate diagrams from India on page 21. With increasing distance from the equator, in other words, there is an increasingly prolonged dry season.

The effect of this seasonal variation in climate on the forest is most apparent in the decrease in the proportion of evergreen to deciduous tree species (see illustration). A deciduous tree, as has been explained, is one that loses all its foliage at a particular time of year, leaving it bare of any leaves. In the Tropics this loss of leaves takes place in the dry season, so clearly a deciduous leaf has certain advantages for a tree under these conditions (see sidebar on page 61).

The semi-evergreen (alternatively called semi-deciduous) forests of the Tropics thus contain a variable mixture of evergreen and deciduous tree species. Moving away from the equator, the deciduous habit first makes an appearance in the emergent and the tallest trees. As has been discussed, it is these trees that are subjected to the greatest water stress in the rain forest, and the increasing degree of drought away from the equator allows deciduous trees to take over this position in the forest canopy. An example is *Cavanillesia platanifolia,* an extremely tall emergent tree found in the forests of South and Central America, from Panama south to Peru. Once a year it loses all of its leaves and then becomes cov-

The forests of the Siwalik Hills in northern India receive rain mainly during the summer monsoon season. The forest is consequently semi-evergreen, as many of the trees lose their leaves throughout the dry season.   *(Peter D. Moore)*

ered in small red flowers, followed by the fruits. This pattern suggests that the dry season can act as a timing mechanism to instigate flowering in the tree as well as causing leaf fall (see "A Time to Flower," pages 132–133).

In regions with increasingly prolonged or severe drought seasons, the proportion of deciduous trees in the canopy increases. Under particularly severe stress, when the dry season lasts for five months or more with only four inches (10 cm) or less rain falling, the forest may become entirely deciduous, in which case it is called tropical deciduous seasonal forest, rather than semi-evergreen (see illustration below). A high proportion of deciduous trees is typical of those tropical areas with a monsoon climate (see "Monsoon," pages 15–16), such as the north of India. In part, the proportion of deciduous and evergreen species depends on the amount of rainfall in the wet season as well

The mountains of the Himalaya foothills in northern India bear forest that is largely deciduous, receiving rain only during the summer monsoon season.   *(Peter D. Moore)*

Why Be Deciduous?

A tree conducts most of its photosynthesis in its leaves, so the leaf is the prime source of energy. Maintaining a large leaf area thus makes energetic sense to the plant and gives it advantages over competitors for the available light. But the pores (stomata) through which carbon dioxide is taken into the leaf also serve as escape routes for water vapor, so a photosynthetically active leaf that has its pores wide open is also a source of considerable water loss to the plant. In the wet rain forests this is not a major problem for most trees because water is abundant, but in regions with a dry season trees may become stressed because of lack of *groundwater* to make up for the canopy water deficit. An evergreen tree can respond by closing its stomata and reducing water loss. However, this also means that the rate of gaseous exchange and thus photosynthesis is reduced, which defeats the object of maintaining a leaf canopy. Some evergreens have evolved with leaves capable of reducing water loss by developing thick, leathery structures with few stomata on the upper surface and a layer of waxy cuticle to reduce evaporation. This type of leaf is called a sclerophyll and is found in some tropical forest trees; it is more frequent, however, in the thorn scrub and savanna woodlands of the drier regions.

The alternative evolutionary strategy is to lose the leaf canopy during the dry season and become deciduous. Such trees effectively shut down their photosynthetic activity during the drought period and become dormant. This pattern is familiar in some temperate forests, where the deciduous habit is adopted because of water unavailability as a result of winter cold, but in the Tropics it is a response to low rainfall causing drought. There are disadvantages associated with being deciduous, however. A lack of leaves during the dry season means that the tree is unable to take advantage of any temporary respite in the drought, which is possible for an evergreen. Also, new leaves have to be constructed each wet season. This process is expensive in terms of energy investment and the nutrient elements, such as nitrogen and phosphorus, which must be absorbed by the roots from the soil. The deciduous habit thus becomes a more efficient option where soils are relatively rich in these elements. The delay involved in constructing new leaves means that the tree may not be able to take advantage of the early part of the wet season for its photosynthesis, which is not the case for the evergreen. The young leaves may also be subjected to excessive grazing by herbivores because it takes them awhile to assemble all the chemical defenses found in older leaves.

With all of these disadvantages it is surprising that the deciduous leaf exists at all. But the fact that it has evolved as a strategy in many tropical trees provides evidence that it is an effective mechanism for drought survival and evidently has competitive advantages over the evergreen leaf in certain circumstances. The deciduous habit may have one further important advantage: Invertebrate herbivores, such as caterpillars, suddenly find themselves deprived of a food source when the leaves wither and fall. When a tree drops all of its leaves at the same time, it is difficult for insects to build their populations to such an extent that they present a serious problem to the tree.

---

as the length of the dry season. Heavy monsoon rainfall can partially compensate for a long dry season.

The structure of the semi-evergreen and deciduous seasonal forest is generally simpler than that of the evergreen rain forest, as shown in the diagram on page 62. The height of the tallest trees is usually lower and emergents are less common. There are normally two main layers of tree canopy, although the better light penetration often leads to a more extensive development of shrubs and herbs at ground level. The diversity of tree species is also lower in the seasonal forests than in the equatorial rain forests, but lianes and epiphytes are still frequent within the canopy. The visual appearance of the forest changes markedly with season because leaf fall occurs synchronously among the deciduous tree species, leaving the canopy at least partially bare. Trees that lose their leaves also need to protect their buds from drought. Buds contain the delicate initials of shoots and leaves that will grow once again in the coming wet season, but they must avoid desiccation. Bud scales are specialized structures that wrap around the developing shoots and leaves and prevent excessive water loss, and they are found in tropical deciduous trees.

Under increasingly severe drought, deciduous seasonal forest is replaced by savanna woodlands or by thorn scrub. Forest can maintain itself under relatively dry conditions, where the annual rainfall is as low as 12 inches (30 cm), but the boundary between the two biomes is often determined by human land use, especially the application of fire as a management tool. Regular firing of scrub and savanna can result in a relatively sharp boundary between the savanna and the forest, for the humidity that prevails in true forest

Profile of a deciduous monsoon tropical forest. Emergent trees are rare and the main canopy is occupied by tall layers of deciduous trees. There is a prominent layer of small shrubs and trees near the ground.

makes it difficult for grass and scrub fires to penetrate. The dry savannas and scrub often exhibit a higher proportion of evergreen species once again, but the evergreen species in this case tend to have small, leathery, sclerophyllous leaves, often accompanied by thorns and other deterrents to mammalian grazers. The relative advantage of evergreen and deciduous leaves is clearly a complex subject in which climate is just one factor to be considered.

■ SECONDARY FORESTS

Forest that has developed free from disturbance (primary forest) is considered to have attained a state of equilibrium with its environment, called the *climax* state. But many ecologists question the whole idea of uninterrupted and undisturbed vegetation, even in the remote regions of the rain forest. Individual trees become old and eventually die, leaving gaps in the canopy that are quickly filled by the growth of younger trees, which may have remained in the lower layers for many decades. But these small gaps and regenerations can be regarded as part of a larger-scale equilibrium. Considering entire watersheds or landscapes, such gaps are part of a relatively uniform whole, so at this larger scale, the forest can be regarded as being in an overall equilibrium, climax state. Indeed, the small gaps that form part of the constant cycle of replacement within mature forest contribute to its biodiversity. Insects are often more abundant within the gaps, and some birds and large mammals prefer these gap habitats, including the Sumatran rhino, which actively seeks them out.

There are many other factors, however, which operate upon the forest and can cause more widespread opening of the canopy. Tropical storms with accompanying winds of over 75 miles per hour (120 km/h^{-1}) can wreak devastation in a forest. Although rare over the continent of Africa, they are known in Madagascar, and are more common in the Bay of Bengal to the east of India and in the Caribbean (see "Tropical Storms," pages 14–15). Hurricane-force winds can cause great damage to emergent trees and may flatten entire forest canopies. Electrical storms with lightning strikes may result in fire within the forest, and tropical regions with high volcanic activity suffer a particularly high fire risk (see "Tropical Volcanoes and Islands," pages 41–43). Although the rain forest is relatively protected against the incidence of fire as a consequence of its high internal humidity, fires are not infrequent and can be extensive. Particularly devastating fires have occurred in recent decades in the tropical forests of Southeast Asia, many of them associated with human activity. The high precipitation associated with equatorial regions may cause excessive runoff water, resulting in slope erosion and landslips. This is especially common where human harvesting of forests on slopes has already made them unstable. Flooding in the forest lowlands is a regular feature of some tropical forests, such as those of the Amazon Basin (see "Tropical Floods," pages 16–19). The resulting changes to the courses of rivers and tributaries lead to constant disturbance of the forest, eroding some areas and depositing new banks of mud and detritus in others. Floodplain forests are thus constantly changing.

Perhaps the greatest disturbance in the tropical forests, however, is due to human influence. Indigenous forest peoples often clear areas for dwellings and local agriculture and then abandon the site as they move on to new locations. This is known as *slash and burn* agriculture, and it leads to patchiness in the forest with areas that have been abandoned for various lengths of time. When the areas involved in the clearance are small, supporting just one or two families, the gaps they open are not much larger than the natural gaps of regeneration within the forest. These can contribute to the diversity of habitats and therefore to the overall biodiversity. More extensive forestry and agriculture has led to much larger areas of forest being cleared, and sometimes these are also abandoned after a crop has been harvested, leaving the forest to regenerate. Large areas of clearing cause much greater change in the local microclimate and soils, making reinvasion by forest species often more difficult, so the secondary forest resulting is usually poorer in species.

All of these different types of disturbance, followed by abandonment, lead to the development of new forest as the process of *succession* takes place (see sidebar below).

Succession

When an area of land becomes accessible to colonization by plants, a process begins that ends in the development of vegetation and a fully functional ecosystem (see "The Concept of the Ecosystem," pages 76–78). Sea shores where sand or mud accumulates provide good examples of situations where new ecosystems, such as tropical mangrove forest, can develop, as do volcanic islands forced up from the seabed by tectonic activity. Previously uninhabited surfaces are invaded by plants, animals, and microbes, and the environment becomes altered by their presence. The growth of a plant changes the microclimate, providing shelter and food for invertebrate animals and leaving dead organic matter that contributes to the soil and supplies energy to microbial decomposers. By changing their surroundings, pioneer plants inadvertently provide new conditions that allow the invasion and survival of a wider range of other plants, so a community of different species develops over the course of time. The development of an ecosystem in this way is called succession.

The course of ecosystem development in any particular region is fairly predictable in its outcome, depending on the overall climate. The precise type of ecosystem that eventually forms may be determined in part by additional factors, such as soil type, aspect, fire frequency, or grazing intensity. Of these factors, the last two are often the result of human management of the environment. When the community attains a steady state, or equilibrium, it is said to have reached its climax.

When an ecosystem develops on a completely new and unvegetated substrate, such as volcanic lava flow, or is derived from the gradual infilling of a lake or an estuary, it is called a primary succession. When the succession takes place following a catastrophe, such as a hurricane or a fire, it is said to be a secondary succession. In the case of secondary succession, some plants, animals, and microbes are already present in the soil or as survivors from the catastrophe that devastated the original community. The starting point of a secondary succession is thus different from that of a primary succession because flora and fauna are already present, perhaps as seeds or insect pupae buried in the soil, and the process is therefore not entirely dependent on invasion from outside. The course of the succession is then partly determined by the initial floral and faunal composition. At the commencement of a secondary succession, the soil itself is also relatively mature, in the sense that it contains organic matter and has already been weathered and leached over the course of time. Tropical forest that has developed following catastrophe by the process of secondary succession is called secondary forest.

A small clearing with tree ferns in montane rain forest on the slopes of Mount Elgon in Uganda, East Africa. *(Peter D. Moore)*

The vegetation resulting from secondary succession is called *secondary forest* to distinguish it from the primeval, primary forest it has replaced (see illustration above), and it differs from primary forest in a number of ways. Abandoned clearings have a very different microclimate from primary forest. Much greater light penetration takes place, in the early stages even lighting up the surface of the soil. The daytime temperature close to the ground is therefore much higher, and the nighttime temperature can be much lower than that previously experienced when the forest canopy was intact. Humidity is lowered, as a result of higher water loss from the soil, higher temperatures, and the penetration of wind to ground level, increasing air movements and evaporation.

Many of the plants and animals that lived close to the ground in primary forest are unable to survive in the new conditions, so the organisms that colonize these new situations are usually pioneer species that were not found in the primary forest. In the forests of Africa, *Musanga* is among the most frequent of the pioneer trees, in Asia it is *Macarnaga* that plays this part, while in Central and South America

Cecropia is the characteristic pioneer. But the growth of these pioneer trees creates a microclimate in which more shade-dwelling species can survive, so the succession usually involves the gradual invasion of forest species, eventually including those of the primeval forest. The pioneers then become shaded, are unable to compete with the new arrivals, and are gradually lost. Some species of primary forest plants and animals, however, find it difficult to invade secondary forest. These may have very specific environmental requirements not commonly found in regenerating forest, or they may have poor properties of dispersal so that they cannot move into the developing forest. As a result, secondary forest may take a very long time or may never succeed in returning to the original condition of the primary forest. This is why ecologists and conservationists are so concerned about the loss of primary forest; it can never be fully replaced.

Secondary forest has a different species composition from that of primary forest. Most secondary forest species have a more extensive geographical range than that of primary forest species, so the secondary forest has fewer rare and localized species. Often one of these pioneer species will completely dominate the forest canopy in the early stages of forest development. Invasive weeds are often found in secondary forest, some of which have been carried around the world by people, such as *Lantana camara,* a shrub of the New World tropics that is now found throughout the tropical world. It is said to be pantropical. Other species, however, may have existed in gaps before the catastrophe that initiated secondary succession in the forest, and become very successful when a natural or human-induced disaster opens up wider areas.

The deep-shade species of primary forest and the rare species that were confined to restricted geographical regions (*endemics*) are usually absent from secondary forest, which is consequently generally lower in diversity than primary forest. In the early stages of development the structure of secondary forest is also very different from the forest it replaced, having a less clearly layered profile, with fewer emergent trees. The density of tree trunks is often greater in the early stages of succession, and the shrub and herbaceous layers are crowded with species, making the secondary forest much more impenetrable than that of the undisturbed areas. Lianes are also more abundant, taking advantage of the additional light to form a tangled mass in the lower canopy. Most of these lianes are also wide-ranging in their distribution, including *Ipomoea* species, relatives of the morning glory.

The density of human populations in the tropical forest is steadily increasing, and the consequent clearance of formerly untouched areas means that primary forest is becoming rarer and more fragmented. Most of the tropical forest regions that are relatively accessible by road now bear secondary forest. The precise outcome of the development of secondary forest in terms of its structure and species composition depends strongly on factors such as aspect and soil.

The continued influence of people can also play a large part in determining the direction taken by the vegetation succession. Where grazing and fuel wood gathering take place, the forest may remain relatively open, and grasses, such as *Imperata* species, or invasive ferns, such as the bracken (*Pteridium aquilinum*), assume dominance.

It is inevitable that in time secondary forest will expand as the remaining fragments of primary forest become increasingly disturbed by human activities.

■ MONTANE FORESTS

Tropical forests are not confined to the lowlands of the Tropics but are also an important feature of the more mountainous areas, where they can be termed montane forest. In some regions, such as Jamaica, virtually all of the lowland tropical forest has been destroyed, so all of the remaining forest is now confined to the mountains. The same is true of East Africa. The structure and composition of montane forest vary with altitude because of the alterations in climate resulting from greater elevation (see

"Tropical Mountains," pages 22–25). Whereas distance from the equator is associated in general with the occurrence of a dry season, altitude brings a gradual reduction in overall temperature and often an increasing exposure to wind. These factors affect the structure of the forest and lead to an altitudinal zonation (see sidebar) of forest types, as illustrated in the diagram on page 24.

In the equatorial rain forest regions of the world, evergreen forest covers the lower slopes of mountains, but this submontane forest differs from the lowland evergreen forest in its lower overall canopy height, simpler structure, and less diverse nature, as shown in the diagram below. Many of the trees associated with lowland rain forest are lost with increasing altitude. In Southeast Asia, for example, trees belonging to the large and important family Dipterocarpaceae are a major constituent of the lowland rain forest, but these

Profile of a montane tropical forest. The architecture of the forest is simpler than in evergreen rain forest or monsoon forest, having only about two layers of main canopy and an understory of small trees or palms.

© Infobase Publishing

Zonation

Plants and animals are rarely dispersed at random in nature, rather, often follow some form of pattern. Sometimes this pattern takes the form of an arrangement of parallel or concentric bands, called *zonation*. The vegetation around the edge of a pool is often zoned in this way, resulting from the decreasing wetness from the open water to the dry land that surrounds the pool. Forest edges may also display a series of zones as light intensity diminishes from the open land to the dense shade of the forest interior. Sometimes zonation results from succession (see sidebar "Succession," page 63), as in the case of the pool, where each band of vegetation, from open-water floating leaves, through emergent aquatics and reeds, to swamp forest, replaces the preceding one by encouraging sedimentation and raising the soil level ever higher in relation to the water level. Here zonation reflects a time sequence as succession proceeds. But this is not always the case. At the forest edge, the zonation is caused by an environmental gradient in the form of decreasing light intensity and its associated microclimatic effects. So, zonation may result from a gradient in an environmental factor that is not related to changes in time. This is the case with altitudinal gradients and their effect on the zonation of vegetation in mountainous regions. Temperature decreases with altitude (see sidebar "Lapse Rate," page 10), and this limits the upward extent of tropical tree species.

epiphytic ferns in a sample plot just 200 feet by 25 feet (61 m by 7.6 m) at an altitude of 5,600 feet (1,710 m). This compared with just 11 epiphytic ferns in a sample of the same size in the lowland forest at an altitude of 1,250 feet (380 m). Mosses, liverworts, and lichens were also more abundant in the higher site. Evidently the cooler (average annual temperature was about 11°F [6°C] lower at the upper site), moister conditions of the higher-altitude forest allowed far more fern and moss species to survive in the forest canopy as epiphytes.

In Africa the tropical mountains lie mainly in regions where human clearance of forest has been extensive, so it is difficult to make comparisons here of lowland and montane forests. In some areas, savanna woodlands and grasslands lie adjacent to the montane forest edge, while in others banana plantations extend into the foothills of mountains. The eastern Congo Basin has the least disturbed transition between lowland and montane forest, where the transition takes place between 3,600 and 5,400 feet (1,100 to 1,650 m). The transition zone has a mixture of lowland and upland tree species but is lower in stature and diversity than the lowland forest (see illustration on page 67). It is mainly evergreen and has a compact canopy with few lianes. Tree ferns are a feature of the montane forest here as in many other parts of the world. As in South American montane forests, epiphytes become increasingly abundant at higher altitudes.

At the upper edge of forest growth in the Tropics there is often a zone of smaller trees, sometimes called elfin forest, which may be subjected to constant mist and cloud, when it can be termed cloud forest. This zone is subalpine and forms a transition between the montane forest and the open conditions of the tropical tundra of very high mountains. At very high altitude in the Tropics the daily range of temperature is often greater than the mean seasonal range. Frost is even possible during the night, so conditions at the upper limit of forest growth are much more severe than in the lowlands, and these conditions determine the growth form of the trees. The height of the trees is much reduced, rarely exceeding 25 feet (8 m), and is sometimes less than 10 feet (3 m) in height. The canopy is often broken and open, sometimes because of the steep and rocky nature of the terrain. The forest composition is completely different from that of the lowland forest and the leaves of the trees are generally much smaller. Some tree ferns and palms may occur in the elfin forest, as in the Caribbean mountains. Epiphytic plants, including orchids, bromeliads, and mosses, are abundant, but generally less diverse than in the montane forest.

On the African tropical mountains, bamboo forests occupy the misty upper regions of the montane forest zone. Bamboo is actually a tall, woody grass, but on the African mountains it forms a dense, almost impenetrable forest, eventually succeeded at higher altitudes by subalpine heath, dominated by members of the heather family (Ericaceae), especially the tree heather (*Erica arborea*). An unusual and

become rarer with height above sea level and are replaced by trees of the beech and oak family, Fagaceae. In New Guinea, the hoop pine (*Araucaria cunninghamii*), a conifer related to the monkey puzzle tree, is one of the largest and most conspicuous members of the lower montane forests, gradually giving way to forests dominated by the southern beech (*Nothofagus* species) with increasing altitude.

Large lianes are generally less common in the higher forests, but epiphytes (plants that grow entirely upon the support structures of other plants), especially ferns, mosses, and liverworts, are often more abundant than in the lowland forest. Overall cooler temperature reduces the rate of evaporation and favors these easily desiccated plants. Specialized flowering plant epiphytes, including orchids and bromeliads (see sidebar "Epiphytes," page 121), are also abundant in the montane forest. In one study comparing montane with lowland rain forest in Ecuador, South America, there were 58 species of

Montane rain forest on the upper slopes of Mount Elgon in East Africa. *(Peter D. Moore)*

distinctive type of woodland sometimes forms the upper forest limit in the East African mountains and extends into the lower parts of the open tundra zone. This is formed by plants that have the appearance of trees but are actually giant herbs, such as giant senecios (related to common groundsel, *Senecio vulgaris*) and giant lobelias. No tropical trees are able to cope with the very cold conditions experienced on the high equatorial mountains, where night temperatures can often fall to 23°F (-5°C), so herbaceous plants have evolved a tree habit with stout, dense stems that take the form of trunks. Compressed stems of this type are called *pachycaul*, and this growth form is widely distributed in the

tropical mountains of Africa and South America, having evolved in a variety of plant groups. The tightly packed leaf bases around the trunks creates an efficient insulating layer that protects the plants from frost.

The altitude to which the forest can grow is ultimately limited by low temperature and wind exposure. Often there is a distinct line of contour where the forest comes to an end and the open grassland of the alpine tundra begins. But local conditions, such as shelter and aspect, can modify this, as can the activity of humans, such as burning and grazing the high open country. The line of forest limit is called the *timberline* (see sidebar below).

Timberlines

The timberline can be defined as the border between forest vegetation and treeless tundra. In the Tropics it occurs at high altitude on mountains, its precise level varying with latitude, wind exposure, and proximity to the ocean. In the equatorial mountains of East Africa, for example, the timberline lies at about 12,000 feet (3,900 m). The boundary may be clear and distinct, tall trees abruptly ceasing and a marked line of dwarf shrub replacing them, but this occurs on tropical mountains largely where intensive grazing or fire has been used in the alpine zone by pastoral farmers. More often, the forest becomes lower in stature as it enters the elfin forest zone. It then forms a ragged boundary where the closed forest is replaced by open patches of trees and

scrub, scattered among the tundra vegetation, especially in the more protected and sheltered sites. Some ecologists distinguish between the forest limit, where uninterrupted forest comes to an end, and the tree limit, beyond which there is no tree growth at all.

Climate, especially temperature, is the major determining factor in establishing the timberline on a tropical mountain. Seasonal variation in temperature is usually relatively small in comparison to daily fluctuations, which can range over 20°F (11°C). Most tropical trees are unable to survive if the annual temperature falls below 41°F (5°C), but some tree-like herbs are able to survive below this threshold and form a type of woodland above the true tree line.

LOWLAND SWAMP AND BOG FORESTS

Many of the tropical rain forest regions of the world lie in the basins of major rivers, such as the Amazon and the Congo. Water reaches the soils of these regions not only from the abundant precipitation but also as a result of floodwater carried down the rivers from the mountains at the heads of their catchments. Periodically the forest may be flooded, and the roots of the trees then lie beneath the water. In the Amazon Basin, many areas of forest are inundated to a depth of 30 feet (9 m) for several weeks each year (see illustration [below]). Some parts of the forest, especially regions close to the river courses, are permanently flooded, forming tropical swamps. The vegetation and soil fauna of these areas are subjected to a number of physiological problems associated with waterlogging (see sidebar [right]).

Epiphytic orchids escape the flooding of the Amazon forest by living on the trunks and branches of the trees.   *(Large-Scale Biosphere-Atmosphere Scientific Conference)*

Waterlogging

Soils in tropical swamps are either temporarily or permanently saturated with water; they are waterlogged. When a soil is completely saturated in this way, all of the spaces in the soil that are normally filled with air become filled with water instead. This means that air-breathing animals in the soil are deprived of their normal means of respiration. Some invertebrates can survive in minute air pockets for long periods of time, and some, such as the springtails, have a hairy covering on their bodies that makes them virtually unwettable, so they manage to exist even in a watery environment. The springtails (Collembola) are soil insects of this type. But swamp conditions are not conducive to the survival of most terrestrial soil invertebrates; they are replaced by species that can cope with a lack of oxygen. The respiration of plants, animals, and many fungi and bacteria depend on a good supply of oxygen, usually directly from the air or dissolved in a thin film of water. Oxygen dissolved in water diffuses very slowly, about 10,000 times as slowly as in air, so organisms that live permanently underwater can easily become short of this vital substance. Plant roots may find it difficult to cope under *anoxic* (low-oxygen) conditions and some have developed sophisticated biochemical pathways to avoid the accumulation of toxic materials, such as ethyl alcohol, in their tissues. Some fungi, such as yeasts, excrete the toxic alcohol when they respire anaerobically, thus ridding themselves of this toxin. But animals, plants, and microbes that cannot adapt to a shortage of oxygen are unlikely to survive under the *anaerobic* (lacking an air supply) conditions of flooding.

There are several different types of swamp forest even within the Amazon Basin. In the regions that border the Andean mountain chain, the floods bring large quantities of suspended silt, much of which is deposited in the upper swamps. These are called white-water swamps, or *varzea*, and the trees of these fertile regions are tall with buttressed roots that help support their great height and weight. Among these are some well-known rain forest trees, including the rubber tree (*Hevea brasiliensis*) and the sand box tree (*Hura crepitans*), that can reach a height of 200 feet (60 m).

In the Amazon floodplain there are also permanently waterlogged sections of the forest, which develop where the river once meandered in a wide curve and are left isolated in a crescent-shaped section called an *oxbow lake* as the river has changed its course. These oxbows are rich wetland habitats that rapidly become colonized by aquatic plants and later by trees to form a swamp forest. The palm *Mauritia flexuosa* is an important member of the swamp community in the Amazon region and supports many species of animal.

Some of the rivers that enter the Amazon floodplain have passed through acidic, nutrient-poor sands, and the drainage waters have a pH of only 4.0–5.0 (see sidebar "Acidity and pH," page 45). The swamps that develop in areas supplied by these waters have lower productivity, biomass, and species diversity than that of the varzeas and are called blackwater swamps, or *igapos*. The waters are dark because the organic matter suspended in them decomposes very slowly.

Swamp forests also occur in the coastal regions of South America north of the Amazon Basin, in Surinam and Guyana. These are similar to the flooded swamps of the Amazon and share many of the same tree species, such as various palms and the kapok tree (*Carapa guianensis*), but generally accumulate more peat in their sediments (see sidebar [right]). In the coastal fringe of South America they merge into the mangrove forest as the salinity of the water rises.

African swamps (see illustration on page 70) are found throughout the rain forest and the deciduous tropical forest zones, wherever water collects among the trees. As in the Amazon forests, palms are important in the composition of African swamp forests, though different types of palms characterize different parts of the continent. In West African swamps, such as those of Nigeria, the raffia palms are important (genus *Raphia*, the source of raffia for matting and basket weaving), while in the more southern swamps, such as the Okavango Swamp in Botswana, trees of the date palm genus, *Phoenix*, are common. These tall, spiny palms fringe the waters of the swamp and are often accompanied by strangler figs (see "The Fight for Light: Upwardly Mobile Plants," pages 125–127). Fan palms and acacias are also present in the Okavango Swamp, but usually in the drier locations that are less frequently flooded.

East African swamps also have *Phoenix* palms rather than raffia palms, and some of them accumulate great depths of peat, suggesting that they have very ancient ecosystems. One such peat deposit in the uplands of Rwanda has been radiocarbon dated to 38,000 years old, showing that it was developing at a time when the high latitudes were undergoing glaciation. As in the case of the montane forests, the trees of these upland swamps are generally shorter and less dense than those of the lowlands.

Indonesia is extremely rich in tropical swamp forests. It has been estimated that 60 percent of all tropical peat lands are found in Indonesia and the coastal regions of other islands in the region. Some of these forests have such a distinctive structure and peat development that they must be regarded as tropical raised mires, or *bog forests*. Wetland ecologists use

Peat

When plants and animals die, or shed part of their bodies, such as leaves, exoskeletons, or pupal cases, these fall to the ground as *litter* and become incorporated into the soil as organic matter. Many of the organic chemicals they contain are rich in energy and provide a food resource for scavenging invertebrate animals, called *detritivores,* and the many bacteria and fungi, the *decomposers,* that devour these materials. In most terrestrial habitats, including the dry-land rain forests, all of the organic matter that falls on the ground is consumed in this way. But under waterlogged conditions, far fewer detritivores and decomposers are able to function because of the lack of oxygen (see sidebar "Waterlogging," page 68). As a consequence, some of the litter is never fully decomposed and accumulates in the soil as organic detritus. In waterlogged environments, therefore, sediments accumulate below the water, consisting partly of the mineral particles eroded from surrounding slope soils and carried into the water body, and partly of the undecomposed organic detritus derived from the local plant and animal life. The relative proportions of these two components varies with the nature of the surrounding catchment, the erosive capacity of the incoming streams, and the local productivity of organic matter, together with the degree to which decomposition has taken place. Soil scientists take a threshold value of 65 percent organic material and 35 percent mineral material to define a sediment as peat. In some wetlands, however, which are fed entirely by rainfall rather than by drainage streams, the organic proportion of the peat can be as high as 99 percent.

the term *bog* in a very specific sense, meaning a wetland in which the development of peat is so rapid and sustained that the resulting peat mass becomes raised above its surroundings to form an extensive dome. A bog of this type can no longer receive drainage water from the ground but is instead fed entirely by rainwater, a condition ecologist term *ombrotrophic*. In this respect they differ from *rheotrophic* swamps fed by flowing water. The tropical raised mires of Southeast Asia, mainly Sarawak, Borneo, New Guinea, and Sumatra, are of this type. A cross section of one of these bogs is shown in the diagram. Since all their water supply comes from precipitation, the peat they form is extremely pure with hardly

Tropical rain forest along the edge of a lake, Lake Nabugabo in western Uganda. The forest along the edge of waterways is particularly dense and rich in lianes. *(Peter D. Moore)*

any mineral content. They lie in the low coastal regions and are often dissected by rivers passing to the ocean, frequently blending in with mangroves around the coast.

Cross section of a coastal forested peat land in Southeast Asia. These coastal forests receive so much rainfall that peat deposits of up to 65 feet (20 m) can develop in the form of extended domes several miles in diameter.

The peat-forming bog forests are among the richest of tropical forests in their diversity of plant and animal life. They are dominated by trees of the family Dipterocarpaceae, and the lower layers are rich in clambering plants, such as the spiny *Pandanus.* There is a distinctive zonation of forest types from the riverine and marginal swamp forests flooded by groundwater, especially during the monsoonal rains, through the sloping edges of the elevated peat dome (often about one to two miles [2–3 km] from the rivers) and

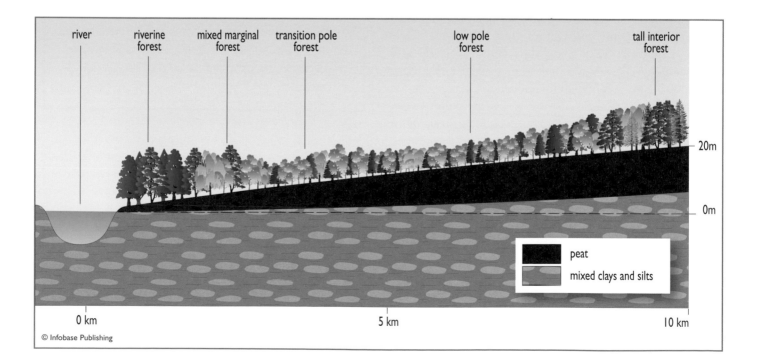

onto the central plateau of the bog. Not only do the species of tree change along this gradient, the size and the density of the trunks also change. The peripheral swamps have tall trees, up to 130–160 feet (40–50 m) in height. The density of large trunks is relatively low and the water level is permanently high. On the sloping edge of the bog dome, however, the water table can fall below the peat surface during the dry season and in this region is a transitional forest of lower stature, only 80–100 feet (25–30 m) in height. On the central bog dome the size of the trees becomes yet smaller, with a maximum around 65 feet (20 m). But the density of the trees is greater in the central part of the bog, producing what has been called a *pole forest,* in which the dipterocarp tree *Shorea albida* is particularly prominent. The diversity of tree species is also lower in the central bog area, with only about half the species density found in the peripheral regions.

The elevated, rain-fed centers of the tropical forested raised mires are very acidic, with a pH often less than 4.0, and sometimes support the growth of the *bog mosses* of the genus *Sphagnum.* The peat surface here may be raised to a level 17–35 feet (5–10 m) above the water table in the surrounding lands. The peat mass itself is almost impervious to the downward drainage of rainwater, so the surface water moves toward the sides of the bog and drains down the surrounding slopes. In many respects these tropical peat forests are the closest surviving ecosystem to the coal-forming tropical swamps of Carboniferous times.

■ MANGROVE FORESTS

Tropical, and some subtropical, coastlines with low energy regimes, that is, gentle wave action and low current velocities, bear a forested wetland vegetation that has been variously called *mangrove* and *mangal.* The main problem with the use of the term mangrove is that it can be applied either to the habitat or to the trees that occupy the habitat, so this can cause confusion. These forests extend to approximately 25° north and south of the equator, reaching up to the subtropical 32°N where conditions are relatively frost-free. Mangrove swamps typically occupy gentle inclines and grow upon fine-particle, silty, muddy sediments. They are subjected to regular tidal movements and occasional tropical storms that can erode the sediment and cause frequent changes in the configuration of the forest.

Mangroves show a distinct zonation in the pattern of their trees, which is especially obvious in estuarine regions, where there may be a strong gradient in the salinity of the water (see sidebar [right]).

The zonation of trees in a coastal mangrove forest largely reflects the salinity tolerance of the different species, although water depth and the process of succession also play a part in determining the pattern. The diagram on page 72

Salinity

Water is an extremely efficient *solvent* and as a result is non-existent in nature in a perfectly pure state. Even rain water contains materials that have been dissolved in the water on its passage through the air. Water that has passed through soils on its way to the sea contains many dissolved elements (*solutes*), and these become more concentrated as the water solvent evaporates. The oceans are rich in dissolved salts that have accumulated over thousands of millions of years, eroded from rocks by the constant recycling of water through the atmosphere and the soil. The concentration of salts, the *salinity,* varies according to geographical location and the effect of freshwater inputs from rivers and icecaps, but the general salt concentration of the oceans is between 33 and 38 parts per thousand (ppt), averaging at 35 ppt, or 3.5 percent. Some inland regions have enclosed water bodies with even higher salt concentrations, the hypersaline lakes with salinity sometimes exceeding 40 ppt. These saline lakes have no exit streams, so the only way water leaves such basins is by evaporation. An example is the Dead Sea in the Rift Valley of the Jordan River, which has a salinity 10 times that of seawater.

Virtually all known elements have been found in seawater, but there are eight elements or ions that account for 99 percent of the total. These are chlorine (1.9 percent), sodium (1.0 percent), sulfate (0.27 percent), magnesium (0.13 percent), calcium (0.04 percent), potassium (0.04 percent), bicarbonate (0.01 percent), and bromine (0.007 percent). All plants and animals need to regulate the content of water in their cells and they normally take up water by a process of *osmosis,* which is driven by the concentration of elements in the aqueous medium of their cells. Water moves out of a low-salinity medium into a high-salinity medium, so the cell must have a higher concentration of salts than the outside environment if it is to take up water. Considering the water molecules themselves, they move from a high water concentration to a lower concentration inside the cells, where there are greater quantities of additional elements. The higher the salinity of the external water, therefore, the more difficult it becomes for organisms to extract water for their own use. Plants adapted to cope with saline waters are termed *halophytes.* The gradient of salinity is the main factor controlling the zonation of trees in a coastal mangrove forest.

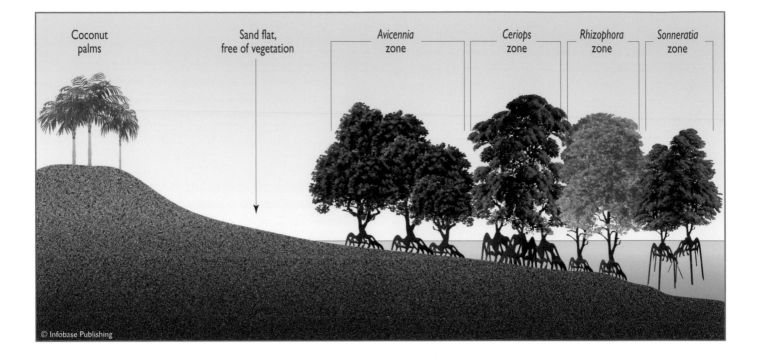

Coconut palms | Sand flat, free of vegetation | Avicennia zone | Ceriops zone | Rhizophora zone | Sonneratia zone

© Infobase Publishing

shows the typical zonation of trees in the mangrove forests of coastal East Africa. There are just 34 species of tree that can be regarded as true members of the mangrove type, the two most abundant genera being *Avicennia* and *Rhizophora*, both of which are found in the mangrove forests of the Old and the New World. Some palms are also characteristic of the mangal habitat, such as *Nypa*, a palm whose fruits have been found in a fossil state in the 60-million-year-old geological deposits east of London, England, showing that tropical conditions extended to these high-latitude regions at that time.

On the seaward side of the mangal fringe lie mudflats, and these are often colonized by sea grasses. The fruits of mangrove trees fall into these muds and germinate, eventually growing and stabilizing the mobile sediments. In the Caribbean, *Rhizophora* is the first colonist, but in northwestern Australia it is *Avicennia* species, together with *Sonneratia*, that first colonize the open mud, so there is no universal sequence that is followed around the world. As the mangrove habitat develops, the complexity of the forest composition also increases. Areas of water may become relatively isolated close to the shore and separated from the open ocean by the zone of mangrove forest. Within such isolated areas, water evaporates quickly in the tropical heat, leading to high salinity close to the shore. Only mangrove species capable of tolerating this raised salinity can survive on the landward side of the mangrove forest. Where mangal develops in the estuaries of large rivers the patterns of salinity can become even more complex. In West Africa the estuary of the Niger River is rich in mangal affected by freshwater flooding down from the continental interior, while salt water pushes into the region with each high tide, and

Profile of a coastal mangrove swamp in East Africa. Different species of trees are best suited to different depths of water and variations in salinity, so they form a zonation pattern. The precise pattern varies in different parts of the world, as does the area covered by the swamp, depending on the steepness of the shore. The entire sequence often covers about a mile from shore to open ocean.

evaporation increases salinity wherever bodies of water are isolated from the flow.

The structure, or architecture, of mangrove vegetation is most complex near the water level. The leaf canopy is relatively simple and single-layered, but the instability of the muddy sediments has resulted in the evolutionary development of unusual root systems. *Rhizophora* species usually have arching prop roots growing from the trunks which act as flying buttresses that support the tree and create tangled tents (see illustration on page 73) built of wooden struts, providing coincidentally shelter for a wide range of fish. *Avicennia* species have roots that grow upward out of the mud, forming vertical structures exposed to air at low tide and acting as organs of gaseous exchange. These erect "breathing roots" are called *pneumatophores*, or *pneumorrhiza*. The roots of *Ceriops*, another genus of mangrove trees, are twisted and knotted, sometimes emerging above the surface of the mud. The complexity of mangal structure is thus found in the rooting systems, which are periodically submerged by the tide and then left exposed as the water recedes.

Lianes, or climbers, are not generally found in the mangal habitat because the rooting conditions in saline mud are too difficult for germination and establishment. Only the canopy

Prop roots, or stilt roots, shown here in a swamp of red mangroves from the Philippines, are organs that provide additional support and stability for spreading canopy branches. *(James P. McVey, NOAA Sea Grant Program)*

■ BIOGEOGRAPHY OF THE TROPICAL FORESTS

Rain forests are found in three main areas of the world (see map on page 2): Central and South America, Africa, and parts of Southeast Asia, the Pacific Islands, and northern Australia (a region that can, for the sake of brevity, be given the name *Malesia*). These three blocks of tropical forests are generally similar in structure but very different in their species composition and their biodiversity.

Tropical forests are among the most diverse ecosystems on the planet. They may well contain more than half of all the species of plants, animals, and microbes found on Earth. As yet, only a small proportion of these species has been discovered and described, so exactly how diverse the tropical forests are is still a matter of speculation. The possible reasons for the Tropics being so much richer in species than temperate regions will be considered later (see "Patterns of Biodiversity" pages 100–103). When comparing the global blocks of tropical forest, however, it becomes apparent that some tropical forest regions are richer than others. In general, the forests of Central and South America and those of Malesia are richer than the forests of Africa.

Tree species are the simplest of the tropical organisms to survey and count, so ecologists have conducted many studies on tree diversity in the tropical forests. A plot of 0.4 acres (1 ha) in the tropical forests of Ecuador, Peru, or Brazil in South America typically carries between 200 and 300 tree species that have a diameter at breast height of more than four inches (10 cm). Studies in Sarawak, Kalimantan, and Malaysia, all in Malesia, show that a plot of the same size also usually bears over 200 species of tree. In West Africa, however, studies in Cameroon and Nigeria show that a plot of that size would only rarely provide a count of 100 tree species. The total numbers of plant species in the three regions can only be guessed at because so many have yet to be described. In Africa, botanists consider that there are at least 12,000 species of tropical forest plants, while in Malesia the number could easily be three times that number. The South American forests are thought to contain three times the number of plant species found in Malesia, so this region is undoubtedly the richest of the tropical forest blocks as far as plants are concerned. The African forests are nonetheless still extremely rich in plant species compared with most other biomes in the world, and many of the plants they contain are restricted to that continent (*endemics*), so their conservation value should not be underrated.

Although the tropical forests can be ranked in the order of plant diversity, with the Americas (Neotropics) highest and Africa lowest, this does not necessarily mean that the same order applies to other groups of organisms. The main problem in conducting this type of exercise is the lack of

close to the shore may support some lianes that are rooted in the drier conditions. Epiphytes are present, but only those tolerant of the salt spray coming from the sea. In general, epiphytes are very sensitive to air-borne chemicals, whether salt spray or pollutants (see sidebar "Epiphytes," page 121), so the range of epiphytes found in mangal is very restricted.

The concept of succession and climax is particularly difficult to apply to mangrove vegetation because the habitat itself is so unstable. Erosion from river flooding and from tidal scour constantly changes the course of water movement through the mangal, although the complex root systems contribute a degree of protection and stability to the sediment. Tropical storms and tidal waves caused by tsunamis can be catastrophic to the mangal, but the destructive energy absorbed by the mangal ecosystem offers some protection to the shoreline habitats, including human settlements, that lie on their landward side.

data for most types of animals, although some groups are relatively well known. In the case of bats, for example, the Neotropics (Central and South America) have 96 species, Malesia has 66 species, and Africa comes last again with 26 species. Butterflies are also quite well recorded, and some tropical groups, such as the swallowtails, show a slightly different rank order. Malesia is at the top with 123 species recorded within 10° latitude of the equator, the Neotropics coming second with 80 species, and Africa still in third place with 58 species. A very different result emerges from the study of primates. Although the Neotropics are slightly ahead with 33 species, Africa is second with 32 species, and Malesia is a poor third with just 11 species. The reason for this distribution is that Africa was probably the center of primate evolution. By 40 million years ago, Africa had become totally separate from South America and the New and Old World primates developed along separate paths. Africa then remained a center of primate evolution and diversity.

On the whole, Africa does appear poorer in species than other tropical forest regions, with important exceptions, such as the primates. A closer comparison of the inhabitants of the Neotropical and African forests reveals that most African tree species are widely distributed throughout the tropical forests of that continent, whereas in Central and South America many tree species have quite restricted distributions. This observation suggests that the New World has more specialist species, while the African forest has more generalists. Very few species of plant (less than 1 percent of the flora) are found in both Africa and the Neotropics, so there is very little overlap in their flora.

It is difficult to explain the differences in biodiversity between the three main tropical forest regions of the world, and biogeographers argue about various theories. The African forests are much smaller than those of the Amazon Basin, only attaining about one sixth of their size, and this is likely to affect overall diversity. There are also ancient geological differences. South America has long been isolated and evolution has proceeded independently on that continent. The plant family Bromeliaceae (the bromeliads), for example, is almost totally restricted to South and Central America. South America also has very extensive areas of lowland, whereas in Africa the lowland regions are more confined. This is important because the climatic fluctuations of the past two million years (the Pleistocene ice ages) have undoubtedly caused extensive disturbance in the tropical forest, with the lowland regions providing the best opportunity for forest survival. Africa is also a more arid continent than South America, with steeper gradients of precipitation running north and south from the equator. Climatic fluctuations in the past would have had a greater effect on the shifting of vegetation boundaries in Africa, leading to more fragmentation and therefore species extinctions. There is much evidence to suggest that the African forests were largely replaced by savanna during the glacial episodes of the high latitudes, and the modern rain forest in Africa is therefore a relatively young biome recovering from times of widespread extinctions. This topic will be explored in greater detail later in the book (see "Tropical Forests in the Last Two Million Years," pages 180–182).

■ CONCLUSIONS

Tropical forests are found both in the lowlands and the highlands of the Tropics, as well as along the coastal fringes and bordering the savanna woodlands of drier regions. The term *forest* is applied when the canopy covers more than 40 percent of the ground; woodland has a less extensive canopy than this. The tropical forests are all dominated by trees, but they vary in their structure, or architecture, and in their evergreen or deciduous nature. The true rain forests of equatorial regions have a canopy that can often be divided into three layers. The upper layer (stratum A) consists of individual emergent trees, forming an open canopy, sometimes reaching heights of 200 feet (60 m) but more frequently attaining a height of up to 140 feet (42 m). The next layer (stratum B) forms a more complete canopy cover, often at a height of about 90 feet (27 m), and below this lies a third layer of smaller trees and younger individuals awaiting a gap in the upper canopy that they can occupy. The third layer (stratum C) lies at a height of approximately 46 feet (14 m). Climbers and epiphytes are often abundant in all these layers, but the ground beneath the forest cover is relatively poor in herbaceous plants.

The penetration of light through these layers of leaves depends on leaf density, leaf arrangements, and the occurrence of gaps. Leaf area index is a measure of how many times an area of ground is covered by leaves, and this index provides an indication of how intensely the plants compete for light in the canopy. The upper part of the forest canopy, where light is more easily available, is called the euphotic zone, in contrast to the oligophotic zone in the darker, lower layers of the canopy.

Evergreen trees predominate in the equatorial zone, where there are few seasonal variations in climate and conditions are always hot and wet. Away from the equator, both to the north and the south, seasonal changes are more apparent, especially taking the form of a dry season, and the occurrence of drought encourages deciduous trees that can avoid water loss by shedding their leaves as conditions become drier. The frequency of deciduous trees in the forest increases gradually away from the equator, and the deciduous habit eventually dominates the vegetation in the tropical deciduous forest of the higher-latitude Tropics. In the monsoon forests, where the drought season is prolonged, the deciduous habit prevails.

Pristine, or primary, forest is found only in areas remote from human habitation. Many tropical forests have been modified by human activity and are recovering from past clearance for forestry or agriculture. The recovery of the forest from catastrophe, whether natural as a result of storms or human induced, follows a course of invasion by pioneer trees and a gradual reestablishment of the full canopy. This process is known as secondary succession and the resulting forest is called secondary forest. Some species are very slow to reinvade disturbed forest, leaving secondary forest generally poorer in its biodiversity than primary forest.

In tropical mountainous regions, the rain forest gives way to montane forest, and there is often a distinct zonation of forest types with altitude. Montane forest is lower in stature and has a simpler structure than the lowland forest. It is generally richer in epiphytes, especially in the high cloud forest where temperatures are often cooler but humidity is constantly high. The high-altitude forests may take on a dwarf form, when they are called elfin forests, or they may be replaced by a forest zone of bamboo, as on some African mountains. The timberline marks the boundary beyond which a closed canopy forest cannot survive and gives way to alpine tundra vegetation. This boundary lies at about 12,000 feet (3,700 m) on East African mountains and is controlled by temperature.

Some lowland regions of the Tropics receive an abundance of water from major rivers draining vast catchments, and waterlogged conditions result in the formation of swamp forests. Only trees capable of survival with their roots frequently underwater can survive in these conditions. In some parts of the Southeast Asian region, coastal wetlands accumulate large quantities of peat that can build up to form extensive raised domes clothed in trees, the bog forests. Along the edge of coasts, however, the influence of saline water from the ocean presents a different set of problems to the vegetation, and a distinctive mangrove forest ecosystem develops. The canopy structure of mangal is relatively simple, but the rooting systems of mangrove trees are complex, with arching and vertical roots protruding from the mud.

These different types of tropical forest, with the exception of the bog forests restricted to Southeast Asia, are found throughout the three major blocks of tropical forest in the world: South and Central America (the Neotropics), Africa, and Southeast Asia, Polynesia, and northern Australia (sometimes called *Malesia*). Of these three regions, however, the African region is generally poorer in species than the other two. The Neotropics are generally the most biodiverse of the three regions, but this may in part be due to their greater area of coverage. African tropical forests may have suffered most from fragmentation and destruction as a result of climate changes over the past two million years, resulting in a high level of extinction and a generally impoverished diversity. Nevertheless, all of the tropical forest blocks are extremely high in their biodiversity in comparison to other terrestrial biomes and in combination contain at least half of the world's living species.

4

The Tropical Forest Ecosystem

Every individual organism on the Earth interacts with its surroundings, and the organisms living in tropical forest habitats are no exception. A terrestrial organism living on the surface of the land is in contact with the atmosphere and the soil, but it will also interact with other members of its own species. An individual organism, especially if it is a relatively advanced animal, will have social interactions with other members of its species, and a collection of individuals of the same species interacting in this way is called a *population*. A collection of macaws living together in a breeding colony along the banks of a South American river is an example of a population. An organism also interacts with other living creatures belonging to different species. Some of these may be its prey, while others may be its predators; some will be harmful parasites, while others compete with it for the same food resources. A collection of different species living together and interacting in all these different ways is called a *community*. Finally, a collection of living organisms coexisting in a community interacts with the nonliving world that forms a setting for all life. Plants are rooted in the soil, from which they absorb the minerals they need to grow, and they take up carbon dioxide gas from the atmosphere, which they convert into sugars in the process of photosynthesis. Animals receive most of the energy and the chemical elements they need from their food, which may consist of plant materials or prey animals. Animals also drink water or absorb water through their surfaces and supplement their mineral intake in this way; some animals may even eat soil if they run short of certain elements. A community of different species of animals and plants living in the physical and chemical setting of the nonliving world is called an *ecosystem*.

■ THE CONCEPT OF THE ECOSYSTEM

The concept of an ecosystem is an extremely useful one to ecologists and conservationists. It provides an approach

to the study of the natural world that can be applied at a range of different scales. A single tree rooted within the soils of a tropical forest can be regarded as an ecosystem. Using this approach, one can study the ways in which the energy and mineral elements contained in the growing plant are consumed by grazing animals or decomposed by fungi and bacteria. The microbes living in the soil beneath the tree are eaten by invertebrate animals, which in turn may be fed upon by carnivores, such as beetles, and these larger creatures attract visiting insectivorous birds, which descend from the canopy and consume the beetle. The bird thus harvests the energy that the beetle contains. Each of these organisms is obtaining energy from its food and uses some of this energy in such processes as growth and movement (see sidebar on page 77), perhaps in seeking more food or in finding a mate. The bird that has eaten the beetle may fly away with its prey, taking it to another part of the forest. When it does this, it is removing energy from one ecosystem, in the case of the single tree, and transporting it elsewhere. In this example, the ecosystem concept is being applied on a relatively small scale to a single tree. It is possible, however, to regard the entire forest as an ecosystem, in which case the single tree is simply a part of a greater whole in which the photosynthesis of all the vegetation is trapping the energy of sunlight, storing it in leaf and root tissues, and eventually providing an energy source for all the grazing herbivores and the assemblage of microbes and invertebrate animals inhabiting the decomposing materials of the soils beneath the forest. The insectivorous bird is now part of the same ecosystem as the beetle, performing its own part in the organization of the whole. Although the ecosystem concept can be applied on a small scale, as in the case of a single tree, it is more often applied at a larger scale, to entire habitats. It is also possible to apply the ecosystem concept to the sum of all the tropical forests of the Earth. At this scale, the term *biome* is usually used.

Although it is possible to use the ecosystem idea at many different scales, all ecosystems have certain features in common. All ecosystems have a flow of energy through them.

Energy

Energy is the capacity to do work. All animals, plants, and microbes need energy for their daily life because they need to grow, or move, or reproduce. When an animal forages for food it is using some of its energy in the pursuit of yet more energy. When a plant absorbs elements from water in the soil it expends energy. Many of the biochemical processes that take place in individual cells also demand the expenditure of energy. Animals need energy to move, hunt, and reproduce. So, all living creatures constantly demand energy.

Energy can be divided into two major types, kinetic and potential. Kinetic energy is the outcome of the motion and the mass of an object. Water running along a stream, a rock falling down a mountain slope, the wind blowing in the trees—all of these exhibit kinetic energy. Potential energy is stored in an object, ready to be released. A rock perched on the top of a cliff has the potential to release kinetic energy if it falls over the edge. Many chemical compounds contain energy in the bonds that hold the molecules together. A sucrose molecule, for example, can be burned to release its component carbon atoms as carbon dioxide and it releases kinetic energy in the form of heat as it does so.

Living organisms can capture the energy they need in either kinetic or potential form. A green plant or a *photosynthetic bacterium* absorbs electromagnetic radiation, which is a form of kinetic energy, when it intercepts sunlight. The green pigment chlorophyll is able to capture this energy and transfer it to chemical potential energy for use in the trapping of carbon dioxide molecules and reducing them to sugar. When a herbivore eats part of a plant, it removes energy-rich chemicals, including sugars and starch, which can then be transformed into animal tissues, or be respired to provide the kinetic energy needed for movement and other activities.

Energy occurs in many different forms and can be converted from one form into another. A burning piece of wood produces light and heat as the oxidation of carbon-containing compounds releases the energy. Falling water can turn a turbine and generate electricity. The movement of energy between its many forms obeys two fundamental laws, the laws of thermodynamics. The first law of thermodynamics is concerned with energy conservation. It states that energy cannot be created or destroyed, so each energy transfer can be described by an energy budget equation that must balance. The second law of thermodynamics states that no transfer of energy from one form to another can be 100 percent efficient. A turbine can never capture all the kinetic energy released by falling water, and the electricity generator will never be able to convert all of the energy released by the rotation of the turbine into electricity. All energy transfers are accompanied by energy losses, often in the form of waste heat. When a person eats a potato the energy contained in the stored starch is not completely transformed into human flesh. Wastage and losses may account for as much as 90 percent of energy intake and only a small proportion of the total energy is captured by the consumer. The second law of thermodynamics explains why the transfers of energy from Sun to plant, from plant to herbivore, and from herbivore to carnivore all involve losses and wastage.

The source of energy for most ecosystems is sunlight, which is made available to living organisms by the photosynthesis of green plants. There are some bacteria that can photosynthesize, and there are some that can obtain energy from nonsolar sources, such as chemical reactions with inorganic materials, including the oxidation of iron, but these are generally of little significance in most ecosystems when compared with the contribution of green plant photosynthesis. Some ecosystems, such as the muddy banks of a tropical river, import energy from other ecosystems. Dead plant materials, rich in the energy derived ultimately from sunlight, are brought into this type of ecosystem from the surrounding forest; these imported sources of energy and chemical elements supply the needs of the animals and microbes that feed upon them within the mud. As energy flows through an ecosystem one can distinguish certain groups of organisms that play different roles. The green plants are primary producers, fixing solar energy into organic matter; they are said to be *autotrophic* in their nutrition, which literally means that they can feed themselves. Some of the energy they trap from the Sun is used in the energy-consuming activities of the plant, such as the uptake of chemical elements against a concentration gradient from the soil. Energy needed for such purposes is released from storage and is liberated by the process of *respiration*. The remaining energy is used in building new plant materials, leaves, stems, roots, flowers, and seeds. Herbivores are primary consumers; they are dependent on plants for energy, so they are said to be heterotrophic, meaning that they need to be fed by others. Predatory animals are also heterotrophic. They too depend ultimately on plants, but indirectly so because they feed on the herbivores or on other animals that eat herbivores. These are secondary and tertiary consumers, respectively. They occupy different positions in a hierarchy of feeding, sometimes referred to as a *food web* in the ecosystem. All of these organisms release kinetic energy according to their immediate needs from stored chemicals, using the process of respiration in their cells.

The waste materials produced by egestion (defecation) and excretion by living organisms, together with the dead parts or dead bodies of those individuals that escape consumption by predators and survive long enough to die a natural death, are used as an energy source by the decomposer organisms. Animals that eat dead plant and animal materials and derive energy from them are called detritivores. But ultimately it is the microbial decomposers in the ecosystem, the bacteria and fungi, that are responsible for the degradation of all the residual, energy-rich materials in the process of *decomposition*. Nothing is wasted, nothing is unaccounted for. All the energy entering the ecosystem is finally used up and is dissipated as heat, released by the respiration of all the various plants and animals involved in the food web.

While energy flows through the ecosystem and is eventually dissipated, chemical elements are recycled around the ecosystem. Carbon atoms, for example, are taken up by plants in a gaseous form as carbon dioxide, and these atoms become incorporated into carbohydrates. They may be stored in this form or as fats, or they may be converted to proteins by the addition of the element nitrogen, derived from the soil. Carbon compounds may also be converted into different kinds of molecules by the addition of other elements, such as phosphorus to make the phospholipids used in membrane construction, or sulfur, used to construct some types of amino acids, the building blocks of proteins. The materials built into plant bodies are consumed by animals and a proportion passes through the body of the consumer to be voided as waste, while some becomes incorporated into the body of the animal. Respiration results in the release of carbon back into the atmosphere as carbon dioxide gas, while the nitrogen and phosphorus, together with other elements, are lost to the organism by excretion or by death, and these elements enter the soil and are gradually released to the environment in the process of decomposition. Chemical elements are thus recycled and can be used over and over again. Energy passes through the ecosystem and is finally dissipated, but chemical elements cycle round and round the ecosystem indefinitely. The constantly turning wheel of element motion is called a *nutrient cycle*.

In addition to the cycle of nutrients, however, there is also usually an import and export of elements to an ecosystem. A stream entering a tropical forest ecosystem will bring dissolved minerals from the other ecosystems in the catchment area, perhaps from the higher regions of mountains. Rainfall will also bring a supply of elements, its richness depending on how close the site is to the ocean. Animals may migrate into an ecosystem and bring elements from outside, such as the songbirds that migrate from North America into the tropical forests of Central and South America each winter, bringing elements they have collected in their northern summer feeding grounds. Plant materials, including twigs and leaves, may be washed or blown into a pond, supplying a source of elements from another ecosystem. But just as these processes bring elements into an ecosystem, they can also take them out. Streams can leave an ecosystem, animals can move out, and plant material may be blown away. So study of an ecosystem involves a consideration of its energy and nutrient imports and exports to understand the balance of supply and loss in both energy and mineral elements. Before an ecosystem can be fully understood, it is necessary to calculate its energy and nutrient budgets.

■ PRIMARY PRODUCTIVITY IN TROPICAL FORESTS

The food webs of the tropical forests are based on plants, mainly trees, which are constantly fixing energy from the Sun in the process of photosynthesis. Energy capture depends on the pigment chlorophyll, and the tissues containing the pigment must be exposed to light to intercept the incoming radiant energy. The most important organ used by most plants for light capture is the leaf, which is often broad, flat, and relatively thin in order to expose a maximum surface area to the sunlight. Stems and branches are used to support the leaves and hold them in an exposed position, but stems and branches can also be used for photosynthetic activity. Many stems, especially in herbaceous plants and small shrubs, are green and assist in the vital process of energy gathering. Even parts of the reproductive organs may be green and photosynthetic, such as the sepals on the outer part of the flower, and the wall of the carpel in the developing fruit. A plant cannot afford to waste any opportunity of using an organ that is exposed to light to boost its food production. There are some organs such as the petals (colored to attract pollinating insects), bark (dead outer cells protecting the water-conducting elements in the wood), and the woody cores of the tree trunks in which other functions are even more important than energy trapping. Roots are important for stability and for taking up water and nutrients from the soil, but their position below the ground obviously precludes any photosynthetic activity. As a plant grows, therefore, it is faced with a strategic problem regarding the various ways it can allocate its growth energy. The plant may allocate materials to the production of more leaves, or it may benefit by placing more emphasis on structural support. Plants have evolved a number of different strategies in their patterns of *resource allocation* (see sidebar [below]).

However the plant allocates its resources, the total amount of energy that accumulates in an area during the course of the year provides the energy resource for the animals that live in that area. Ecologists need to measure this total rate of energy collection, or *primary productivity,* so that they can gauge the efficiency of an ecosystem's vegetation as a support system for its animal and microbial life. Measuring the primary productivity, however, is difficult. The most commonly used method for terrestrial vegetation studies is to harvest sample areas of the plant community at different times during the growing season. The researcher selects an area of ground, in the case of grasslands often about a square yard in area, and completely crops of all its vegetation down to soil level. The next step is to sort the

harvested material into the different species contained in the vegetation and then to dry the samples in an oven. The drying process is used to remove all water so that the results are not affected by the degree of wetness of the samples. Care must be taken, however, not to let the temperature rise too high, because this might cause the evaporation of other components of the plants, such as oils, which may constitute important reservoirs of energy. For detailed studies of energy flow it is useful to conduct one additional analysis to obtain the energy content, or calorific value, or the cropped material. This is achieved by reducing the dried samples to ash in a bomb calorimeter, a piece of equipment in which a known weight of material can be ignited to release the energy it contains. The rise in temperature of the calorimeter is directly related to the energy released, so the energy content of the sample can be calculated.

As a result of this type of study, scientists can calculate how much new plant material is added to a given area of the ecosystem each year, and this amount can be expressed

Resource Allocation

As a plant gains in weight, the newly acquired material must be moved to appropriate parts of the plant body. In the early stages of plant growth, the new products of photosynthesis may be directed toward the construction of new leaves, in which case all of the investment is used for the development of productive organs in which more photosynthesis can occur. But root development must keep pace with the growth of the shoot so that the plant remains stable in the soil and a supply of water and mineral nutrients is maintained for continued transpiration and growth. As the shoot continues to grow it needs extra support, so instead of making photosynthetic organs, such as leaves, the plant must expend an increasing amount of its energy on support structures, including stems and branches. In the case of shrubs and trees, support tissues include wood, which holds the plant relatively rigid and assists in the development of a complex canopy, enabling the leaves to reach the light and continue their productive function. Competition for light in the tropical forest means that the rapid construction of tall support is essential if the plant is going to survive. The allocation of new material to different parts of the plant is called *resource allocation.*

As plants grow, they must devote an increasing amount of their energy to nonproductive support tissues, especially roots and rigid stems. In its early stages of growth, a plant invests energy almost at a compound

rate of interest. This means that every portion of new material is reinvested in production, so the growth is potentially logarithmic. But the older the plant becomes, the lower the proportion of new production (the "interest" gained) can be reinvested in this way. As more support tissues are needed, the growth rate falls from a logarithmic one and becomes slower, eventually reaching a plateau as the plant attains its maximum size.

Eventually the plant begins to reproduce, and the investment of new material is directed away from both productive and support tissues into the development of flowers and then fruits. In many annual plants the proportion of energy invested in reproduction is very high because such species rely heavily on seed production for survival into the next growing season. Long-lived perennials devote proportionally less energy to seed production. To them the maintenance of a robust body and long-term survival of the individual are more important than producing masses of young seedlings. Most tropical forest trees adopt this strategy.

Grazing animals, which rely on plant products for their food, thus have a number of options for feeding. They may attack leaves directly or may feed on roots or timber. Alternatively, they may gain energy from nectar, or pollen, or the seeds subsequently produced by the plant. Resource allocation by plants presents consumer organisms in the ecosystem with a range of feeding strategies.

either as dry weight per unit area per year, or as energy (calories) per unit area per year. There are several problems with this type of study that need to be considered. The first problem is that of sampling. Almost all types of vegetation are extremely varied and nonhomogeneous. Selecting relatively small samples as representatives of the entire ecosystem can therefore be misleading. This is particularly true of the tropical forests, where individual plants (trees) are big, so very large areas are needed to achieve a representative sample of the vegetation, and replication of sample plots becomes even more challenging. A square foot of a tropical forest would not provide a representative sample; several hundreds of square feet would be needed. The destructive nature of the harvesting technique also becomes a major consideration when large areas of a fragile habitat are concerned. Cropping all the vegetation from a square foot of grassland creates relatively little damage and can be repeated many times before there is a significant threat to the ecosystem in general. But harvesting all the vegetation from several hundred square feet of tropical rain forest is not an appealing prospect for ecologists concerned with habitat conservation.

The problem facing ecologists in this situation is precisely the same as the one facing foresters in a managed forest. Timber is a valuable crop, and measuring its rate of growth (productivity) by periodically felling large samples is not an economically sensible approach. Instead of wholesale cropping, foresters use a technique called *dimension analysis*. The forester takes an easily obtained measurement from a living tree and from this calculates the total volume of timber it contains, or the total weight of the tree. The measurement usually taken is the diameter of the tree trunk at a height of 4.25 feet (1.3 m) above the ground. This is known as the *diameter at breast height* (DBH). The forester then fells a series of individual trees of various DBH sizes and measures their timber volume or their weight. When the DBH is graphed against weight, a close relationship is usually revealed, so the weight of a tree can be estimated on the basis of this relationship and can be obtained without the need to cut down large areas of forest. In tropical rain forests, therefore, ecologists can estimate the total weight of the trees without having to fell them, and successive measurements can provide an indication of the rate of growth of the forest.

An additional problem for ecologists studying energy flow in a tropical rain forest is that it is almost impossible to sample the plant growth taking place below the surface of the soil. Roots are very difficult to separate from the soil in which they grow, and very fine roots have a short life span, often only a few days, so it would be well nigh impossible to estimate all the productivity that goes on beneath the ground.

One further problem is the loss of energy from the ecosystem even while the productivity is taking place. Some of the energy fixed by plant photosynthesis is used by the plant in its own activities and metabolism, so what accumulates at the site over time is actually the excess of productivity over and above the needs of the plant. This is called net primary productivity, in contrast to gross primary productivity, which is the total amount of energy fixed in an area in a given time. In a natural ecosystem under field conditions, there are additional drains on the energy that is fixed in photosynthesis. Grazing animals, both large and small, take their crop of energy from the plants even while the experiment is proceeding. In a tropical rain forest it is difficult to fence an area to exclude even large grazers, such as tapirs, but keeping out all invertebrate grazers, from caterpillars to leaf-cutting ants, is impossible. A full study of productivity therefore requires further work on the losses due to these various grazers. Only with this knowledge can the entire productivity of the forest be calculated. It is also possible that some plant material will die during the experiment. Old leaves may wither and fall, providing a source of energy for the detritivores and decomposers. One way of checking this source of loss is to place trays within the vegetation that collect any falling plant tissues. In the tropical forest even these trays may be raided by herbivores, so the experimenter has to make frequent collections of fallen litter before it all disappears.

As a result of all these complications to the study of plant productivity, the results available from studies are quite variable. The following table on page 81 shows a range of productivity values from a number of different types of tropical forest habitat, as well as data from some other ecosystems for comparison. These data are expressed as pounds dry weight per square yard per year (or kg m^{-2} per year) and it is important to remember that productivity is a *rate*, not a quantity, so time measurements are involved. Harvesting a sample of vegetation at a given point in time provides information about the quantity of vegetation present per unit area, called the *biomass*. This is measured as pounds dry weight per square yard (or kg m^{-2}). The productivity can only be obtained by measuring changes in the biomass over the course of time. Dry weight figures are used rather than wet weight because the amount of water in different tissues is extremely variable and will change according to the amounts of water available in the habitat.

As can be seen from the table, the lowland tropical rain forests generally have a high level of productivity, although this is variable, depending on the structure of the forest and its position within the successional sequence. Some of the highest productivity figures come from regenerating secondary forest just prior to its reaching a stable equilibrium or climax point. The most productive of tropical rain forests exceed all of the other biomes listed in their capacity to fix the energy of sunlight, although still higher productivity figures have occasionally been obtained by researchers exam-

Primary Productivity Values for a Range of Tropical Forests and Other Ecosystems

(pounds per square foot per year [figures in brackets are in kg m⁻²y⁻¹] in dry weight)

TROPICAL FOREST ECOSYSTEMS	NET PRIMARY PRODUCTIVITY
Tropical lowland rain forests	0.5–1.0 (1.7–3.5)
Tropical montane forests	0.2–0.7 (1.0–2.5)
Tropical seasonal forests	0.3–0.7 (1.1–2.5)
OTHER ECOSYSTEMS	
Hot desert	0.003–0.06 (0.01–0.2)
Moss tundra	0.02–0.022 (0.06–0.07)
Temperate grassland	0.06–0.44 (0.2–1.5)
Savanna grassland	0.06–0.59 (0.2–2.0)
Boreal evergreen forest	0.11–0.59 (0.4–2.0)
Temperate deciduous forest	0.17–0.73 (0.6–2.5)

ining tropical marshes and swamps, and even from some temperate reed beds and wetlands. But tropical rain forests are clearly highly productive habitats and their great energy-fixing capacity allows them to support the extensive range of heterotrophic organisms, including animals and microbes, living within them.

Tropical montane forests and tropical seasonal forests are also very productive, but they tend to have lower overall primary productivity levels than the lowland rain forests. There is much overlap in the data, however, and some temperate forest ecosystems are as productive as the poorer tropical forests. Tropical savanna woodlands and tall grasslands also overlap with the tropical forests in their productivity values, but temperate grasslands, deserts, and tundra tend to have lower levels of productivity. The figures that researchers have obtained in their various studies of these habitats in different parts of the world are bound to be variable when one considers the relatively small samples of the ecosystems that are actually measured and taken to be representative of whole biomes.

Biomass is the total amount of above-ground material (mainly vegetation) present in an ecosystem. Biomass includes all living and growing tissues as well as the non-living materials that are intimately associated with living organisms, such as wood in the case of trees, and hair and claws in the case of mammals. It does not include the dead remains of plants and animals, consisting of shed leaves, fallen twigs and flower, shed antlers of deer, exoskeletons of insects, and so on. These items are included in the *litter*, which is the accumulation of all such dead and shed

parts upon the surface of the soil. Considering the entire ecosystem, there is often a general relationship between total biomass and primary productivity. Lowland rain forest above-ground biomass is usually about 11 pounds per square foot (40 kg m⁻²), while montane forest has an above-ground biomass of around nine pounds per square foot (30 kg m⁻²). So the ecosystem with the higher biomass (the lowland forest) also has the higher productivity in this case. This situation makes sense in a forest, as the more biomass above the soil surface, the greater the abundance of leaf tissue and the greater the potential productivity. In the lowland tropical rain forest, the canopy usually forms a series of three layers of trees, and this has the effect of concentrating the leaves into canopy zones, as shown in the diagram below. In this diagram the percentage cover of leaves is shown at different heights and the main layers of the tree canopy are clearly visible. This layering of leaves in the forest canopy is evidently an efficient way of organizing the leaves to ensure a maximum trapping of light energy and thus enables the forest to maintain its high primary productivity.

The relationship between high biomass and high productivity found in the tropical rain forest does not hold

The leaf cover, equivalent to the canopy biomass, in a tropical forest forms a series of layers at different heights above the ground. Here there are three or possibly four layers within the canopy.

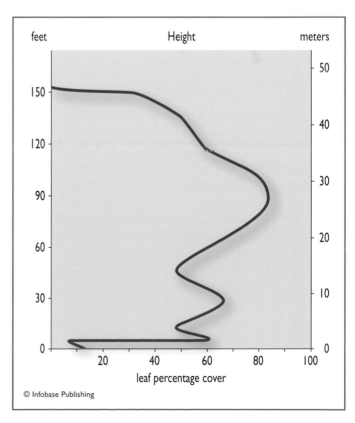

true for all ecosystems, however. In the tropical savanna biome, for example, especially in areas entirely dominated by grasses, the biomass may be no greater than the annual productivity, because all the above-ground vegetation may die down during the dry season. So, if the efficiency of production is measured by comparing how much productivity is achieved per unit biomass, the savanna is more efficient than the rain forest, where productivity is less than 10 percent of the biomass. This efficiency of production, coupled with the palatability of grasses to large herbivores, including domesticated ones, explains why farmers find it economically attractive to clear rain forest to make way for grasslands (see "Forest Clearance," pages 213–216).

As has been explained, measuring biomass and productivity in a forest is extremely difficult, partly because so much takes place below the surface of the soil, where observation and measurement are almost impossible. In one study of an Amazonian rain forest, scientists found that the total above-ground biomass was 11 pounds per square foot (36 kg m^{-2}), of which almost 90 percent consisted of trunks and branches of trees, just 2.5 percent was leaf tissue, and the remainder was largely epiphytes and lianes. The researchers carried out a very careful separation of all the roots within the soil of their sample plot and found a below-ground root biomass of 3.6 pounds per square foot (12 kg m^{-2}). So this study revealed that about 25 percent of the tree is situated below ground in the form of roots. In a savanna ecosystem, approximately 50 percent of the plant biomass is underground, so the ratio of above-ground to below-ground plant biomass evidently varies between ecosystems, with the tropical forests carrying a substantially greater proportion above the soil surface.

The biomass of an ecosystem can be thought of as a reservoir of energy. It can be divided up into a series of compartments, most of which reside in the vegetation, such as trunk wood, leaves, branches and twigs, roots, flowers, and fruit. There are also other, smaller reservoirs of energy, such as the invertebrates, birds, mammals, reptiles, and amphibians—in other words, the animals of the ecosystem—which also contribute to the biomass, as do the microbes, including the fungi on the forest floor. In addition to the living biomass there is also nonliving organic matter that represents an additional reservoir of energy in the ecosystem. Dead trunks of trees may remain upright for a long while, sometimes referred to by ecologists as the "standing dead." Litter, consisting of dead plant and animal remains, lies on the surface of the soil, and beneath the surface is another reservoir of energy, the soil organic matter that is in the process of decaying. All of these different energy reservoirs form the food base for the heterotrophic organisms that occupy the ecosystem, and these animals and microbes are linked to one another in complex webs of interaction.

FOOD WEBS IN THE TROPICAL FORESTS

The primary productivity of tropical forest forms the energy input for the entire ecosystem. The energy fixed in photosynthesis and stored in plant tissues ultimately provides the source of food for all the heterotrophic organisms found within the ecosystem. There are two major pathways along which this energy derived from primary production can travel. Energy-rich plant tissues may be consumed by herbivorous organisms and subsequently pass on to predators, or the plant organ may escape the attentions of grazers until its eventual death, when it will enter the detritivore and decomposer system. But the track of energy movement is actually more complex than this because energy may move between these major pathways, crossing from one to the other as it proceeds through the ecosystem. Fungi are important decomposers, but they may be eaten by tiny soil invertebrates, such as springtails, which may then form the prey of beetles, themselves a source of food for insectivorous birds, and so on. Energy has thus moved from the decomposer system back into the herbivore–predator system. Similarly, the insectivorous bird may die and fall to the floor, where it is invaded by bacteria and its energy returned to the decomposer system.

It is possible, therefore, for energy to move on complex tracks through the ecosystem, but it cannot circulate indefinitely, because some energy is lost at each exchange (see sidebar "Energy," page 77). Eventually the energy is used by an organism to produce work and is then lost as heat in the process of respiration. If it were possible to follow the course of movement of one unit of energy subsequent to being trapped by a plant in photosynthesis, it would pass through a series of stages as it is transferred from one organism to another. A forest tree might put some of its energy into the production of fruit, which might then be consumed by a herbivorous animal, such as a tapir. The tapir might in turn fall prey to a jaguar. Energy has thus moved along a food chain, eventually being released by the respiration of the predator as it conducts its energetic hunting, or perhaps then entering the decomposer system when the predator dies. But the complex possibilities for energy movement in an ecosystem mean that the range of possible food chains is extremely large, and ecologists prefer to think in terms of a *food web*. A highly simplified example of a tropical forest food web is shown in the diagram on page 83.

When the feeding relationships of an ecosystem are simplified in this way, it is possible to detect a series of feeding levels, called trophic levels, in the food web. The first trophic level (to the left of the diagram) is occupied by the primary producers, the plants of the tropical forest. The primary con-

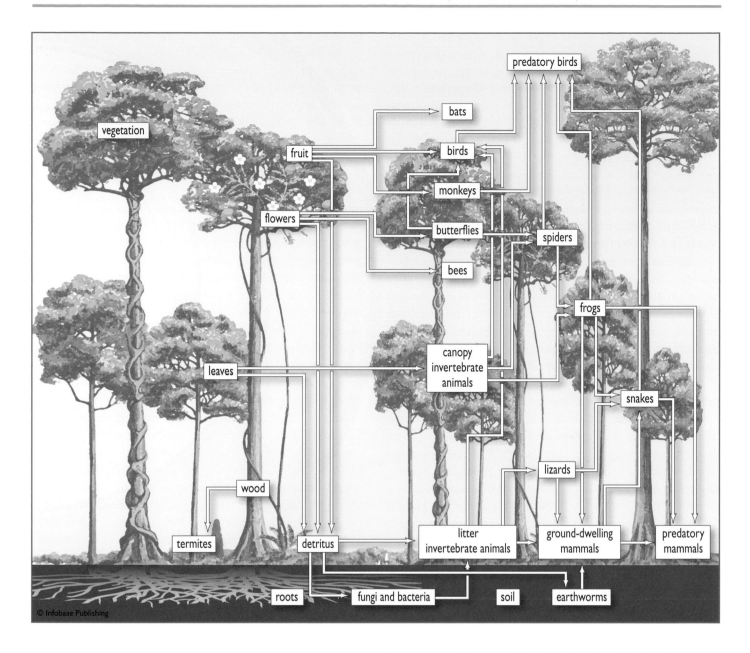

The food web of a tropical forest. Photosythesis by the trees forms the basis of all food webs in the tropical forest, and the fall of leaf litter to the forest floor provides energy for a range of animals and microbes that live there. Herbivores feed at different levels within the canopy and are preyed upon by carnivorous animals ranging from spiders to snakes and predatory birds. All the plants and animals in this flow sheet release energy by respiration and deposit litter or excreta, and eventually their dead remains, on the forest floor.

sumers, whether invertebrates, birds, or mammalian grazers and fruit eaters, form the second trophic level, and this is followed by one or two layers of predators, belonging to the third and fourth trophic levels, respectively (on the right-hand side of the diagram). For example, an invertebrate grazer, such as a caterpillar (the second trophic level), could

be consumed by a tree frog (the third trophic level), only to be eaten in turn by a snake (the fourth trophic level).

As energy moves from one trophic level to the next, however, some is inevitably lost and the efficiency of transfer is often only about 10 percent. One consequence of the inefficiency of energy transfer in ecosystems is that the amount of energy available gradually decreases as it moves up from one trophic level to the next. As has been seen, there is a larger quantity of biomass of vegetation in a tropical forest than there is of herbivores. Similarly, if one could gather together all the herbivores in the forest, from caterpillars to capybara, their combined mass would be several times greater than that of their predators. The loss of about 90 percent of the energy at each stage of transfer means that the biomass present within each trophic level declines with each step up the trophic levels of the food web. By the time the energy resources reach the top predator level (such as the jaguar), there is only a very small quantity of living material present per unit area of the forest. This arrangement is known as a *pyramid of biomass*. The energy pyramid is broad at the base, supported by the mass of vegetation, and becomes smaller at each trophic level until the top predators form the very small amount of biomass supported at the summit. This energy structure is common to most terrestrial ecosystems, including the tropical forests.

The potential length of food chains and the complexity of food webs are therefore limited by how much energy is available at the base. In other words, the rate of primary production and the quantity of plant biomass sustained in an ecosystem have a profound effect on the range of different organisms supported and the complexity of their interactions (see "Patterns of Biodiversity," pages 100–103). With its very high primary productivity, the lowland rain forest is able to support relatively long food chains, containing five or more links, and hence a very complex food web. High productivity is therefore one important element in explaining the biodiversity of the tropical forests.

Because there is so much plant biomass in the tropical forests compared with the biomass of herbivores, it is extremely unlikely that a herbivore will build such large populations that it threatens the survival of its plant food. The diversity of species also helps here. If grazing becomes locally intensive, perhaps depleting the fruits of a particular tree species, then the herbivore soon discovers that it is expending more energy in finding the fruit than it gains when it eventually eats its prize. The fruit-eating animal will then turn its attentions elsewhere and seek alternative fruits, which are usually abundant in a biodiverse forest. Similarly, a predator usually has a wide range of alternative prey animals to hunt, so if one species becomes infrequent the predator can concentrate on another, giving the first prey species a chance to recover its former population levels. This is known as prey switching. It is a form of shuffling the food web pattern when one link starts to become weak, and some ecologists believe that the diversity of species available within the food web is a major contributor to the stability of this complex ecosystem (see "Ecosystem Development and Stability," pages 91–94).

■ ELEMENT CYCLING IN THE TROPICAL FORESTS

Just as energy is transferred from one trophic level to another, so are the elements that make up the food. Most of the nutrient elements needed by chimpanzees, macaws, and jaguars, for example, are obtained from their food, though they may supplement this with their drinking water and, in conditions of nutrient scarcity, may even lick rocks or eat soil, an activity that zoologists call geophagy. An animal loses elements when it urinates, defecates, sheds hair or other body parts, or dies. The elements within the organic remains of the animal enter the soil, where they are released by microbial decomposition and are available for uptake by the vegetation once more. The nutrients in the tropical forests, as in all ecosystems, thus circulate in a nutrient cycle.

The nutrient cycle of the tropical forests, like the food web, is very complex. The existence of several trophic levels in the forest ensures that some elements do not necessarily pass straight from a grazing animal to the soil, but can move along a series of alternative routes. For example, a capuchin monkey feeding on the leaves high in the canopy in a South American rain forest may fall prey to a harpy eagle (*Harpia harpyja*). An atom of nitrogen the monkey has taken in from the leaves it has eaten may then pass into the gut of the eagle and may subsequently fall to the ground when the bird defecates. Or the atom may be absorbed by the eagle and built into its proteins, only falling to the ground for recycling when the eagle dies. The movement of elements along such a chain, however, does not follow the same rules as energy. There is no loss of nutrients to the ecosystem as a consequence of transfer between trophic levels. Nutrients can move out of ecoystems, for example, if the eagle flies off to a different habitat, but they are never entirely lost to living organisms in the way that energy is lost. This is why ecologists speak of nutrient cycling as distinct from energy flow.

As in the case of energy, however, the movement of nutrient elements around the ecosystem is determined to a degree by the activities of the organisms present. The faster the rate of primary productivity, the higher the rate of nutrient uptake by plants from the soil. The more intensive the grazing and predation, the faster nutrients are circulated between trophic levels. The speed of decomposition of dead materials also determines how fast elements are liberated and made available for plants once more (see "Decomposition in Tropical Forests," pages 89–91). Warm temperatures throughout the year and a constant supply of water in the lowland rain forests ensure that the nutrient cycle proceeds uninterrupted throughout the year. In seasonal forests drought may cause changes to the rate of cycling as the movement of elements from forest canopy to the soil surface take place in a single season rather than being spread throughout the year.

Although elements cycle around an ecosystem, they can also enter and leave it. It is important to understand the pattern of gains and losses to ecosystems, especially if a harvest, such as forest logging, is to be made sustainable over a period of time. There are several different ways in which elements enter and leave the tropical forest ecosystem, which can be classified into geological, meteorological, and biological mechanisms. Geological sources of elements involve the breakdown of rocks. In the weathering processes ancient elements trapped for millions of years are released back into circulation (see "Tropical Rock Weathering," pages 43–46). The constant movement of slightly acidic water through the soil, and the chemical consequences of microbial activity in the soil, all assist in the degradation of rocks and the liberation of elements into the forest soil environment. Some of the elements released into the soil are leached by the drainage water and pass into streams and rivers that transport the dissolved elements out of the forest ecosystem and eventually into the ocean.

The atmosphere is also a source of nutrient imports to ecosystems, both directly in supplying the gases carbon dioxide and nitrogen and indirectly through the meteorological processes within the atmosphere. Precipitation in the form of rain is rarely pure. As water droplets form in the atmosphere and as they fall toward the ground, they collect dust fragments and dissolve chemical atoms and molecules that they encounter on their descent. Some of these atmospheric components result from human pollution, often many thousands of miles away, other molecules are entirely natural, derived from wind-eroded soils or from the spray of the oceans. Rainfall in the tropical forest is thus a route by which mineral elements can enter the ecosystem, and some members of the forest community, such as the epiphytes growing on high branches, far from the ground and its mineral supply in the soil, depend entirely on this source for their inorganic requirements. The quantities of nutrient elements brought by rainfall are extremely variable, ranging between five and 18 pounds per acre (6–20 kg ha^{-1}) for nitrogen, 0.4 and 15 pounds per acre (0.5–17 kg ha^{-1}) for phosphorus, and between six and 16 pounds per acre (7–18 kg ha^{-1}) for potassium in a year. Oceanic sites are generally richer in calcium and magnesium than those farther from the coast.

The rain that falls on a tropical forest can thus bring elements from the atmosphere and the oceans, and it can remove elements from the soil in the process of leaching. The movement of rain through the ecosystem also plays an important part in the cycling of nutrients between the tree canopy and the soil below (see sidebar [right]).

Element movement from the tree canopy to the soil can thus take place by a variety of routes. Leaves may fall to the ground and decompose. Invertebrate and vertebrate grazers consume some elements and pass them into food webs, ultimately delivering them to the ground as excreta or dead remains. Leaves may also be leached of their chemical contents by rainwater passing through on its way to the ground (see illustration on page 96). Ecologists can compare these different pathways experimentally. They place a rain collector above the canopy or outside the canopy in a clearing, and can then analyze the chemical content of the water arriving at the upper level of the ecosystem. They also place a rain

Canopy Leaching

When raindrops collide with the leaves of a forest canopy they can cause physical and chemical changes to the leaves. The physical force of falling rain can strip away the waxy layer of the upper surface of the leaves, the cuticle, leaving the underlying tissues exposed to desiccation and to infection by fungi and bacteria. This is especially true of rainfall that has been made acidic by industrial pollution, but that problem is more frequently associated with the temperate forests growing close to regions of major industrial activity than with the relatively unpolluted forests of the Tropics. But even unpolluted rainwater can lead to physical damage to the leaf.

Water landing on the leaf canopy drains over the surface of the leaf and is then shed to lower layers and ultimately to the soil. There are two main drainage routes for water running off leaves. Either the water may drip directly from the leaf tip, or it may run down the leaf stalk, the petiole, and then along twigs and branches until it reaches the main tree trunk. Water passing along this route moves down the trunk in streams; in heavy storms it may even cascade in torrents. Water moving down the trunk is termed *stem flow,* as distinct from the canopy *throughfall* that drips directly from the leaves.

Water moving through the leaf canopy and descending to the ground by one or other of these two routes picks up elements from leaf surfaces as it flows. All leaves are slightly leaky and release small amounts of cell exudates onto their surfaces. These materials are washed off by the moving rainwater and carried down to the ground. Rainwater passing through the canopy thus acts in a similar way to water draining through the soil—it leaches out elements from the upper layers of the forest and delivers them to the soil where they are available for recycling by the trees. This process is called canopy leaching.

collector beneath the forest canopy and analyze the chemistry of water that has passed through the layers of leaves above. In detailed studies, they may also collect stem flow water passing down the trunks and determine its chemical constitution. The difference between the concentration of elements in the above- and below-canopy collections is the quantity removed from the leaves by the passing water. To measure the movement of elements by leaf fall and animal

Rain forests in Sarawak, Malaysia, recycle the water that falls on them. As water passes out of the trees by transpiration and into the atmosphere above the canopy, it condenses into mist and clouds, retaining the humid atmosphere over the forest and collecting on the foliage as occult precipitation. *(Kenneth McIntosh)*

detritus, they set out trays beneath the forest canopy and collect the materials that fall from above, including dead leaves and twigs, dead insects, fecal material, and other falling fragments. Invertebrate animal detritus is called *frass*. The detritus can be separated into its different components, such as plant and animal remains, but often ecologists combine all the material under the title litter.

Researchers have carried out experiments concerning nutrient movements in different rain forests around the world and the results have proved quite variable between sites and between different elements. In one experiment from the rain forests of Ghana, Africa, the rainfall contained 16 pounds per acre per year of potassium (18 kg ha^{-1}y^{-1}).

The water from below the canopy contained 211 pounds per acre per year (237 kg ha^{-1}y^{-1}), so a total of 195 pounds per acre per year (219 kg ha^{-1}y^{-1}) had been leached from the leaf canopy by the passing rain water. The litter fall in a year contained 61 pounds per acre (68 kg ha^{-1}y^{-1}), so canopy leaching accounted for 76 percent of the total movement of potassium from the leaf canopy to the soil. Falling leaves and animal detritus accounted for only 24 percent of the total. The conclusion from the study of potassium movement and that of other elements was that canopy leaching is a very significant process in the nutrient cycling pattern of the tropical rain forest. There is, however, considerable variation in the data obtained from different sites and for different elements, as shown in the accompanying table on page 87.

The amount of different elements contained in the rainfall varies between oceanic and continental sites, but the leaching of potassium from leaves as rain passes through is a common feature of all sites. More detailed studies show that the leaching of leaves is most apparent when they are very young because the protective wax cuticle is incompletely formed and potassium leaks more freely than when

The Importance of Throughfall as a Mechanism for Nutrient Movement from Canopy to Soil in Tropical Forests

	K	P	N
Ghana	76	37	5
New Guinea	69	27	21
Venezuela	30	–	9

Figures are expressed as percentage of nutrient transfer accounted for by throughfall for a range of elements.

the leaf is mature. There is a second leaching peak when the leaf is very old and the tissues are beginning to break down. Cells then become infected with fungi and cell contents often leak into the surroundings. At all sites canopy leaching is an important, often the most important, mechanism by which potassium is transferred from the canopy to the soil. Nitrogen is also leached from the canopy in this way, but the quantities carried by rain are relatively small in comparison to the litter fall, which accounts for over 90 percent of nitrogen transfers in all sites. Phosphorus behaves more erratically. In the Ghana and New Guinea sites throughfall is an important pathway for phosphorus loss from the canopy; no figure is given for the Venezuelan site. This is because the water collected below the canopy in this experiment contained more phosphorus than the original rain water. In other words, the leaf canopy had absorbed phosphorus from the rainfall. Phosphorus is a relatively scarce element in nature, but is one that plants and animals need in considerable quantities for their maintenance. The leaf canopy is evidently acting as an absorptive organ for the trees, taking up the scarce phosphorus from the rainfall and using it as a kind of foliar fertilizer. Similar results have been found in many of the world's forests, both tropical and temperate. Canopies are thus not always leached by rainfall, they can reverse the process and use the rainfall as a direct supply of scarce elements. Epiphytes have no contact with the ground, so they must receive all of their nutrient input from rainfall and local throughfall.

Some of the nutrients that are leached from the canopy or fall in the litter never reach the ground. The tangled complexity of the layers of branches results in some organic detritus being caught in the angles of branches, or even on their horizontal surfaces. As organic matter builds up in crevices and crannies, some of the leached elements are adsorbed and enrich the quality of this aerial soil, just as organic matter in the soil acts as a reservoir for nutrient elements. These pockets of soil in the canopy are an important microhabitat for epiphytic plants (see sidebar "Epiphytes,"

page 121) and for invertebrate animals that can dwell in soil, yet never touch the ground. Some tree species even manage to steal nutrients from these suspended pockets of soil by sending their roots up the trunks of neighboring trees to tap into the reserves of their canopy soils. The roots of plants are normally positively geotropic, which means that they can detect the force of gravity and respond by growing downward. But canopy-tapping roots are negatively geotropic and grow upward instead, following the course of stem flow water up to the source of their nutrient enrichment in the soil pockets. Ecologists have conducted experiments in which they placed animal manure on platforms at the top of tall poles and allowed the leachate to leak down the poles to the ground. Roots from neighboring trees immediately started to grow up the poles, seeking the source of the nutrients at its summit.

The cycling of nitrogen in ecosystems is more complex than that of most other elements. Certain microbes possess the ability to reduce atmospheric nitrogen to ammonia and thus bring it into the nutrient cycle of ecosystems (see sidebar on page 88), and these constitute an important biological source of nitrogen to many ecosystems, including tropical forests.

Nitrogen fixation within an ecosystem is difficult to measure because it takes place wherever the appropriate microbes are able to live and conduct their biochemical activities. Much nitrogen fixing goes on in the soil or in plant roots within the soil. Ecosystem studies in Costa Rica have revealed that almost 20 percent of the forest trees belong to the pea family, Fabaceae (previously called Leguminosae), and have nodulated roots in which symbiotic bacteria fix nitrogen. Free-living cyanobacteria and other fixing microbes live on soil surfaces, tree trunks, pockets of water or aerial soil, and even on the surface of leaves. Microbes that find a home on leaf surfaces are said to be *epiphyllous*. The nitrogen fixed by epiphyllous microbes is quickly assimilated by the leaves of the host plant, so the trees can derive great advantages in having these microbes on their surfaces. One other surprising source of nitrogen in the tropical forest ecosystem is found in the guts of termites. These colonial invertebrates, whose main activity is the consumption of dead wood in the ecosystem, have symbiotic bacteria that are able to fix atmospheric nitrogen in their guts, which operate rather like the nodule-inhabiting microbes of leguminous plants. Because of these numerous sources, all of which are difficult to monitor, estimates for nitrogen fixation in tropical forests are quite variable. Most of the available estimates of nitrogen fixation in tropical forests fall between 36 and 100 pounds per acre per year (40–110 kg ha^{-1}y^{-1}).

The cycling of nutrients within a tropical forest is summarized in the accompanying simplified diagram on page 88, where the main reservoirs and pathways of nutrient movement are set out. The boxes show where nutrients are to be found at any given moment, and the arrows illustrate

Nitrogen Fixation

Atmospheric nitrogen occurs in the form of the dinitrogen (N_2) molecule, which has a very strong set of bonds between the two nitrogen atoms and is consequently a relatively inert material. But certain microbes are able to break this bond and to reduce nitrogen gas into the ammonium ion NH_4^+ and subsequently to an organic form. Among these are some free-living organisms, such as the photosynthetic *blue-green bacteria* (*cyanobacteria*) and non-photosynthetic microbes such as *Azotobacter* and *Clostridium.* Sometimes the cyanobacteria are found in a symbiotic relationship with plants, such as the water fern *Azolla,* and the cycads, a tropical group of gymnosperms. Other nitrogen-fixing microbes are found only in association with higher plants, including *Rhizobium* and *Bradyrhizobium,* a group of bacteria commonly known as *rhizobia,* found in the root nodules of members of the pea plant family (Fabaceae), and others associated with such plants as alder (*Alnus* species), bog myrtle (*Myrica gale*), sweet fern (*Comptonia* species), and mountain lilacs (*Ceanothus* species). The process of fixation consumes a large amount of energy, provided by the molecule *ATP,* which is why so many nitrogen-fixing microbes are found in association with photosynthetic plants. The plant produces the energy and the bacteria do the nitrogen fixing. So when a plant develops a symbiotic relationship with a nitrogen-fixing bacterium it takes on a very expensive partner in energetic terms. But the payoff is worthwhile, because the plant is able to enhance its protein supplies.

Vegetation containing plants of the pea family enhances the nitrogen levels of the soil, thus acting as a fertilizer. The ancient Greeks knew about this means of soil improvement and used beans to improve their agricultural soils. Clover (*Trifolium* species) is often used in the same way by modern farmers, who find that up to 270 pounds per acre (300 kg ha⁻¹) of nitrogen can be added to the soil each year by this means. Some calculations suggest that 150 million tons of nitrogen are fixed from the atmosphere each year by natural biological fixation over the surface of the Earth. Members of the Fabaceae plant family are common as trees in tropical forests, and many epiphytes have associations with nitrogen-fixing microbes.

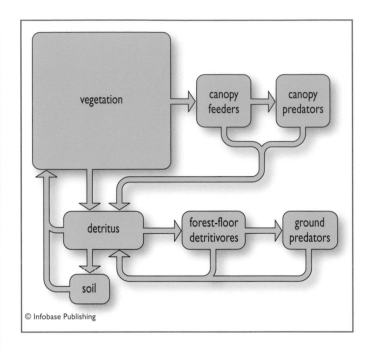

© Infobase Publishing

The simplified nutrient cycle of a tropical forest. The bulk of the nutrient capital of the forest lies in the vegetation, with only a relatively small reservoir in the soil. Decomposition of leaf litter releases the nutrient content, which is immediately recaptured by the tree roots in the surface layers.

the direction of nutrient flow. The sizes of the boxes give some indication of the quantities of nutrients present within each of the main element reservoirs of the ecosystem; it can be seen that the vegetation biomass is the largest of all the nutrient reservoirs. In comparison, the nutrient contents of the animals, the detritus (litter), and the soil are all relatively small. In this respect, the tropical forests differ from most temperate forests, where the soil nutrient capital is usually comparable to that of the vegetation biomass. The reason for this difference between temperate and tropical forest is the faster rate of decomposition in the Tropics, leading to a rapid breakdown of litter and soil organic matter, forcing the trees to compete for the released elements in the soil before they are leached out by the constant flow of water through the soil.

As described in chapter 3, there are many different kinds of tropical forest, and they differ considerably in their nutrient richness. Some forests, called *eutrophic* forests, are rich in nutrients and have a high level of element concentration in their biomass and their soils. Others are relatively poor in nutrients, depending on the rocks on which they are developed and the quantity of new nutrients brought into the ecosystem by streams, rivers, and rainfall. Generally, the tropical forests of Costa Rica and Panama, together with the montane forests of the Ivory Coast in Africa, are eutrophic forests with foliage relatively rich in elements such as phosphorus and nitrogen. The montane forests of Puerto Rico and the dipterocarp forests of Malaysia, by contrast, are relatively poor in nutrients. They are said to be *oligotrophic* forests. The forests of the Amazon Basin are extremely variable, depending on their location and the nature of the rivers that feed them. The white-water rivers bring relatively nutrient-rich waters to the forest, whereas the blackwater rivers are extremely poor and support oligotrophic forests.

■ DECOMPOSITION IN THE TROPICAL FORESTS

Many chemical reactions are strongly influenced by temperature, proceeding more rapidly when the temperature rises. Most *microbes* (bacteria and fungi) are affected by this because they obtain their energy through a process of external digestion. Rather than ingesting food material and soaking it in a solution of digestive enzymes, they secrete the enzymes into their environment and absorb the breakdown products through their cell surfaces. If the conditions in the environment are warm, the chemical reactions involved in organic matter digestion take place faster and the microbial populations respond by reproduc-

ing faster. This is why food materials are best protected from microbial attack by keeping them at low temperature in a refrigerator. In the equatorial regions, the constantly high temperature and high humidity are ideal for microbial activity and growth. Most microbial digestive enzymes operate most effectively in a temperature range of 75°–85°F (25°–30°C), so organic matter decomposition takes place very rapidly in tropical soils (see illustration below).

The external digestion employed by microbes is essentially similar to the internal digestion used by higher animals. Complex molecules containing energy in their chemical bonds are broken down by reactions involving enzymes, proteins that form a temporary link with the substrate and weaken their bonds, allowing the reactions to take place even under relatively low temperatures and pressures. As the complex molecules are broken down, energy is released. Starch, for example, is released from the dying cells of a fallen leaf and is attached by microbial enzymes,

Fungi lack chlorophyll and depend on organic materials such as dead wood and leaf litter for their energy. They are abundant on the forest floor, where light levels are low but organic detritus is plentiful.   *(fotosav)*

splitting the complex molecule into its glucose components. Glucose can then be absorbed by the microbe and used as an energy source during the further breakdown into carbon dioxide and water. Similarly, proteins are broken into their constituent building blocks of amino acids, which contain nitrogen atoms, bound up with hydrogen to form amino groups. These amino groups can also be oxidized by microbes to release energy and result in an oxidized form of nitrogen called the nitrate ion. The process of oxidation of nitrogen atoms in the soil is particularly important because it influences the availability of this vital element to plants. The process is known as nitrification (see sidebar [right]).

Nitrate ions, unlike ammonium ions are not retained very efficiently by soils, so unless they are rapidly taken up by plants they are in danger of being leached out of the ecosystem. In the waterlogged soils of the tropical swamps there is an additional cause of nitrogen loss from the soil, namely denitrification. Oxygen is in short supply in the stagnant water of saturated soils, and some microbes, such as members of the genus *Clostridium*, use nitrate as an oxidizing agent in their respiration instead of oxygen. As they decompose organic materials they produce carbon dioxide and nitrogen gas, thus depleting the soil further of its nitrogen resource.

Nitrate ions are thus extremely valuable commodities to the trees of the tropical forests, but they are also rapidly lost, so plant roots compete to collect these vital nutrients before they disappear. There is one method of competition for nitrogen that shortcuts the nitrification process, which involves tapping the resource while it is still in its ammonium ion form. Relatively few species of plants have developed the capacity to absorb nitrogen from the soil in its ammonium form, but there is a clear advantage in doing so. Since ammonium ions are positively charged, they are more easily retained in soils than the negatively charged nitrate ions, but most tropical soils are relatively poor in clay minerals with their negative charges that hold on to ions, so even ammonium ions must be taken up quickly if they are not to be lost to the ecosystem. However, some rain forest trees have the capacity to absorb the ammonium ions directly, and there is some evidence to suggest that such trees also actively inhibit the process of nitrification. The likely mechanism they use is production of toxic chemicals in their leaves that inhibit the activity of *Nitrosomonas* and *Nitrobacter*, resulting in the accumulation of ammonium nitrogen in the soil and the starvation of other plant species awaiting the production of nitrates. This active competition is known from a range of temperate ecosystems poor in nitrogen, such as bogs, heathlands, and pine forests. This competitive technique ensures a supply of nitrogen for the specialist plants and leaves other species weakened in the struggle for survival. The use of chemical warfare by plants in this way is called *allelopathy*. Evidence is accumulating for the existence of such chemical

Nitrification

Nitrogen is present in living plants, animals, and microbes in a number of different forms, but most frequently as a constituent of amino acids, the components of proteins. Proteins are present in the cell either as enzymes, chemical catalysts that are responsible for regulating all the metabolic processes within the cell, or as having structural functions in some tissues, such as hair and claws. When an organism dies, microbes use proteins among other substrates as a source of energy and they release the energy by breaking the bonds in complex organic molecules. The decay of proteins first involves breaking the large molecule into amino acids and then a process called ammonification, or nitrogen mineralization. In this reaction, the nitrogen component of amino acids is removed from the carbon skeleton and released in the form of ammonium ions, NH_4^+. A wide variety of microbes are capable of this part of the decay process.

Ammonium ions are retained by soils because of their positive charge. Clay particles in the soil are negatively charged, so they adsorb these cations and hold them in a loose bond. Most plants, however, are unable to absorb nitrogen in this form, so the ammonium nitrogen is unavailable to much of the vegetation. There are other microbes in the soil, however, that are capable of oxidizing the ammonium ions to nitrate ions.

$$2NH_4^+ + 3O_2 \rightarrow 2NO_2^- + 4H^+ + 2H_2O$$

$$2NO_2^- + O_2 \rightarrow 2NO_3^-$$

This conversion is called *nitrification*. It is carried out in two stages, as shown here, and involves two different groups of bacteria. The fist stage is conducted by bacteria of the genus *Nitrosomonas,* and the second by those of the genus *Nitrobacter*. These microbes are chemosynthetic autotrophs. Like photosynthetic autotrophs they can feed themselves, but they obtain their energy not from light but from inorganic chemical reactions. The oxidation of ammonium ions and nitrate ions releases energy, so the microbes that conduct this process are making a living by regulating this chemical reaction. The outcome of nitrification is the production of an oxidized form of nitrogen, nitrate ions, NO_3^-, and most plants use this form of nitrogen as their major source of the element.

Mycorrhizae

The word mycorrhizal literally means "fungus root," and it describes an intimate association between a fungus and the root of a higher plant. In most of these associations, the fungus penetrates the cells of the host plant and then divides into a series of fine branches, called *arbuscules,* and sometimes produces swellings called *vesicles.* This type of mycorrhizal association is thus called vesicular–arbuscular (V/A). The fungus obtains energy from the plant in the form of sugars and other products of photosynthesis and plant metabolism, but in return it supplies to the plant an enhanced system for gathering water and mineral elements from the surrounding soil. The free mycelia of the fungus in the soil gather elements such as phosphorus, zinc, manganese, and copper and transport these to the plant root. Seedlings become infected with the mycorrhizal fungus early in life and usually fail to survive if infection does not take place. The relationship is thus a true symbiosis, with both partners benefiting from the association. Perhaps 80 percent of plant species are infected with V/A mycorrhizae, and many of the remainder have a less invasive association called ectomycorrhizae, rather than the intimate endomycorrhizae of the V/A type. Ecotmycorrhizae occur in the form of a sheath of fungal tissue outside the plant root, and these are found in many types of trees, including members of the beech, oak, birch, pine, and eucalyptus families. The ecological importance of mycorrhizae is considerable, enabling plants to survive under harsh conditions and providing an additional competitive mechanism in soils with low nutrient content. The fungal mycelium greatly increases the surface area over which water and nutrient elements can be absorbed and passed on to the plant root. It is possible that elements may proceed directly from decaying organic matter, through the fungus, to the host plant without ever having entered the soil. Many ecologists support this idea of a closed cycle of nutrients in the tropical forests, although it has yet to be conclusively demonstrated. Mycorrhizae may also play an even more subtle role in population ecology because a fungus can invade the roots of neighboring plants and they then form a link between individuals in a population.

techniques of plant competition both against soil microbes and in the suppression of seed germination, especially of potential competitor species.

The roots of plants, foraging for nutrient elements in the face of scarcity and competition, have an additional secret weapon, the assistance of symbiotic fungi. Many, perhaps most, of the tropical forest trees have roots that form close associations with soil fungi to form mycorrhizae (see sidebar [left]), which are of great value in assisting the roots to gain access to the limited nutrient resource.

Overall, therefore, the rapid rate of decomposition in tropical soils leads to intense competition between plant roots, ensuring that elements released by microbial activity are rapidly taken up by the trees. The demand for nutrients is so strong that these valuable commodities are not left for long within the soil and if picked up by mycorrhizal fungi may never enter a dissolved state in the soil water (see sidebar). The whole ecosystem is thus a highly conservative one, geared to seal possible pathways of nutrient loss and keeping the bulk of the required elements tightly stored in the plant biomass. At times of biomass growth, such as during the regeneration that takes place in a forest following a catastrophe, the demand for nutrients becomes even greater because a growing ecosystem needs more resources than a stable one.

■ ECOSYSTEM DEVELOPMENT AND STABILITY

Ecosytems grow almost like organisms. Areas denuded by tropical storms and river flooding are gradually colonized by pioneer plants and animals, and the microbial populations of their soils actively decompose any detritus left by the catastrophe. The ecosystem steadily increases in its diversity and complexity (see sidebar "Succession," page 63) over the course of time. As the ecosystem develops, changes occur in both its energy flow patterns and its nutrient cycling systems.

A young ecosystem is dominated by its nonliving components, such as the fragmented rocks of its primitive soils, and the atmosphere. In the case of ecosystems developing from devastated land, there may be a residue of seeds and soil animals remaining, which form the basis for the new development. Additional living organisms arrive, often carried in the air as spores and seeds or, in the case of mobile animals, flying, so the living component of the ecosystem begins to expand and assume greater importance. Blue-green bacteria and lichens arrive as spores in the air, colonize rock and soil surfaces, and begin to take in solar energy and carbon dioxide from the atmosphere to form organic compounds. Bacteria and fungi also arrive by air and feed on the dead remains of the photosynthetic microbes. Mosses and ferns are widely dispersed by air as spores and colonize primitive soils, especially where damp conditions prevail, and they add more energy to the expanding biomass of the ecosystem and more organic matter to the soil (see "Soil Formation and Maturation," pages 46–50). Invertebrate animals may also arrive by air, such as tiny spiders, which spin webs that act as parachutes and carry their living passengers for many miles. Vertebrate animals, including birds, wander into the open clearings to feed upon the rich vegetative growth in the open, unshaded conditions, and they may defecate while present, leaving behind seeds from the forest that have passed through their guts and are now available to grow and outcompete the pioneer species. In Madagascar, the black lemur is an important transporter of seeds from mature forest to clearings, and in many tropical forests monkeys, parrots, and hornbills play an important role in the spread of forest trees and hence the healing of gaps in the canopy.

The gap-filling process of succession in a tropical forest. (A) When an opening occurs in the forest, either by the death of an old tree, wind blow, lightning strike, or human clearance, the process of recolonization begins almost immediately. (B) New tree seedlings invade and (C) grow rapidly, succeeding one another as (D) light-demanding species are replaced by shade-tolerant species. Patches of regenerating forest add to its overall biodiversity.

The outcome of these invasive events is that the composition of the ecosystem changes. There is a buildup of energy, residing within the living biomass in the form of organic chemicals. The changes that take place in succession may be predictable, in which case the process is said to be *deterministic;* or the succession may be influenced by *stochastic,* chance events that deflect development from the expected path. There are some general features, however, that all successions share. Biomass continues to grow until climate or some other factor, such as shallow soil or high wind, prevents further growth. The biomass then stabilizes at its maximum level, as shown in the diagram on page 92. Biomass growth in a clearing or devastated area accelerates to a maximum as the trees begin to establish and grow, and then slows down as the gap in the canopy is filled. Studies in the rain forests of Venezuela have shown that maximum biomass growth takes place in the first 40 years of regeneration, after which the rate of growth in biomass slows down and finally achieves a degree of stability after about one or two centuries. This final, equilibrium state is called the *climax* of the succession. At this stage, the energy budget of the ecosystem is balanced. All of the energy fixed in primary production is being used for the respiration of the plants, animals, and microbes within the system, so there is no excess energy that can accumulate as additional biomass. Ecosystem growth has terminated.

Just as energy builds up in the living biomass during succession, so do nutrients. The expanding biomass can be considered a living reservoir of elements, especially carbon, nitrogen, phosphorus, calcium, and so on. At the start of the succession all of the chemicals were contained in the nonliving (abiotic) part of the ecosystem, mainly in the soil, but at climax the biomass itself forms the major nutrient reservoir, especially in tropical forests. Locked up within living tissues, elements are less likely to be leached from the ecosystem and lost, but they are released gradually by canopy leaching and litter fall leading to decomposition, so they can be recycled. At climax, the nutrient budget of the entire ecosystem is in equilibrium in the same way as the energy budget. The continued gains of nutrients to the ecosystem, by rock weathering, atmospheric deposition, stream inputs, and so on, are equaled by the losses from the ecosystem, especially by soil leaching. The ecosystem is now in nutrient balance unless conditions change and the biomass starts to grow again, which is possible if there is a shift in climate or some other form of disturbance.

Because the climax ecosystem is in equilibrium, it is also regarded as being stable. Stability is an important ecological concept, especially in the modern world where human disturbance and modification of the environment is so widespread and ecosystem conservation is thus a matter for great concern. But the concept of stability is also difficult to define and demands careful consideration (see sidebar).

The Concept of Stability

Stability is difficult to define. If a situation exhibits no change over time, it can be considered stable. An ecosystem in equilibrium with its physical environment will remain in its current state and will not alter over the course of time unless there is a change in the prevailing conditions, such as in climate. Such an ecosystem can be regarded as stable. But the word stability implies more than simply a lack of change. There are two additional ideas contained within the concept of stability: One is the ability to recover rapidly from any disturbance, and the other is the capacity to resist such alteration in the first place.

Rapid recovery from disturbance helps an ecosystem sustain itself in the face of modifying forces. Ecologists give this property the title *resilience.* A resilient ecosystem can cope with perturbation by returning to its original state rapidly without being altered by the experience. An ecosystem capable of healing itself effectively in this way can be regarded as stable.

Resistance to change is called *inertia.* All ecosystems are subjected to disturbing forces, such as extreme conditions of weather or human pollution events. Some ecosystems are easily damaged by such perturbations and can thus be considered relatively unstable, while others are more resistant; the possession of such inertia protects them from harmful effects.

A stable ecosystem, therefore, is one that is strong and unmoved by disturbing pressures, and one that recovers rapidly and returns to its original state if it is damaged. Such an ecosystem continues unaltered in the long term despite the forces that tend to deflect it from its present condition.

A stable ecosystem possesses both inertia (resistance to change) and resilience (the ability to recover from damage), but what features of an ecosystem can supply these qualities? One helpful way of approaching this important ecological question is to use the analogy of a bank. Before depositing money in a bank, an investor usually asks questions about the bank's stability and in this respect is well advised to check out the assets of the institution. A large bank with extensive assets is more likely to be a safe place in which to deposit money. If there is a sudden demand on its resources, it should be able to take the strain without injury and any money invested remains safe. A small bank with low assets, on the other hand, could be severely damaged in the event of any sudden large demand on its resources. Invested money is therefore less secure. To summarize, the stability of a bank depends on the size of its assets in relation to the demands placed on it. The same reasoning can be applied to ecosystems. The assets of the ecosystem are its biomass and any other nutrient reserves, such as the soil, and the kind of pressures exerted on those reserves range from heavy grazing (possibly by an insect population explosion) to pollution events.

Tropical forests are characterized by their very large biomass, contained within their trees. This biomass is a great store of energy available to the grazers or to humans if they wish to harvest it for timber. The biomass is also a reservoir of nutrient elements; indeed, it is the major nutrient reservoir of the entire ecosystem because the soils of tropical forest are generally rather poor in these elements. Studies of nutrient cycling in tropical forest usually reveal that the quantity of nutrients in circulation from such processes as canopy and soil leaching is small in comparison to the massive reserves of the forest biomass. For example, the biomass of a tropical forest in Thailand contains as much as 2,670 pounds per acre (3,000 kg ha^{-1}) of the element potassium. The annual loss of potassium from the soil by leaching is about nine pounds per acre (10 kg ha^{-1}). So the losses to the system represent only 0.3 percent of the "capital," and this is actually compensated for by the arrival of potassium dissolved in the rainwater. So the movement of nutrients through the ecosystem is small in comparison to the reserves held within the biomass. As in the case of the bank with large assets, such an ecosystem is well protected against an unexpected calamity, such as an excessive loss of potassium in a storm or as a result of an insect plague. If, however, the entire biomass of the ecosystem were removed, either by natural catastrophe or by human harvesting, the disturbance would severely damage the ecosystem and recovery would take a long time.

Tropical forests are thus protected from damage to a certain extent by their large biomass. Another feature of the forests that might enhance their stability is their biodiversity and the resultant complexity of their food webs (see "Food Webs in the Tropical Forests," pages 82–84). The great range of plants and animals found within the forest means that energy and nutrients can pass along many different paths, depending on who eats whom. If a grazer or a predator becomes abundant the population of its prey may become reduced, but in the tropical forest there are many alternative sources of food. Once its initial prey becomes difficult to find, the predator is likely to move its attention to an alternative food source, a phenomenon called prey switching. In a highly complex system, like the forest, prey switching is easy because of the large range of alternatives. No one is likely to go hungry in a well stocked restaurant. Similarly, if a plant or animal species becomes excessively

successful and expands its population, it is likely that it will attract the attention of a predator, parasite, or pathogen that will quickly bring it under control. Diversity and complexity in an ecosystem, therefore, should bring stability in its wake.

This line of reasoning dominated ecological thinking throughout the early part of last century because of its logical attractiveness, but in the 1970s it was brought into question. The development of computers led to an increase in research using mathematical modeling, and the model ecosystems constructed behaved in a way that differed from expectation. Complex ecosystems with many species and numerous interactions proved less stable than simple ecosystems. They were more easily disrupted by perturbations and failed to recover to their former state, so they lacked both inertia and resilience. These findings led to a prolonged debate about the relationship between complexity and stability in ecosystems that continues to the present day. Researchers are now using elaborate experiments to test the relationship, constructing artificial ecosystems with known numbers of plants, animals, and microbes, and subjecting them to stress to observe their degree of inertia. Although disagreement still prevails, ecologists are beginning to return to the idea that complexity does indeed impart stability to an ecosystem, so the argument is gradually turning full circle. This is the way science develops.

CYCLES IN THE FORESTS

Stability is a complicated concept, but it is important to remember that a stable ecosystem does not have to be static. The idea of dynamic stability is particularly important in natural ecosystems. The tropical forest, rather like science, can go round in circles and yet stay in approximately the same place. Small openings in the forest, created by wind blow, age, lightning strikes, and so on, heal by a process of succession, or gap regeneration. Observing rain forests from the air, they often look patchy and patterned as a result of the numerous small blocks in different stages of regeneration, yet the whole forest system maintains its overall stability. Scale is thus important in the consideration of stability. Each patch of forest is constantly changing, but the forest as a whole is in equilibrium. When dealing with hundreds of square miles of the ecosystem, the budgets for nutrient inputs and outputs, biomass, and energy transfers show no net gains or losses; the system is in equilibrium. This is an important concept when considering the value of tropical forest in relation to global carbon cycling (see "The Forest as a Carbon Sink," pages 209–211).

Time scales are also important in the concept of stability, as are spatial scales. Everyday conditions change in the forest with the coming of day and night, but this type of

r- and K-selection

The population growth curves of different organisms follow a similar basic plan. The simplest and fastest type of population growth can be observed among bacteria, and since they reproduce by simple cell-splitting (binary fission), these organisms provide an ideal model for population growth studies. Suppose a bacterium divides in two every 20 minutes. If a single bacterial cell is introduced into an ideal growth medium and commences its process of division, then within an hour there would be eight cells. In three hours the population would be 512, and in just six hours the population would have reached 262,000 cells. Mathematically, this is known as a logarithmic growth curve. If it continued indefinitely, then the Earth would be packed full of bacteria within a matter of weeks. In fact, there are limitations to population growth that may take the form of resource limitation, predation, parasitism, and so on. In natural populations, mortality resulting from such limitations keeps most populations under control for much of the time; only occasionally are there population explosions because of unusually high availability of food or lack of predation. A population growth curve over time thus generally shows a rapid expansion in its early stages, accelerating to a maximum rate and then gradually declining until it becomes level. In these latter stages the limitations to growth hold back any further population growth and the population has reached its maximum extent.

Population biologists use two terms to describe the particular shape of the population growth curve of an organism. The letter K represents the carrying capacity, which is given by the maximum number of individuals of the species that can be sustained by the environment. This is the density at which there is no additional population growth. The letter r represents the initial rate of population increase, and it relates to how rapidly a species can reproduce. A bacterium is very efficient at rapid reproduction, whereas an elephant is not. The bacterium has a higher r value. Some organisms are very good at expanding their populations but do not survive very well when placed under competitive stress. They are typical of open, noncompetitive environments, where resources are not limited, such as plowed arable fields or gaps in a forest canopy created by a minor catastrophe. Such species are said to be r-selected. The evolutionary pressures on them over many millennia have led to their adaptations as pioneers and invaders. They are the opportunists of the plant and animal worlds. K-selected species, by contrast, are not very good at invading new habitats. They may take a long while to reach an appropriate location, but once they establish themselves they can build up their populations and outcompete many of the pioneer species that have preceded them. The K-selected species is thus best adapted for long-term survival under stable conditions.

cycle does not interrupt the overall stability of the system. Seasonal cycles may also occur. In the equatorial regions, the climate varies very little between the seasons, but away from the equator a dry season becomes increasingly likely (see "Tropical Climates," pages 9–14). Although the seasonal change in climate causes alterations in the energy flow and nutrient-cycling patterns of the tropical forests, the annual cycles are part of the general stability of the system.

The most obvious effect of seasonal drought on the ecosystem is the occurrence of leaf fall during a specific period of the year. In the monsoon forests of northern India and the Himalaya foothills, for example, leaf fall takes place mainly in early summer from May to June, just prior to the monsoon rains. Approximately 80 percent of the litter falling to the floor of this forest takes place in this early summer season. Decomposition processes are then fastest during the monsoon rains when conditions are warm and moist. The rate of decomposition is slower at higher altitude in the Himalaya

Mountains, as might be expected. In the lower levels of the mountains the tropical seasonal forest litter decomposes as rapidly as that of the true rain forest during the wet season, and elements are released into the soil from the decaying leaves. The renewed growth of leaves following the monsoon rains then encourages the uptake of the soil nutrients by the tree roots. The seasonal cycle thus results in a regular pattern of leaf fall and maximum decomposition rates, followed by renewed growth and enhanced nutrient uptake on the part of the trees. Seasonal changes can also act as important cues for reproduction in the plants and animals of the forest (see "A Time to Flower," pages 132–133). But this seasonal cycling in ecosystem function does not detract from long-term stability.

Some cyclic processes in the tropical forests are not related to season and can be unpredictable. The populations of some animals, for example, may expand and contract seemingly irregularly. Sometimes it is possible to relate such cycles to environmental causes, as in the case of amphibian populations that vary with the availability of small water bodies at the time needed for amphibian reproduction. In the forests of the Ivory Coast, for example, there is a great deal of variability in pool frequency from year to year, and this influences amphibian populations. Species that succeed in unpredictable habitats often have a type of breeding system that enables them to cope with occasional failure and yet expand their populations rapidly once more when conditions improve. They are said to be *r-selected* (see sidebar on page 95). Species that are adapted to more stable and predictable conditions are said to be *K-selected*.

One general characteristic of r-selected organisms is that they put much of their energy into reproduction. The amphibians of temporary pools in the forests of the Ivory Coast produce very large number of eggs in their spawn. Mortality among the tadpoles is usually very high, but if conditions are unusually favorable, the amphibian population expands rapidly to take advantage of the opportunity. They thus cope well even in a catastrophic and unpredictable environment. Some plants adopt the same strategy. In Mexico, the tree *Cecropia obtusifolia* invades gaps and grows best under well-lit conditions. It produces large numbers of seeds that are spread by fruit-eating animals into the gaps in the forest. In forest gap regeneration, therefore, there is often a cycle of r-selected species gradually becoming replaced by K-selected species.

Flowering and fruiting among trees is often unpredictable in the tropical forests. Some years a tree species will produce large numbers of fruit, and other years, very few. Animal populations dependent on particular types of fruit may also vary according to food availability, or may migrate from one region of the forest to another to obtain the required food. The wild pig (*Sus barbatus*) of the Borneo forest migrates in this way to obtain fruits of the dipterocarp trees (family Dipterocarpaceae), which are rich in oils but are produced in abundance only during certain years, called mast years. The reproductive behavior of the pigs depends on the abundance of this food source, and pig populations expand following the mast years of the dipterocarps. Several of the bird species found in the tropical forests of Australia and New Guinea are similarly nomadic, seeking out areas where their preferred food is in greatest abundance.

Insect population explosions can cause defoliation of tropical forest trees, as in the case of the moth *Zunacetha annulata* in South America, which feeds upon trees of *Hybanthus prunifolius* in seasonal forests. This tree flowers in abundance when rain falls after a dry spell in the forest, but it is occasionally defoliated by outbreaks of caterpillars. The insect population explosion does not usually persist for long, however, because the abundance of eggs, larvae, and pupae is often followed by an increase in fungal infections leading to widespread death of the immature insects, so they do not survive to breed. The complexity and biodiversity of the tropical forest once again operates to damp down erratic fluctuations in population numbers.

Cycles of populations in the forest may thus be seasonal and predictable or triggered by unpredictable events, such as short periods of drought or rain. Some cycles, like the mast years of certain trees, are still not understood. But whatever the cause of cyclic events among the plants and animals of the forest, the overall stability of the ecosystem is rarely disturbed.

■ CONCLUSIONS

Ecosystems are a fundamental unit of ecology in which the nonliving and living components interact with one another. Energy, trapped from sunlight by photosynthetic plants, is transferred to herbivores, then to carnivores, or to the detritivores and decomposer pathways. All energy transfers are inefficient, so the number of energy transfers is limited, but the possible paths of energy movement through the complex food webs of the forest are very numerous. Chemical elements form the bodies of living organisms but are exchanged with the nonliving world. Plant roots are responsible for most of the nutrient absorption by living things, and microbial decay results in their recycling in a nonliving form.

Net primary production is the rate of matter accumulation by plants as a result of their trapping of the Sun's energy and using it to build energy-rich organic materials from inorganic carbon dioxide and elements from the soil. In lowland wet tropical forests, net primary production reaches some of the highest values recorded on Earth. The multiple layers of leaves ensure that much of the incident sunlight is captured, and the consistently warm and wet conditions of the equatorial regions allow year-round growth. Productivity is slightly lower in the deciduous seasonal forests and in the higher altitude montane forests. Much of the material accumulated by the plants is used in construction of the elaborate support systems for their leaf canopies, hence resources are directed into trunks, branches, and roots.

The biodiversity of the tropical forest results in very complex food webs, so both energy and nutrient elements can move along many different paths. But whereas energy is ultimately dissipated as heat, the elements are retained within the ecosystem. New supplies of elements arrive in the forest through rock weathering and rainfall. As rain passes through the canopy, however, it leaches some elements from the leaves, so water falling beneath the canopy is richer in most elements than the rainfall collected above the canopy. Potassium, in particular, is leached from the canopy leaves in large quantities. Indeed, more potassium may move from canopy to soil by rainfall leaching than by litter fall, though this varies from one forest to another. Phosphorus is in particularly short supply in the tropical forests generally, and the leaf canopy sometimes removes it from rainfall, leaving the water falling from the canopy depleted in this element. Nitrogen cycling in the tropical forests, as in all terrestrial ecosystems, is complicated because of the biological fixation of nitrogen gas from the atmosphere. Both symbiotic and free-living microbes within the tropical forest acquire nitrogen by fixation.

Decomposition is rapid in the warm and wet soils of the tropical forests, so elements are rapidly converted into inorganic form and are equally rapidly taken up once more by plant roots. Fungal associations with roots, mycorrhizae, are extremely important in assisting the trees as they compete for soil nutrients, and in return the fungi receive organic materials from their hosts. Competition for the short-lived pool of nutrients in the tropical soils can lead to a form of chemical warfare between plants. Chemicals produced in the leaves of some plants can suppress the activity of certain soil microbes, especially those involved with the conversion of ammonium ions to nitrate ions in the soil (nitrification). When this suppression is accompanied by absorption of the ammonium ions directly rather than waiting for the generation of nitrate ions, the normal route is bypassed and the tree gains earlier access to the vital nitrogen resource.

The competition for soil nutrients results in the plant biomass becoming the major reservoir for most elements within the ecosystem, rather than the soil. It can be argued that this lends stability to the ecosystem because the gains and losses for the entire system are very small in comparison to the amount of chemical elements bound up in the biomass. The biomass grows during the course of succession, placing greatest demand on the soil's resources while it is growing most rapidly, usually the first 40 years of growth. The biomass, together with the energy and nutrient budgets, then stabilizes as the ecosystem attains its equilibrium state, or climax.

The complexity and biodiversity of mature forest are probably contributors to its stability. Stability involves resistance to disturbance (inertia), together with the ability to recover quickly from perturbation (resilience). Complex food webs and the opportunity to switch between prey species enable herbivores and predators to resist major disruption when the forest is put under stress or if a population goes into unexpected decline. Major disruption, however, such as that caused by the removal of the bulk of the ecosystem's biomass and nutrient capital during logging, can still cause immense damage to the forest, and recovery takes centuries.

The seasonal forests of the Tropics, such as the monsoon forests and seasonal deciduous forests, exhibit regular and predictable cycles of change. Productivity, leaf fall, reproduction, decomposition rate, and vegetative regrowth are all related to the time of year. Cycles also take place that are not seasonal in nature. Gaps created by tree death and forest regeneration are frequent in the forest and provide opportunities for invasive, r-selected species to maintain themselves in the forest, acting as pioneers in the gaps and commencing the recovery process. They are eventually replaced by the K-selected species that reproduce more slowly but come to dominate the mature forest. Other cycles are even more unpredictable, such as the population explosions of animals that may follow unusually lush production of fruit. These events are usually quickly dampened down by the expansion of parasites, disease, or predator populations, thus maintaining the overall stability of the forest.

The forest ecosystem is highly complex and integrated, which is in part a consequence of its large biomass, elaborate physical structure, and its biodiverse plant and animal communities.

5

Biodiversity of the Tropical Forests

The Earth is a remarkable planet because it teems with life, while, as far as humankind currently knows, the remainder of the universe may well prove to be lifeless. Life makes the Earth special, thus it is reasonable that people should have due regard and respect for the richness of living things that exist here, the Earth's biodiversity. Of all the biomes of the Earth, the tropical forests are by far the richest in living organisms; about half of all the species on the surface of the planet find their home here. There are so many species that many, perhaps most, have yet to be discovered and described. Any assessment of the Earth's biodiversity must begin here, in the tropical forests.

■ THE MEANING OF BIODIVERSITY

The word *biodiversity* is currently used both by scientists and journalists, but it is not easy to define. The main component of biodiversity, and the one that immediately comes to mind, is the abundance of species present in an area, the *species richness*. Assembling species lists is therefore the first step in determining the biodiversity of an area. But diversity involves more than the number of species present. Consider a collection of 100 colored balls and suppose that they come in 10 different colors. The collection could have 91 balls of one color and one each of all the remaining nine colors, or it could have 10 balls of each color so that there was an even distribution of colors among the balls. The same could be true of a natural ecosystem with 10 species and 100 individuals. The individuals may be evenly spread among the different species, or one species might dominate the system and take up the bulk of the community. Faced with this situation, the ecosystem with the more even and equitable distribution of individuals among the species should be regarded as more diverse than the ecosystem dominated by one species. Ecologists therefore make a distinction between richness (number of species in an area) and diversity (a combination of richness and evenness).

Biodiversity, however, is yet more complex than this. All of the individuals in a population, unless they are vegetative clones, differ slightly in their genetic constitution. In human societies only identical twins have exactly the same genetic makeup, all others differ to some extent. This variety makes society much more interesting and flexible; the thought of identical people behaving in precisely the same way is the basis for disturbing science fiction, as in Aldous Huxley's *Brave New World*. Sexual reproduction involves the shuffling and recombination of the genetic attributes of parents, leading to totally new arrangements in the offspring, making them into distinctive and unique individuals. This is true of humans and plants, mammals and microbes.

The outcome of genetic recombination is genetic diversity. A population of organisms in which breeding takes place between unrelated individuals is said to exhibit *outbreeding*. New genetic characters are constantly being added to the population and new combinations of characters lead to greater genetic variability. This, in turn, equips the population to cope with new challenges, including adaptation to changing climate, pollution pressures, predation, disease, and so on. A small population in which breeding takes place between related individuals is deprived of new genetic input and new genetic combinations of characters and is said to demonstrate *inbreeding*. There is a danger that genes leading to physical weakness, normally recessive to more dominant healthy genes, will accumulate and make their presence felt in the population. Hemophilia in humans is an example of the type of recessive gene that shows itself under such circumstances of continued inbreeding. An organism that has outbreeding populations and a diverse genetic constitution is therefore more likely to survive in a changing world. It is said to have a wide *gene pool*.

Biodiversity is a concept that includes the breadth of the gene pool within the populations of organisms that constitute its species lists. Genetic diversity is a part of biodiversity because it is a component of the range of variation found within a particular habitat or ecosystem.

Biodiversity also includes the wealth of different microhabitats found within a landscape. This is closely related to species diversity because a diverse landscape, consisting of patches of many different habitats and microhabitats, contains more species than a uniform landscape that lacks variety. A varied geology and topography in a region develops a range of different soil types that in turn may support a variety of vegetation types with different species compositions and physical structures. Within these vegetation types, many different animals and microbes are able to survive, leading to high biodiversity.

Conservationists lay much stress on biodiversity. They regard each species as an irreplaceable combination of genetic material, sorted by the long and painful process of natural selection and honed to a high degree of fitness. Regions rich in biodiversity are therefore ranked highly in terms of conservation priority.

HOW MANY SPECIES ARE THERE ON EARTH?

Even the first component of biodiversity, species richness, is still a mystery. When asked the question "How many species are there?" biologists may give estimates ranging between 10 and 30 million. At present the total number of species of all living things that biologists have discovered and described comes to about 1.8 million, so all scientists are agreed that the actual number is very much greater than the current total. The following table shows the breakdown of this figure among the various plant, animal, and microbial groups, together with an estimate of what proportion of the Earth's total is presently known to science.

Some groups of organisms are well known, such as the vascular plants and vertebrates. Each year a few more plants, reptiles, birds, and mammals are found, even among the primates, which are among the best known groups of animals. The lemurs of Madagascar, for example, were thought to number 47 species, but two new species have been added in the last few years as a result of new surveys. Both are species of mouse lemur, *Microcebus lehilahytsara* and *Mirza zaza,* the former being about the size of the palm of a human hand and the latter as big as a squirrel. The Amazon Basin also continues to yield new primate species, such as two monkey species belonging to the genus *Callicebus.* But the total number of plant and vertebrate species is unlikely to increase significantly in the future. Apart from the discovery of new species, sometimes taxonomists (those scientists concerned with the classification and evolutionary relationships of living things) decide that what was once considered one species is better regarded as a number of species. The herring gull (*Larus argentatus*), for example, is a very widely distributed species with many different races, and ornithologists now feel that some of these races are in fact worthy of being considered species in their own right. This means that the number of bird species has increased without any actual "discovery" of new species. On the other hand, some species have been crossed off the list, either because they are now regarded as variants of a single species or because they have been separately described in different parts of the world, when they are actually the same species.

Certain groups of organisms are poorly studied, very small, or difficult to identify, and these groups are poorly known. As the table shows, bacteria, viruses, fungi, and nematode worms are all in this category. Less than 10 percent of theses groups have yet been described and there are undoubtedly many species that remain to be discovered. Estimates of microbial biodiversity are difficult to achieve,

GROUP OF ORGANISMS	NUMBER OF KNOWN SPECIES	LIKELY TOTAL	PERCENTAGE KNOWN
Insects	950,000	8,000,000	12
Vascular plants	250,000	300,000	83
Arachnids	75,000	750,000	10
Fungi	70,000	1,000,000	7
Mollusks	70,000	200,000	35
Vertebrates	45,000	50,000	90
Crustaceans	40,000	150,000	27
Protozoans	40,000	200,000	20
Algae	40,000	200,000	20
Nematodes	15,000	500,000	3
Viruses	5,000	500,000	1
Bacteria	4,000	400,000	1

but the extraction of DNA from soil can provide some idea of the range of microbes present. Microbiologists who have used this technique estimate that 0.035 ounces (1 g) of soil may contain a million different species of bacteria. The task of listing the world's bacteria is therefore daunting. Other groups of organisms, such as the insects, are also extremely numerous, so they remain poorly studied. The beetles (Coleoptera) alone comprise an enormous group of creatures, and scientists will take a very long time to describe them all.

The possible total number of species of insect is really an informed guess, and several techniques can be employed to help to achieve such an estimate. One entomologist, Terry Erwin, used a system of extrapolation from a limited set of observations. He studied just one species of tree in the forests of Panama and collected all the beetles from a number of examples of this tree by "fogging" them. This is a technique for collecting insects that involves producing an intoxicating smoke beneath the tree canopy and catching the falling insects in trays. When he had treated just 12 trees of a single species in this way, Erwin had collected over 1,200 species of beetle. If this is typical of the relationship between beetles and trees, Erwin calculated that there could be over 30 million species of insects in the world. Other studies of this sort have now been conducted, and the general consensus among scientists is that this is a high estimate, but eight million insects would not be unreasonable. Calculating the numbers of fungi on Earth has also been challenging. The best approach is to examine the data for an area that has been intensively studied, such as the British Isles, and extrapolate from those data. In Great Britain, there are about five or six species of fungi present for every species of flowering plant. If this is true worldwide, then there should be about a million fungi, although the assumption that this ratio holds true on a global scale is unproven.

Another approach to estimating the total number of species on Earth has been to use body size. On the whole, small species of animals are much more numerous than large ones, and the larger species are very naturally better known than the very small ones. For example, there are fewer than 10 species of animals that have a body length greater than 16 feet (5 m), including elephants, but there are several hundred with a length of 3–16 feet (1–5 m), such as antelopes and horses. About 1,000 species have a body length of 20–36 inches (50–100 cm), including monkeys and foxes, while there are about 10,000 species of rat-sized organisms 4–20 inches in length (10–50 cm). Most of the animals that fall into these size ranges have probably been described and are well known, but smaller creatures are less well known, so here is where the problem begins. If the relationships between size and diversity hold for the smaller organisms, then one would expect increasing numbers among the small species. The known curve seems to be approximately logarithmic in character, so extrapolating back from this into the murky world of the very small organisms the total number of species on Earth can be calculated, coming to approximately 10 million. This number falls in the same order of magnitude as that obtained by other methods, so the general conclusion obtained is that the Earth contains somewhere between 10 and 30 million species.

These methods are rather crude and simple, but they represent the best approach currently available to estimate the total number of living creatures on the planet. One fact is evident from all of these studies: The number of species currently known is only a very small proportion, perhaps only 10 percent of the world's total richness.

■ PATTERNS OF BIODIVERSITY

Biodiversity is not evenly spread over the face of the Earth. The Tropics contain a much higher diversity of species than either the temperate or the polar regions. Alaska contains 40 breeding species of mammal, for example, while British Columbia holds 70 species, California 100 species, and Costa Rica 140 species. Much of the gradient in mammal species is due to bats, which become much more diverse in the Tropics. At latitude 70°N in North America only four species of swallowtail butterfly are found, while there are 21 species at latitude 40°N and 80 species at the equator (0°) in northern South America. Costa Rica, in Central America, has 660 species of breeding birds, the region of Mexico around Mexico City has 340 species, Oklahoma has 140 species, the southern part of Hudson Bay has 110 species, and the North Slope of Alaska has 60 species. So for many different groups of animals there exists a latitudinal gradient of species richness, which reaches its peak in the Tropics.

Biogeographers have expended much effort and thought into explaining the latitudinal gradient in species richness. Climate seems the most obvious cause of the pattern, but how might climate operate in controlling biodiversity? Climate has an immediate effect on ecosystems by determining the productivity of vegetation (see "Primary Productivity in the Tropical Forests," pages 78–82). Vegetation is most productive when it is supplied with abundant water and warm temperature, which is usually the case in the equatorial regions. If either of these two requirements is lacking then productivity falls. Overall primary productivity in North America, for example, is greatest in the southeastern region, including Florida, Alabama, Georgia, and South Carolina. To the west and north of that area productivity falls. Altitude also has a marked effect, with productivity being low on the

Rocky Mountain chain and rising again on the West Coast; the general trend is a declining productivity away from the Southeast. Primary productivity lies at the base of almost all food webs, so a low productivity results in proportionally shorter food webs and fewer trophic levels. Consequently the ecosystem may be expected to support fewer species, and this certainly seems to apply to trees. The southeast region contains 180 species of trees and shrubs of 10 feet (3 m) height or more. New England and British Columbia each have about 60 species, declining steadily to zero in the High Arctic.

Although productivity is certainly an important factor in explaining gradients of diversity there are also other considerations. Warm, moist climates are the most suitable for the growth of trees, thus the vegetation structure, that is, the organization of the ecosystem's architecture, is more complex. This in turn affects microclimates and the variety of microhabitats available to all kinds of organisms, including mosses, invertebrates, epiphytes (plants that use other species for support), birds, microbes, and mammals. Ecologists have recently made careful studies of the epiphytic ferns that abound in the canopies of rain forests and have found that one large fern in the canopy can contain as many invertebrates as the whole of the tree crown. A complex structure therefore provides many opportunities for different organisms to survive and make a living. This relationship between architectural complexity and diversity may also explain why there are so many very small creatures that can take advantage of a complex environment.

Bird life, by contrast, does not precisely follow the predicted pattern based on productivity and architectural complexity. There are only about 100 breeding land bird species in southern Florida, which is similar to the number at Hudson Bay. In the mountain regions of the West, however, such as Idaho, the number of breeding birds can rise above 200 species. The likely explanation for this exception to the latitudinal and productivity gradients is that the landscapes of the mountain country are far more diverse than in the Southeast, and this diversity provides additional niches for birds. This principle could equally apply in some areas of the Tropics, where the mountain regions, such as the Andes in South America, are also richer in species than the lowlands. Landscape diversity can thus create additional complexities in the arrangement of biodiversity over the Earth's surface besides the general latitudinal relationship.

Another possible explanation for the latitudinal gradient in species richness is that evolution has progressed faster in the Tropics; perhaps this is where new species are more rapidly generated. Although it is difficult to test this idea experimentally, the assembling of fossil data can be of assistance. In the case of flowering plants, for instance, their fossils first appeared in the Tropics in late Jurassic times, over 140 million years ago (see "Geological History of Tropical Forests," pages 175–178). They spread out from the Tropics, but it took at least 30 million years before they reached the high latitudes, the total numbers there remaining lower than that close to the equator. About 20 percent of modern flowering plant families are mainly temperate in distribution, while 50 percent are mainly tropical, with the remainder being widespread. Some biogeographers argue that this implies a tropical origin for most families. It is clear that the latitudinal gradient in diversity has always existed, but this could still have been controlled by climate rather than rates of evolution. There is currently no experimental evidence to suggest that evolution proceeds faster at low latitudes.

The climate of the Earth has not been stable over the course of time, and the high latitudes have periodically been subjected to the advance of ice sheets, burying much of the land and part of the oceans beneath deep layers of ice. Consequently all living things have been eliminated from parts of the high latitudes and from high altitudes at various times in the past, but particularly in the last two million years (see "Climatic and Biological History of the Earth," pages 174–175). Apart from the scattered high mountains, the Tropics have not been glaciated in this way, but they have suffered from climate change, so they have not totally escaped the effects of climatic instability (see "Tropical Forests in the Last Two Million Years," pages 180–182). Much of the temperate zone has been covered by ice or converted to tundra during the ice ages, so the plants and animals of this region have been partially eradicated or even brought to extinction in some cases. It can be argued that the high latitudes have suffered most from the fluctuating climate of recent times and that this accounts for the low biodiversity. But this fluctuation does not explain the fossil evidence, which suggests that the high latitudes have always been less rich in species than the low latitudes, even before the onset of the recent ice ages.

Ecosystems develop and grow eventually reaching maturity, but even the final stages often pass through a constant cycle of disturbance and renewal (see "Ecosystem Development and Stability," pages 91–94). During ecosystem development, different species replace one another in a fairly predictable manner in the process that ecologists term succession (see sidebar "Succession," page 63) This process is driven by the interaction of species leading to successive replacement of one set of organisms by another until eventually some degree of stability is attained. The invasive pioneer species must be capable of growing in a relatively severe environment but does not have to be an aggressive competitor to settle in the area because there are few other species demanding space and other resources. The presence of an organism in a habitat has an effect on the physical and chemical nature of its surroundings, so as soon as the pioneer has established itself, the environment inevitably begins to change. The growth of a reed in a pond slows the flow of water and leads to higher rates of sedimentation; the growth of a grass on a sand dune slows the movement of air and leads to the deposition of airborne sand grains; the establishment of tree seedling in a forest clearing creates a shady microclimate and a source of organic matter for the soil. As the pioneer inadvertently changes its surroundings, it makes a habitat less severe and creates conditions in which other species are able to invade. This process is called *facilitation* and is one of the major driving forces of succession. Forces driving succession from within the ecosystem are said to be *autogenic.*

Facilitation does not imply that pioneer organisms are behaving in an altruistic way toward other species. The changes they make to an environment are inevitable consequences of their presence, and the arrival of new species is rarely beneficial to them. As new species arrive, the next stage in the process begins and *competition* becomes an important force in the succession. Once established, later arrivals may prove more competitive than the original pioneers, and the outcome of the competition may well be the loss of the pioneer species. In this way, one species replaces another, and its continued presence alters the environment further, leading to new invasions, more competition, and additional local extinctions. Even when a final stability is attained, the equilibrium is a dynamic one. At any given location, there is a turnover of species as individuals grow old and die. The death of a tree, for example, can lead to a cycle of successional events on a miniature scale as invaders once again struggle for supremacy in the gap.

Diversity changes over the course of succession. On the whole, more mature ecosystems contain more species of plants, animals, and microbes than immature ones during succession, especially when the final, or climax, ecosystem has matured into a mosaic of patches each in different stages of recovery and renewal (see "Cycles in the Forest," pages 94–95). Disturbance can enhance biodiversity because it offers an opportunity for pioneers and early successional species, together with light-demanding species such as tree ferns (see illustration below), to commence the process again, so that these species are not totally lost to an area. Fire

---

Tree ferns are characteristic of montane tropical forests, especially in open clearings where light penetrates but humidity remains high.   *(Ben Ryan)*

and flood on a local scale can diversify a landscape and create just the kind of complex arrangement of different types of habitat that ensures maximum diversity. Tropical storms, fire, the raging torrents of rivers flooding though the forest all add to the disturbance and the diversity of the forest landscape, and these features contribute to the biodiversity of tropical forests. Forces driving succession from outside the ecosystem are called *allogenic.* Disturbance alone cannot account for the latitudinal gradient of diversity, but it is important to bear in mind that stability is not always the means of maximizing diversity.

Apart from short-term disturbance and forest regeneration, the tropical forests are changing in response to long-term environmental trends. Ecologists studying the forests of the Amazon over the past 20 years have found that the tree composition is slowly but constantly changing. Slow-growing tree species are gradually giving way to faster-growing species, some of which are canopy emergent trees, but these are not pioneer trees, so the process cannot be explained by physical disturbance. It is possible that the forests are responding to the gradual rise in carbon dioxide in the atmosphere, faster-growing trees being the first to exploit this new resource and increase their biomass. These findings, if confirmed, will have considerable implications for forest structure, microhabitat complexity, biodiversity, and the role of the tropical forests in acting as a sink for atmospheric carbon in the global carbon cycle (see "The Forest as a Carbon Sink," pages 209–211).

BIODIVERSITY AND COMPETITION

Charles Darwin developed his theory of natural selection on the basis of two observations. First, there is a tendency for all organisms to produce more offspring than can be supported by their environment, and second, there is an inevitable struggle for existence in which the fit survive and the unfit are eliminated. This struggle entails *competition,* which can be defined as an active demand by two or more individuals for a single resource that is in short supply. Precisely what constitutes a resource varies from one species to another. For an insectivorous bird, for example, the resources it requires include insect food, space in which to hunt that food, a mate, and a nesting site. For a tropical forest tree, the required resources include access to light, rooting space, water, nutrients, pollination vectors, and seed dispersers. The greatest intensity of competition for any individual is likely to come from other members of its own species because these have precisely the same resource requirements. This is known as *intraspecific* competi-

tion. This process is extremely important in the course of Darwinian natural selection because it determines which individuals are most successful and consequently generate the greatest numbers of offspring.

Competition also occurs between individuals of different species, called *interspecific* competition. Such competition results from the fact that the resources required by different species may coincide; all green plants need light, and all insectivorous birds need insect food, for example. The process of natural selection operating over long periods of time encourages specialization of species, which reduces the degree of competition. Different species of insectivorous birds, for example, may concentrate on different sizes of insect prey, and various species of plant may develop the capacity to cope with different levels of light intensity. This is known as *niche* specialization (see sidebar on page 104) and the outcome is that interspecific competition is reduced and more species can coexist in the same habitat.

One of the most distinctive features of the tropical forests is the degree to which different species are specialized in their requirements and the ways in which they live. Some flowers can be pollinated by only one species of insects, and many fruit-eating birds are able to consume a very limited range of fruit sizes. Some frogs will breed only in the hollows of rotting trees, and some primates are very fussy about the height of the canopy they occupy. Specialization is thus a means of overcoming interspecific competition and develops as a result of natural selection that favors those individuals that exhibit specialist behavior.

Ecologists have devoted a great deal of time and research to the question of whether the high biodiversity of a tropical forest makes it more or less stable (see sidebar "The Concept of Stability," page 93). One argument maintains that a complex food web allows organisms a range of choices, so that when one species population declines its predators can turn to alternative sources of food and allow the threatened species to recover. Diversity would thus provide the ecosystem with a degree of stability and the ability to resist change (called inertia). But if the degree of specialization is very high, then the range of choices available to a predator (or parasite, or grazer) becomes limited. In this case the decline or local extinction of one species could cause a cascade of extinctions in its wake as those species dependent upon it also decline. So, it is possible that a highly diverse system, such as the tropical forest, may actually be fragile as a result of niche specialization and high levels of interspecies dependence. The argument and the ecological research will undoubtedly continue, but until the ecosystem is better understood, conservationists are wise to adopt a cautious attitude and avoid putting undue pressure upon the forest. Unfortunately, this is not an easy policy to implement (see "Forest Clearance," pages 213–216).

The Ecological Niche

Ecologists have long pondered the question of how so many different species of plants and animals manage to live alongside one another in a reasonable degree of harmony and stability. Some species clearly have very different ways of living, so the problem does not arise. In a tropical forest a toucan and a tapir can coexist because they have very different modes of life. Both eat fruits, but one feeds in the canopy and the other on the ground. But trees all engage in the same basic activity, fixing carbon dioxide from the atmosphere by trapping and using the energy of sunlight, so how can they coexist? Or how can two species of hummingbird, both dependent on the nectar from flowers for their food, live together without one or the other species declining to extinction? Close examination of two similar species that successfully occupy the same area usually reveals that there are subtle differences in either the resources they require or the manner in which they obtain those resources.

Trees in the tropical forest occupy different levels above the ground and trap their light at different intensities; they may also root at different depths in the soil and thus avoid direct confrontation with their potential competitors. Hawks and owls may feed upon the same small mammal prey, but one taps the resource by day and the other by night, so the resource is effectively divided between them and they can coexist. The sum of an organism's requirements, together with its manner of obtaining its needs, is called its ecological niche. The niche can be considered the role played by an organism in the community, where it is found and how it makes its living.

The ecologist G. Evelyn Hutchinson developed the concept of the ecological niche by considering two possible forms that it can take. If an organism is allowed to live in the absence of other potential competitors that demand the same resources, it expands to its full potential. It occupies more space and feeds on a greater range of prey species than is usual in a natural, multispecies environment. Hutchinson called this broad potential niche, which is bounded solely by the physical limitations of the species, its *fundamental niche*. In the real world, where competitors are present and make demands on the same resources, a species is more constrained in the space it can occupy and the food available to it, so its observed niche is smaller than its fundamental niche and is called its *realized niche*. If two species are to coexist in a habitat, they must differ in some aspects of their realized niches. Their fundamental niches may be similar, but they will prove more efficient than their competitor in some specific area, perhaps by catching prey of a particular size or from a particular microhabitat more effectively, thus developing a different realized niche.

Species thus give the appearance of dividing resources among themselves, a process known as *resource partitioning*. This should not be regarded as a friendly affair in which agreement is achieved by negotiation. Resource partitioning results from intensive competition in which each species can survive only by a degree of efficiency and fitness in some aspects of its niche that prevent its extinction by other organisms. Where resources are abundant resource partitioning can proceed in a manner that allows many species to be accommodated within the ecosystem, known as niche packing. In tropical forests, where conditions are appropriate for high levels of productivity, individual niches can be narrow and tightly packed, leading to high levels of biodiversity.

■ SPECIES/AREA RELATIONSHIPS

One very simple rule in ecology is that the larger the area of ground an observer samples, the more species of plants and animals will be found. It is very easy to test this concept. A sample of one square foot in the middle of a region of relatively uniform grassland contains a certain number of species of plant, perhaps about five. When the area is doubled, additional species can be found, perhaps an extra four. Doubling the area sequentially adds more and more species, but the numbers of new species added tends to decline as the area becomes very large, so a graph of species number against area shows a rapid initial increase followed by a gradual falling off in the steepness of the gradient. In a perfectly uniform landscape, the graph would eventually become completely level once all the species in the ecosystem had been observed, but in reality landscapes are rarely uniform and new minor habitats may be encountered, such as damp or shaded areas, in which new species may find a suitable place to grow. This simple relationship between species number and area of ground is universal for all habitats and ecosystems and can be described in mathematical terms. If the logarithm of the species number is plotted against the logarithm of sample area, a straight-line relationship results.

Biogeographers have examined this relationship between species richness and area in relation to islands, where larger islands are usually richer in species than smaller islands, provided other factors, such as climate and soil variability, are equal. In this situation, it is possible to understand the ecological processes that underlie the species/area curve. A large island is more likely to be encountered by wandering organisms, either carried by air or water from the mainland, so the rate of immigration of new species is higher than that of a small island. On the larger island there is more room and therefore more opportunity to establish a colony for the immigrant species without succumbing to competition from those organisms already established on the island. The rate of extinction on a large island is thus lower than that of a small island. With higher rates of immigration and lower rates of extinction, a large island is thus more likely to settle down to an equilibrium in which the total resident species is greater than that of a small island.

A complicating factor is that some islands may be far from the mainland whereas others are close by, and this discrepancy affects the rate of arrival of immigrant species. So the distance between islands or their distance from the mainland affects the immigration rate and hence the equilibrium level of species number. The combined effects of island area and the distance from other islands provide the basis for a predictive model describing the number of expected species on any particular island. This model has proved relatively robust when put to the test and has become known as the theory of island biogeography.

There are many factors that can complicate the simple model described in the theory of island biogeography. Landscape diversity within an island, such as the presence of many different microhabitats, can increase the capacity for an island to support species. Intense disturbance, either by humans or by tropical storms, on the other hand, can reduce the potential of an island for supporting different species. Newly formed volcanic islands are in an early stage of succession, so they will not have reached the equilibrium required for the island biogeographic theory to apply. An island sinking below the waves, by contrast may contain more species than is predicted by the model because it is effectively "supersaturated" with species and will equilibrate downward as some species become locally extinct.

The theory of island biogeography has proved extremely useful to biogeographers and to conservationists in helping to explain why some areas are rich in species. The principles on which it is based apply not only to real oceanic islands but also to habitat "islands," such as fragments of forest surrounded by agricultural land. One of the main problems facing the tropical forests is an increasing fragmentation as a result of first development of roads through the forest and then ribbon development of agriculture and villages (see "Forest Clearance," pages 213–216). Conservationists can

use the theory of island biogeography to predict the outcome of such fragmentation on species richness within the remaining forest islands. Some ecologists have been able to test the application of the island biogeographic model by observing the rates of extinction within isolated forest fragments, to see whether small fragments do show the expected higher rate of local extinctions.

In one such experiment in the 1990s in Brazil, researchers observed forest fragments of different sizes following their isolation by general forest clearance. They concentrated on bird species because these are most easily observed and monitored in survey work. The ecologists determined how many species of birds bred in each of the areas prior to disturbance and then documented the number breeding one year following the fragmentation. In an isolated area of forest 3,500 acres (1,400 ha) in extent, the number of breeding birds fell by 14 percent during the first year. In a similar block of forest just 625 acres (250 ha) in area 41 percent were lost, and in an area of 53 acres (21 ha) 62 percent of birds had disappeared after one year. The general principle relating species number to area thus seems to apply in this experimental situation and provides a basis for estimating the impact of forest clearance on species diversity in the tropical forest. It also helps conservationists evaluate potential conservation areas, because large blocks of forest are likely to be more valuable than small ones when seeking to conserve a maximum number of species (see "The Problem of Evaluation," pages 199–200).

As in the case of the true islands, the model should be applied to forest fragments with caution. The model is based on the assumption of equilibrium being established between immigration and extinction, and periods longer than a year are probably needed for this equilibration to take place. Some of the fragments in the Brazilian experiment may still be supersaturated following isolation and extinction may continue for some years. On the other hand, the effects of severe disturbance may have caused excessive emigration on the part of mobile organisms such as birds, and some additional immigration may take place as the level of environmental disturbance lessens.

The presence of landscape diversity in the form of varied microhabitats is also an important consideration and can strongly influence the survival of organisms following fragmentation. An example from the Amazon Basin illustrates the importance of microhabitats in determining the survival of biodiversity. There are 39 species of frogs in the primary forests of the central Amazon Basin living in a wide range of habitats, from the forest floor to the leafy canopy. For their breeding, however, all frogs need to have access to pools of water, and the different frog species of this region have varied requirements. Seven species prefer narrow streams, less than 10 feet (3 m) wide, while a further three species like wider streams. Two species seek out pools in the hollows of decaying tree stumps in which to breed, while a total of 11 species use temporary flooded pools created by the wallowing of the peccary, a wild pig that lives on the forest floor. Overall, about half of the frogs of the jungle have very specialized breeding requirements. If an area of forest is to support all these species of frogs, it must contain all the microhabitats that the amphibians need, so the conservationist must not only be concerned with extensive areas of forest but also ensure that the selected region supplies all the requirements of these demanding animals. A large proportion of the frogs is evidently dependent on the activities of the peccary, so maintenance of frog diversity demands the conservation of this pig. In this situation, the peccary is acting as a keystone species, which supports a number of other species as a result of its manipulation of the habitat as it creates wet wallows. As a general rule, larger areas of forest maintain higher numbers of species, but local factors, including the presence of keystone species and a wide array of microhabitats, are also important in the conservation of biodiversity.

■ BIODIVERSITY HOTSPOTS

The Tropics are generally richer in species than any other part of the Earth's surface (see "Patterns of Biodiversity," pages 100–103), but there are gradients of biodiversity even within the Tropics. The land areas close to the tropics of Cancer and Capricorn, especially on the continental landmasses of Africa, India, and Australia, are arid desert regions. Their biodiversity is generally lower than that of the moist equatorial regions and gradually increases along a gradient approaching the equator. The biodiversity of the African equatorial region is lower than that of Southeast Asia and South America, so here there is a longitudinal gradient of biodiversity as well as a latitudinal one.

Some regions of the world are exceptionally rich in species and have become known as biodiversity hotspots. When the term was first coined, just 10 such hotspots were recognized; ecologists now agree that there are about 25 global hotspots, many of which lie within the tropical forest zone. These include Central America, the tropical Andes, the Caribbean, the Brazilian Atlantic forests, Western Ecuador, the Brazilian Cerrado, Madagascar, the East African mountains, West African forests, the islands of Southeast Asia, the Philippines, Southern India and Sri Lanka, Indo-Burma New Caledonia, and Polynesia. The remaining hotspots are concentrated in those parts of the world with a Mediterranean climate, including the Mediterranean Basin, California, Chile, the Cape region of South Africa, and Southwest Australia. Other hotspots occur in the Caucasus mountains of Asia and in the mountains of South and Central China. Together, these hotspots cover only 12 percent of the Earth's land surface area but contain well over half of the world's biodiversity, so they are clearly areas for special conservation effort.

Why are some parts of the world so very rich in species? This question is even more difficult to answer than that of latitudinal gradients in diversity. The two questions are, in fact, closely related because there is a very clear concentration of hotspots in the Tropics and there are no high latitude hotspots at all. Are they areas of undisturbed wilderness that have suffered least at the hands of humans? The answer to this is clearly negative. The Mediterranean Basin, the mountains of China, and the Central American regions have been centers in which human civilization has developed and agriculture flourished. Currently 16 of the 25 hotspots have a higher than average human population density of 109 people per square mile (42 people per km²), and in 19 of the hotspots the population is growing faster than world average (0.3 percent per year). Although some hotspots lie in wilderness areas, this is certainly not the case for most of them.

Another possible explanation for biodiversity hotspots is that they are centers of endemism (see the following sidebar). Some hotspots, such as California and Madagascar, have very high numbers of species that are confined to that particular region and are found in no other part of the world. In the case of California, many plant species seem to have evolved in this region in geologically recent times and have not spread beyond the confines of that state, perhaps in part because of the barriers of mountains and deserts in the east. In Madagascar the causes of the richness in endemism go back much further in Earth history. Over 100 million years ago Madagascar was part of the continent of Africa, but by 90 million years ago a rift had formed that split the island off from the mainland and the two began to drift apart. Since that time, evolution has proceeded independently on the island and the mainland, leading to the development of a distinctive flora and fauna on Madagascar, many of which are confined to the island and contribute to its species richness.

There is no single universal explanation accounting for some locations being so rich in species. Hotspot areas do not always prove rich in all groups of organisms. In East Africa, for example, the richest area for endemic plants lies in the eastward projection of Somalia and Ethiopia, while the richest region for endemic birds is in the Rift Valley and the region surrounding Lake Victoria. So it is probable that the causes of richness differ for the two groups. Biogeographers still need to conduct a great deal of research into the development of hotspots if their richness is to be fully understood. The first step in this process, identifying the biodiversity hotspots, has now been completed and this will prove extremely useful in establishing conservation priorities as ecologists seek to maintain the Earth's biodiversity.

Hotspots are important areas for the conservation of biodiversity and should receive priority in the selection of

Endemism

An endemic organism is one that is native to a particular region of the world and is confined to that area. There are two main causes for endemism among plants and animals. If a species evolves in isolation from other similar habitats and lacks the capacity for long-distance dispersal into other regions with similar conditions, it becomes effectively imprisoned in the region where it evolved. The lemurs of Madagascar are an example of a mammal group that evolved on the island and are endemic to it. Indeed, all of the 101 species of mammals native to Madagascar are found nowhere else on Earth. These species are called palaeoendemics because they evolved long ago and have failed to disperse and establish themselves elsewhere. The second cause of endemism is the continued generation of new species in an area, where their confinement is mainly due to lack of time for dispersal to new regions. Such species are called neoendemics. Many of the plant groups of California, such as the genera *Aquilegia* and *Clarkia,* have recently radiated there in their evolutionary development, but many species are still confined to the region.

Some regions rich in endemics, such as the Cuatro Ciéngas Basin in Mexico, which is part of the Central American hotspot, owe their high levels of endemicity to long-term climatic stability. This site in Mexico appears to have had a long history in which the climate has varied much less than in those parts of the world affected by the climatic fluctuations of the Pleistocene. Stability and isolation have permitted the development of a flora and fauna with a high proportion of endemics.

reserves for protection. The transitional regions between these hotspots and surrounding areas, however, must not be neglected. In these border areas, species are subjected to the problems of survival in suboptimal habitat conditions, and, as a consequence, a wide range of genetic diversity is often found within them. A population of a species under stress experiences strong pressures of selection in which particular survival traits are favored. Transition areas around the edges of hotspots are therefore often sites of rapid evolutionary advancement and the development of new characteristics that will ensure the long-term survival of a species.

The tropical forests play a crucial role in the maintenance of the world's biodiversity. They occupy only about 6 percent of the land area of the Earth, yet they probably contain over half of the world's species of plants, animals, and microbes. Tropical forests, therefore, form the heart of the planet's biodiversity.

■ CONCLUSIONS

The Earth is unusual, perhaps unique, in the universe because of its support of life. The abundance of living things on the planet represents its greatest heritage and is often referred to as its biodiversity. Biodiversity means more than just the number of different organisms on Earth; it also includes the genetic variability within different species and the range of habitats and ecosystems present within a landscape. The word diversity also implies more than a simple list of the species present within an area; it encompasses the degree of evenness within the collection of populations of a community. A diverse system has many species and each species contributes a similar proportion to the entire community.

At present, biologists have described only a very small percentage of the total number of species of plants, animals, and microbes found on Earth. They can only speculate how many species there may be in total, with opinions ranging from 10 to 30 million. Some groups of organisms, particularly the large and prominent ones, such as mammals, birds, and flowering plants, are relatively well known and most of the species present on Earth have now been described, but even among the primates new species are found each year. Other groups, including fungi, bacteria, insects, and nematode worms, are poorly described and much work will be needed before the lists for these organisms are complete. It is likely that many will have become extinct even before they are recognized. The total numbers in these lesser-known groups has to be estimated by various processes of extrapolation on the basis of limited studies, or by methods using body size or the ratio of unknown groups to well documented ones.

Biodiversity itself is therefore poorly understood over the face of the planet, but the observations that scientists have made so far suggest that biodiversity is not evenly distributed, being far greater in the tropical regions. Studies of mammals, birds, amphibians, butterflies, and flowering plants have all shown a general latitudinal gradient in species richness, falling from high levels in the Tropics to low levels at polar latitudes. Biogeographers have put forward a number of possible explanations for this gradient. Diversity is often highest in areas where primary productivity is greatest, so perhaps the availability of energy to support lengthy food chains and complex food webs lies at the heart of the latitudinal gradient. Interruptions in the general gradient of diversity decline can be detected, such as the deserts around the tropics of Cancer and Capricorn, where diversity dips to lower levels. This phenomenon adds weight to the suggestion that productivity, which is also low in deserts, may underlie biodiversity.

Associated with productivity are biomass and the architectural complexity of ecosystems. The spatial structure of an ecosystem affects the availability of microhabitats and microclimates, providing a range of locations in which small organisms can live. High biodiversity is often found in ecosystems with a complex structure, such as tropical forests. Landscape complexity can similarly add to the opportunities for species to survive, leading to high biodiversity in regions with mountains, valleys, rivers, and wetlands.

Climatic stability has often been proposed as contributor to biodiversity, but some regions of high diversity, such as the tropical forests, occupy regions that have not been as climatically stable as was once thought. Even the tropics experienced fluctuations of climate during the ice ages. Perhaps evolution proceeds faster in the Tropics. Fossil evidence suggests that many great developments in evolution, such as the rise of flowering plants, took place in the Tropics, but there is no evidence to suggest that evolution is faster in tropical climates. Change is a feature of all ecosystems, whether it is disturbance caused by natural disasters, or cyclic changes as individual organisms die and are replaced; the process of recovery and development within ecosystems also contributes to biodiversity.

Biodiversity is partially determined by how many different ways an organism can make a living within a habitat. This concept of having a role in a community is called the ecological niche. Highly biodiverse ecosystems, such as tropical forests, have many opportunities for niche specialization, where each species has a very closely defined microhabitat as well as food and other physical requirements.

Niche packing is a feature of a biodiverse ecosystem; different organisms can minimize the degree of competition they experience from other species.

Spatial area is also an important determinant of diversity because large areas generally contain more species than small areas. Simple mathematical expressions describe the relationship between area and richness, and such studies have made an important contribution to conservation biology because they predict that the reduction of the area of a habitat will involve local species extinction. Experimental studies have confirmed such predictions and have shown that the biodiversity of any habitat, including tropical forests, will be reduced if the area is fragmented into smaller units. Because large areas are more likely to contain a full range of microhabitats it is important to take size into consideration when selecting areas for the representation and conservation of particular types of ecosystem.

Some parts of the world are extremely rich in species and are thus termed biodiversity hotspots. Of the 25 hotspots that are currently recognized, most fall within the Tropics and include tropical forest locations. Biogeographers have not been able to determine why some regions are so rich, but one factor is likely to be endemism. Endemic organisms are native to an area and geographically confined to that area. Some locations seem particularly rich in endemics, possibly for reasons of geological or climatic history, isolation, or unusual climatic stability. Whatever the cause of hotspot development, these areas must clearly form the focus of future conservation efforts.

Of all the world's ecosystems, the tropical forests are the most biodiverse. Containing more than half of the Earth's living species, they occupy a critical position in the sustenance of the biodiversity of the planet.

6

Biology of the Tropical Forests: Plants

Life can exist only within a certain set of physical limits, such as temperature, and where particular resources, such as water and food, are available. Most parts of the Earth's land surface provide the necessary conditions for life, with the possible exception of glacial ice and volcanic craters, but in many regions conditions are far from ideal. Very high or very low temperatures or a lack of water or scarcity of a particular element needed for life can make survival difficult for living organisms. The regions occupied by the tropical forests are generally well supplied with all that is needed for living things to thrive, accounting for the high productivity and biomass of the forests (see "Primary Productivity in Tropical Forests," pages 78–82). The climate is warm and frost is virtually unknown except in the high altitudes (see "Tropical Climates," pages 9–14), and precipitation provides an abundance of water in the equatorial regions. Drought, however, can become a factor that limits plant growth and animal survival in the higher latitudes of the Tropics. The mineral elements needed to construct the bodies of living organisms are rapidly recycled because of the fast rate of decomposition in tropical forest soils (see "Decomposition in Tropical Forests," pages 89–91), so even those elements that are relatively scarce, such as phosphorus, are conserved within the living biomass of the ecosystem. The physical conditions for life in the tropical forest regions are as close to ideal as can be found on the surface of the planet.

It is thus not surprising that life is abundant in the tropical forests. This abundance leads to the one kind of stress that living things have to cope with—namely, competition for the available resources. Survival for a species can be achieved only if it takes up a particular role in the community, a particular and specialized way of making a living (see "Biodiversity and Competition," pages 103–104). Although the physical conditions within the forest present little challenge to its inhabitants, the biological stress from competi-

tion with other organisms is a problem that all inhabitants of the tropical forest have to face.

■ TEMPERATURE AND FOREST LIFE

All living organisms are affected by temperature. If the temperature becomes too high or too low they die. The chemistry and the physics of individual cells are strongly affected by temperature and cell processes can operate only within certain temperature limits. As a general rule, the activity of a chemical reaction roughly halves with every 18°F (10°C) drop in temperature. All life processes are believed to cease when the temperature falls to -94°F (-70°C) or when it reaches the boiling point of water 212°F (100°C). At extreme temperatures the physical structure of the cell can also change, as when water freezes or oils solidify into fats. Living things have evolved in a world where water is usually available in a liquid form and where these extremes of temperature are rarely encountered.

Cell metabolism, that is, the general working and functioning of the living cell, is largely controlled by the activities of specialized molecules called enzymes. Enzymes are proteins and are large molecules built from chains of smaller units called amino acids. These chains are usually not straight but twisted and curled, the folds and bends being held together by weak bonds. Their contorted structure is important for the actions they perform in the cell. Enzymes are chemical catalysts, assisting in chemical reactions, but are themselves unchanged at the end of the reaction. Thus they can be reused time and time again in the living cell. An example of an enzyme is nitrogenase, which is used by some microbes for the fixation of atmospheric nitrogen (see sidebar "Nitrogen Fixation," on page 88). In this process the nitrogenase enzyme acts as a catalyst in the reduction of nitrogen gas to ammonium ions, which is a reaction that can be carried out in the absence of the enzyme only by applying very high temperature and pressure to this relatively inert gas. Enzymes thus facilitate chemical reactions that would otherwise be extremely difficult and energy consuming. They achieve this by forming a loose combination with the substance they operate on, called a substrate, forming an enzyme–substrate complex. When they combine in this way, they distort the substrate molecule, straining its chemical bonds and thus allowing it to undergo chemical changes more easily; this distortion is partly due to the complex structure of the protein molecule.

Enzymes each have their own optimum working temperature that varies from one enzyme to another and, indeed, differs for the same enzyme in different species or even different populations of organisms. The optimum temperature for activity is under genetic control and can be modified in the course of evolution. Most enzymes are inhibited at very high and very low temperatures because their structure begins to disintegrate and they no longer function effectively as catalysts. Relatively few enzymes in the natural world operate effectively above a temperature of about 105°F (40°C) because at this temperature the protein molecule begins to unravel; a denatured protein that has lost its structure can no longer operate as an effective catalyst. At low temperatures an enzyme also loses efficiency because all chemical reactions are slower when the temperature drops. Plants generally assume the temperature of their surroundings, so when the air temperature falls their shoot temperature is also likely to fall with it. If the Sun is shining, however, any dark pigments in the plant, including chlorophyll, absorb some of the radiant energy and can raise the temperature of leaves and stems. Invertebrate animals and some vertebrates, such as amphibians and reptiles, are unable to control their body temperature and, like plants, assume the same temperature as their surroundings. They are said to be *poikilothermic,* or cold-blooded. Like plants, they are able to take advantage of the radiant energy of sunshine. They also have the advantage of mobility, so they can seek out suitable microclimates, such as illuminated, sheltered spots, where they can maximize their energy absorption. By deliberately raising body temperature in this way, a snake or a butterfly can enhance its cell metabolism by encouraging enzyme activity, enabling it to perform its functions such as movement in hunting or escaping predators more quickly and effectively. When overheated, poikilotherms can regulate their temperature to some degree by losing water and dissipating some energy as latent heat of vaporization.

Mammals and birds are warm-blooded, or *homiothermic.* They control the temperature of their cells quite closely by maintaining a constant blood temperature. This might seem to be the answer to living in very cold conditions,

and it does indeed offer many advantages, but an organism that maintains its temperature well above its surroundings inevitably suffers rapid energy loss. Newton's law of cooling states that a body cools more rapidly if there is a large temperature difference between the body and its environment. So a warm-blooded animal needs to expend more energy to keep its higher temperature, so it has to eat more to sustain these energy losses. In the warmth of the tropical forests, the advantages of homiothermy over poikilothermy is not as obvious as in a cold environment such as the tundra, but it can be valuable where night temperature falls, such as at higher altitude. Overheating is a more common problem for tropical homiotherms, and some tropical mammals have developed mechanisms for heat loss, such as the large, thin ears of elephants, which can be used to disperse heat from the blood and help control the internal temperature (thermoregulate). Spraying the body with water is an alternative approach to thermoregulation. But overheating is more often associated with conditions in the tropical grasslands (savannas) and the desert regions than with forests, where shade provides relief from the strong sunlight.

Just as enzyme activity is destroyed at high temperature, so it can be severely curtailed at very low temperature. At temperatures below about 50°F (10°C), enzymes operate progressively less efficiently and eventually lose their structure, becoming denatured. This means that the activity of the cell begins to slow down as temperature falls, but the relationship between temperature and enzyme activity varies greatly between different species of plant and animal. Some species are much more tolerant of low temperature than others, due to temperature optima of their enzymes, which is in turn controlled by the genetic constitution of the cell. The cells of sensitive organisms, both plants and animals, may even die at low temperature before the freezing point of water is reached. They are said to demonstrate chilling injury; most of the inhabitants of the tropical forest fall into this group.

Some tropical organisms begin to show signs of chilling injury if their temperature falls below 50°F (10°C), and many are affected below 41°F (5°C). In the case of aquatic organisms, which live in a less variable environment than terrestrial ones, chilling can occur at quite a high temperature. Some tropical fish, for example, may die if the temperature drops below 68°F (20°C). Close inspection of a cell that is undergoing chilling injury reveals a series of changes that eventually prove fatal. At low temperature the mobile cytoplasm of a cell becomes more static; cytoplasmic streaming is reduced. The respiration rate of the cell then increases, due to the mitochondria that control the energy relations of the cell becoming uncoupled. The energy-regulating molecule of the cell, *adenosine triphosphate (ATP),* is not properly linked up to the oxidation of substrates, which is a process that takes place within the mitochondria. The mitochondria then run out of control. Membranes of chilled cells then begin to malfunction. Cell membranes are layered structures of proteins and phospholipids, and they are temperature sensitive. The lipid component, as its name implies, is a fat, and, like all fats and oils, it alters its state according to the temperature. As the temperature falls, the state of the membrane changes, becoming more stiff and solid, and it no longer effectively controls the passage of chemicals into and out of the cell. In other words, the cell becomes leaky. At this stage, the whole structure of the cell collapses and the damage is irremediable. In some cold-tolerant (psychrophilic) fungi the fats in the membranes are largely unsaturated, resulting in their being more fluid even at low temperature.

The effect of cold on the cell is greatest if the temperature drops suddenly. If a cell is gradually exposed to an increasing degree of cold, it may have a limited capacity for adjustment. This is called cold acclimation. Some plants are better able to cope with chilling if they are first exposed to drought. Perhaps they close their stomata and effectively enter a more dormant state, leaving themselves less susceptible to injury.

Once the temperature falls below the freezing point of water, cells become liable to frost damage, which is quite different from chilling injury. When water freezes it increases in volume by about 10 percent. If the contents of a cell freeze, considerable disruption of the structure, particularly the membranes, is likely to result. Often this damage reveals itself not during the freezing but when the cell subsequently thaws, especially if the thaw is rapid. When a household water pipe becomes frozen and bursts, the resulting leak is not evident until the thaw comes, as with a frozen cell. The presence of a solute lowers the freezing point of water, so cell contents, which are rich in solutes, are unlikely to freeze at 32°F (0°C). In plant tissues, however, relatively pure water is between the cells, outside the main cell wall. Ice crystals thus form and grow between cells and their formation gradually extracts water from the living cells, causing them to desiccate. The plant may recover from this condition, but if ice crystals form within the cell, cell death is inevitable. Freezing injury within the cell is thus caused by membrane disruption, protein denaturation, and dehydration. Frost acclimation in plants is familiar to gardeners, and a plant exposed to progressively lower temperatures often develops the capacity to withstand frost more effectively, becoming frost hardy. Most plants of tropical forest are never exposed to frost and are immediately killed by the experience. When they thaw, the leaves droop and turn black and they never recover. In the dry regions that border the forests and in the high-altitude regions of some montane forests, however, occasional frost is possible; the plants of these regions are moderately tolerant of such low temperatures.

Animal cells can survive freezing, especially if both the freezing and the subsequent thawing are rapid, which is quite different from plants. Animal cells do not have a vacuole, which tends to be the focus of freezing injury inside the plant cell, so they are better able to cope with low cell temperature. At low temperature, the whole cell becomes solidified with amorphous ice. This, however, is not a likely event for a tropical forest inhabitant.

■ PLANT LIFE-FORMS IN THE TROPICAL FORESTS

The different biomes of the Earth are characterized by their vegetation; the prairies are mainly covered with relatively low-growing herbs; the boreal forests have evergreen trees; the tropical savannas are clothed with grasses and low-growing trees, and so on. Botanists have tried to devise systems for describing the general types of plants that contribute to the vegetation types of the different biomes, classifying plants according to their form rather than their taxonomic and evolutionary relationships. One of the most successful and widely used systems for classification was constructed in 1909 by the Danish botanist Christen Raunkiaer (1860–1938) (see sidebar).

Using the classification of plants on the basis of their perennation characteristics, Raunkiaer found that different vegetation types could be described according to the proportions of the various life forms they contained. He called this their *biological spectrum* (see the following sidebar). The biological spectrum of the tropical forests is dominated by phanerophytes because the physical environment of the regions occupied by this biome does not suffer from excessive cold or drought and plants are able to expose their growing points throughout the year, maximizing biomass and productivity. Trees, therefore, are by far the most important element in the vegetation of the tropical forests, comprising over two thirds of the total plant species present. The can grow without interruption by frost and are usually supplied with an abundance of water from the high levels of precipitation. They accumulate high levels of biomass (see "Primary Productivity in Tropical Forests," pages 78–82) but the height of the canopy does seem to have its limits. Water appears to be the factor that limits the height of trees in tropical and other types of forest, but not because it is in short supply in the environment. Tree height is actually limited by the physical attributes of the water molecule.

■ TREES AND WATER

Water is a remarkable material that is essential for all living organisms. In chemical terms it is an oxide of hydrogen that exists in liquid, vapor, and solid form on the face of the Earth. Each atom of oxygen is combined with two atoms

Plant Life-forms

For several centuries plant geographers have tried to describe and classify plants according to their structure in a way that would relate to their adaptations to climate. It has long been appreciated that the vegetation of the world can be divided into major types, such as tropical rain forest, savanna, deciduous woodland, and tundra, and that the general structure, or *physiognomy,* differed between these various types. Regardless of the taxonomic relationships, vegetation types could be described in terms of the general *life-form* of its components; rain forest consists largely of trees, tundra of low-stature plants, and so on. The problem was how to classify plants in a way that described this form and expressed the way in which the form relates to climate. One solution that has been widely accepted and used was devised by the Danish botanist Christen Raunkiaer, who felt that the most important feature of a plant in terms of climatic adaptation is how it survives from one year to the next. For many plants this depends on how it manages to protect its delicate growing points, the buds. In a wet tropical environment, survival is made relatively easy for the plant because there is little seasonal variation and no extremes of temperature. The buds of tropical plant species can therefore be held high in the air and do not even need protection from drought or frost by the development of bud scales. Some tropical environments have a dry season that results in stress to the plant's growing point, meaning that buds may not be held quite so far from the ground and have to be covered by protective scales. In deserts the situation is even more testing for plants, as many can only survive by becoming dormant as seeds.

Raunkiaer therefore developed a classification scheme for plants that bore no relationship to their evolutionary origins but to their structural strategies for survival. The following is a classification of the plants, in particular on the basis of where they held their perennating organs.

Phanerophytes. Plants holding their buds more than 10 inches (25 cm) above the soil surface, including all trees and shrubs. The group can be subdivided according to height into mega-, meso-, micro-, and nanophanerophyte, the last subgroup having its buds between 10 inches and 6.6 feet (25 cm and 2 m) above the ground.

Chamaephytes. Plants with their perennating buds less than 10 inches (25 cm) in the air but above the soil surface. This group includes the dwarf shrubs and herbs that grow in dense cushions.

Hemicryptophytes. Plants that hold their buds at the level of the soil surface. Many temperate herbs fall into this category, including many grasses, together with plants like stinging nettles.

Cryptophytes. Plants with their perennating organs below ground (geophytes) or beneath the surface of water (hydophytes). This group includes plants with bulbs, corms, tubers, and rhizomes, together with aquatic plants.

Therophytes. Plants that survive through unfavorable periods of cold or drought as dormant seeds. These have no vegetative perennating organs and rely entirely on seeds for survival. The group includes annual plants, together with ephemerals, plants that have such a short life history they can produce more than one generation in a growing season.

Raunkiaer found that the vegetation of the major biomes of the world differed in the proportions of these various types of plant life forms present within them. Climate, he felt, determines which of the life-forms is most appropriate as a survival mechanism for each geographic region.

---

of hydrogen, but there are additional weak bonds, called *hydrogen bonds,* that link the molecules of water together in a fluid whole. When liquid water is heated, the molecules move faster and eventually the hydrogen bonds between the molecules are ruptured, leading to the liberated individual molecules becoming separated from the liquid mass and emerging as vapor. Disruption of the hydrogen bonds, however, requires an input of energy, called *latent heat,* so when a film of water evaporates from a surface it takes heat from the body on which it lies, cooling that body in the process. In tropical environments the evaporation of water is an important mechanism for temperature control.

Water is also a very effective solvent for many inorganic and some organic compounds. Chemicals as varied as nitrates, phosphates, sugars, and enzymes can dissolve in water, a process that supplies living cells a means of moving compounds around. Water movements through the entire body of an organism can also provide a means of transport

Biological Spectrum

Having defined the different types of life forms found among the plants of the world, Raunkiaer analyzed the major vegetation types (now called biomes) to establish whether there was a pattern to the global distribution of life forms, and whether this pattern could be explained by variation in climate. He took the full list of the flora growing in selected sites within the different biomes and determined the proportions of each life form within that flora. The outcome was a biological spectrum depicting the life form assemblages of the various biomes. Some generalized results are shown below. The figures are the percentage of the flora that fall into each life form category. The dominant life form is indicated in bold.

Predictably, the tropical rain forest flora is dominated by trees, but temperate deciduous forests and the coniferous boreal forests actually have a higher proportion of hemicryptophytes in their flora than they have trees. The grasslands of the world, both tropical (savanna) and temperate, have high proportions of hemicryptophytes, but also have a large contingent of annual therophytes, which take advantage of open gaps in the turf and benefit from any disturbance by trampling animals. Deserts have an even greater proportion of these opportunistic therophytes, and the flora is dominated by them. The tundra, both polar and alpine, is dominated by hemicryptophytes and chamaephytes, having a higher proportion of chamaephytes in their flora than any other biome.

	TRF	SAV	DES	TDF	TG	BF	PT	AT
Phanerophytes	**69**	18	3	**36**	5	13	0	0
Chamaephytes	8	16	16	3	6	17	**23**	**25**
Hemicryptophytes	13	**27**	19	**42**	**56**	**50**	**66**	**68**
Cryptophytes	3	13	5	14	10	13	11	9
Therophytes	7	**26**	**57**	5	23	7	0	3

Key to Biomes:

TRF = tropical rain forest; Sav = savanna; Des = desert; TDF = temperate deciduous forest; TG = temperate grassland; BF = boreal forest; PT = polar tundra; AT = alpine tundra. **Bold type** indicates the dominant life forms in each biome.

---

for chemical products from one part of the body to another. All cells need a supply of water for their continued function, and the loss of water by evaporation must constantly be compensated for by the arrival of a new supply derived from drinking, eating water containing foods, absorption through the body surface, or, in the case of plants, absorption through the roots.

Plants have a particular problem with water supply because the loss of water from leaves (transpiration) is associated with the essential process of photosynthesis (see sidebar on page 114). In the case of tall trees and extensive creepers, the water must be conveyed long distances against the force of gravity to supply the leaves of the upper canopy. Almost all of the water entering a plant is lost from the leaves by transpiration; only a very small proportion becomes involved in photosynthesis, where it is a source of electrons and protons, generating free oxygen as a byproduct.

Transpiration may seem a wasteful process, an unfortunate but inevitable consequence of the plant needing to keep its pores open in order to photosynthesize. But there are useful functions served by the transpiration stream up the trunk of the tree. In addition to water, plant roots absorb dissolved elements, many of which are vital for building cell materials. These include nitrates, phosphates, calcium, potassium, magnesium, and many others needed in smaller quantities. The dissolved molecules and charged ions diffuse through the water and are carried upward in the transpiration stream, reaching all parts of the plant, including growing points where they are incorporated into the plant biomass.

The vapor losses from the leaf of a plant may also serve the valuable function of keeping the leaf temperature cool as it provides latent heat of evaporation for the water molecules it releases. Canopy leaves in full sunlight may become extremely warm, possibly even hot enough to impair the function of their cell enzymes, so the loss of energy to the leaf as water evaporates is an effective way of thermoregulation.

Carrying water from the ground to the high canopy without the use of pumps is a remarkable achievement, one that has caused much close scrutiny by physicists, who have found the cohesion-tension theory of water ascent difficult to believe. The theory is based on the concept of hydrogen bonding that is so strong it resists the dangers of air bubbles forming, called embolism, and the water column collapsing, called *cavitation*. Many tropical forest trees grow to heights of over 160 feet (50 m), and trees from temperate rain forests, such as the giant redwoods, grow to heights exceeding 330 feet (100 m). Climbing plants, or lianes, must have even longer columns of water extending through their twining stems, reaching over 800 feet (244 m) into the forest canopy. The dead xylem elements that carry the water columns must be narrow enough to provide an adequate surface tension grip on their walls, and wide enough to provide a good movement of water, so there is a tradeoff between these two requirements. Physicists calculate that the optimum diameter for an element should be between 10 and 500 μ (1 μ is 0.00004 inches). Lianes tend to have wide xylem vessels, suggesting that advantage is gained by allowing extra water flow in these very long stems.

Cavitation remains a serious problem for a plant and any air bubble can cause the collapse of that water column.

Transpiration, Water Potential, and Wilt

Photosynthesis involves trapping energy from sunlight and using that energy to reduce atmospheric carbon dioxide to organic molecules. To obtain the carbon dioxide gas opening pores (*stomata*) in the surface of leaves are required so that the gas can enter the plant, and the open pores allow the diffusion of water vapor out of the leaf and into the atmosphere, called transpiration. Inside the leaf there are small chambers below the pores, surrounded by the cells that absorb the carbon dioxide gas, and these cells are constantly losing water as vapor when the pores are open. Such losses would rapidly cause the leaves to wilt if they were not provided with a renewed supply of water, and this is achieved by setting up a negative gradient of water pressure through the plant. As the cell lining the stomatal chamber loses water the cell contents become more concentrated in their solutes, and the walls of the cell may even be put under a negative pressure, all of which create a demand for water, or, to use the technical expression, a reduced water potential. A cell with a low water potential exerts a strong pull on the water in surrounding cells that have a more abundant supply of water (a higher water potential). They draw water from these cells, and the demand for water is thus conducted across the leaf to the vascular bundles, the veins through which water passes up the plant stems from the roots. In the vascular bundles are xylem vessels, which are dead, empty cells that form a continuous system of small tubes running right from the leaves, through the trunk, and into the roots of the tree.

Water is thus removed from the vascular bundles of the leaf, which effectively pulls water in a continuous column directly from the roots, which act as absorptive organs in the soil. Water molecules have the property of cohesion, binding together through their hydrogen bonds so effectively that a very strong pulling force can be exerted without disrupting the water column. This is known as the cohesion-tension theory of water ascent. Although many physicists and physiologists have questioned its validity, the theory remains the best available explanation of how water is transported from the soil to forest canopy.

As water is pulled upward from the vascular tissues in the roots, a negative pressure is exerted on the surrounding cells, removing water from them and placing them under a water demand that then extends across the roots to the root hairs. When a root hair cell becomes short of water, it has only one remaining source, the water in the soil outside the plant. If the soil is moist, water is relatively easily available because its water potential is high, but if the soil is dry, water is held more tightly by the capillarity of the fine pores between the soil particles and the water potential of the soil is thus low. In saline conditions, the soil water potential is even lower because of the effect of the dissolved solutes, making it even more difficult for the plant root to extract water.

In prolonged drought conditions, the soil water potential may fall to such a low level that plant roots are unable to extract the remaining, tightly bound water. The entire plant continues to lose water without being able to compensate by root absorption, and eventually the turgor of the cells is lost and the plant begins to wilt. The leaves will droop, no longer being held rigid by water tension. There is some relief during the night, when photosynthesis ceases, stomata close, the temperature falls, evaporation rates drop, and the water stress on the plant is reduced to some extent. When water arrives as precipitation or water flow within the soil, the wilting may not prove fatal, and recovery of the plant takes place. But if the drought persists, the plant will enter a condition called permanent wilt, and at this stage even a renewal of water supply cannot prevent plant death.

Wood is composed of large numbers of xylem elements, so the tree or creeper can afford to have some columns failing without total disruption to the water supply at the crown. Plant physiologists have attached acoustic devices to the trunks of trees and can detect the occurrence of cavitations by the sounds they generate. Insects that bore into wood and fungi that invade tree trunks can create cavitations and loss of water transport, which is why some pest infestations of trees results in wilting and leaf loss in the crowns.

It takes a long time for water absorbed by the root to reach the top of a tall tree. In the case of the world's tallest trees, the redwoods, it may take as long as 24 days. In times of drought this delay in water reaching the canopy could prove serious and result in the shutting down of stomata and the cessation of photosynthesis, but there is a store of water available to trees in their sapwood. This is a layer of wood in the outer layers, between the heartwood and the bark. It makes up only about 2 or 3 percent of the total wood volume but contains a reserve of water that tall trees draw upon when they are under water stress. Overall, the height to which trees can grow is limited by the physics of water, especially the cohesive forces of the water molecules. Throughout the history of tree growth on Earth, even back in the days of the Carboniferous coal swamps, plant height has never greatly exceeded that of modern redwoods and some Australian eucalyptus trees.

A cross section of a tree trunk shows that timber is largely composed of xylem elements, which are continually added to by the dividing cells, or vascular cambium, that lies close to the bark on the outer edge of the wood. The oldest elements, the heartwood, are therefore close to the center of the trunk and the youngest, the sapwood, are nearer the outer circumference. Old xylem elements in the heartwood become nonfunctional as water-conducting vessels because they are more likely to have experienced cavitation, but they often become filled with chemical products that make them distasteful to pests and parasites. It is possible that these toxins are waste products of the living parts of the plants and are simply deposited in the old wood as a means of disposal, but it is more likely that their presence ensures that the old heartwood is not subjected to rot. Although the center of the wood is no longer functional for water transport, it remains a very valuable means of mechanical support, so it has to be protected from the ravages of insects and disease. This is a particular danger in the tropical forests, where many different types of organism feed on dead wood.

The cross section of a temperate tree shows a series of concentric rings of darker and lighter wood that result from annual surges of growth each spring. Generally, the new growth consists of wider xylem elements with larger intervening cells, giving the wood a light appearance. As the growing season continues, the size of the cells decreases and the wood looks darker. Annual rings, of course, develop only

Mahogany

There are three species of tropical trees that are a source of timber given the general name "mahogany," all belonging to the genus *Swietenia* (family Meliaceae) and originating in the New World Tropics. The Indian, or Cuban, mahogany (*S. mahagoni*) is a native of Florida, the Bahamas, Cuba, and parts of the West Indies. Honduran mahogany (*S. macrophylla*) has a wide distribution from Mexico to Venezuela and throughout Amazonia. A Mexican species (*S. humilis*) that occurs on the Pacific coast has a twisted growth form, so is of less commercial interest and value. All are fast-growing trees that achieve reproductive maturity soon after 10 years, but it is also a large tree, reaching up to 150 feet (50 m) in height and with a basal diameter up to 10 feet (3 m). It is capable of living for several hundred years but is often harvested well before that. The base of the trunk has buttresses that provide additional support for its great height.

Mahogany grows in the wet rain forests, but is also capable of survival in regions with a dry season. Although it has a period without leaves associated with the dry season, even in the perpetually wet equatorial regions it loses its leaves completely for a few weeks. At the end of the dry season it produces a flush of new leaves, together with clusters of small, yellow, sweetly scented flowers, which are visited by a number of small insects, including butterflies. The fruit is a capsule that splits to release winged seeds that flutter to the ground, assisted by any wind in the canopy, but that rarely travel more than 300 feet (100 m) from the parent tree. The seeds will germinate even in dense shade but remain small and slow growing until the canopy opens, as a result of either a tree death or damage by wind or river action. Mahogany is thus one of the many tropical forest trees that benefits from small-scale disturbance. Large-scale damage, such as that caused by severe hurricanes, damages the tops of the mature seed-producing trees, so this has a negative effect on mahogany regeneration.

The tree has been exploited and conserved by local peoples throughout human history of the area since the times of the Mayan civilization. The timber is easily worked, of high quality, and is very resistant to insect and fungal attack because of the natural preservatives that the tree uses to protect its tissues.

The roots of many tropical trees are relatively shallow, but they often form buttresses that add stability to the tall trunks. *(Ben Ryan)*

when there are seasonal differences in climate and therefore seasonal changes in tree growth. In equatorial conditions, there is little change in climate during the course of the year, so trees from these regions do not display annual growth rings in their woods. Some trees show signs of periodic growth surges that leave a record in the wood, but these are not usually annual, so they cannot be used to calculate the age of a tree in the manner used for temperate trees. At higher latitudes within the Tropics, in locations where there are distinct annual wet and dry seasons (see "Tropical Climates," pages 9–14), trees vary in their growth rates, usually growing fastest in the wet season, and can thus produce wood with growth rings.

Tall trees produce large volumes of timber, much of which serves a support function as well as the conduction of water to the canopy, and this timber has made the tropical forests the focus of much attention as sources of a timber harvest (see "Timber Production and Forestry," pages 200–202). Mahogany (see sidebar on page 115) is typical of the trees that have provided timber for human consumption.

The height of tropical trees, coupled with the constant danger of high winds and tropical storms, results in considerable mechanical stresses on their trunks. Many trees in the forest enhance their stability by developing buttress roots. These are massive triangular structures of wood that lead from the lower parts of the trunk out into surrounding areas, eventually merging into the root system (see illustra-

tion above). These large wooden flanges on the trunk greatly assist the stability of a tree when subjected to high wind. In some species the buttresses are hollow below, becoming "flying buttresses," or stilt roots, which are also very effective in adding to trunk stability. These stilt roots, or prop roots, are frequent in the wet and flooded forests, especially coastal mangroves (see illustration on page 73).

The trunks of trees, both tropical and temperate, are covered by bark. Bark consists of dead cells produced by a specialized layer of living and dividing cells, the cork cambium. This lies outside the main vascular cambium, which produces xylem, water-conducting elements to the inside, and phloem, food-conducting cells (sometimes called bast tissue) to the outside. The cork cambium is concerned solely with the production of bark, both to the inside and the outside. The outer cells are impregnated with a rubbery material, suberin, which makes them impervious to water, and this inevitably causes their rapid death. Although dead, these cells remain attached to one another and form a layer of protective tissue around the trunk of a tree. Being impervious to water, the bark prevents the loss of water from the tree trunk but also, especially in a very wet environment, prevents the tree trunk from becoming saturated with water from the environment and therefore exposed to disease. Many types of bark contain chemical that deter insects and other organisms from feeding on the tree, including tannins and various aromatic compounds. The cinnamon tree (*Cinnamomum verum*), for example, contains a spicy material, and eucalyptus species and sandalwood (*Santalum album*) contain aromatic, fragrant materials that are appealing to human senses but deter hungry insects.

Bark may also protect the tree from fire, and some of the thickest tree barks belong to species of trees that live in fire-prone habitats. The cork oak (*Quercus suber*) of Mediterranean regions and the giant redwood (*Sequoiadendron giganteum*), for example, have thick, fibrous barks that are resistant to flame penetration. Fire is occasionally experienced in the tropical forests, but it is not a regular feature except in the drier regions and those bordering the tropical savanna woodlands. Thick bark is rare among tropical trees. Most tree trunks have thin bark coverings that do not need to protect the tree from excessive evaporation, fire, or frost.

In a study of tropical forests in India, botanists found that the trees with the thickest bark are those associated with disturbance, occurring mainly in clearings. These are also trees with the highest levels of production of gums and resins in their bark, thus trees from disturbed sites need to be better protected from drought, fire, and predation by grazing animals. The bark is clearly an important material for affording such protection. Many species of palms are

protected by thick, fibrous leaf bases that persist long after the main photosynthetic frond of the leaf has died and fallen to the floor. Palms are found largely in the Tropics, as shown in the map below, and are trees often associated with challenging habitats, such as drought-prone sites and beaches.

The root systems of tropical trees are extremely variable, and often they contribute to the stability of the trunk, forming buttresses or stilts for additional support. Adventitious roots, formed directly from stem structures, sometimes descend from branches and add to overall stability, or they may tap in on the soils of the epiphyte communities (see sidebar "Epiphytes," page 121). Below the soil, roots of tropical trees are often shallow. In the wet rain forests, water supply arrives from above and is taken up by roots in the upper layers of the soil. Nutrient cycling also takes place rapidly in

The global distribution of palms. The palms are among the most diverse and widespread of tropical trees. There are about 2,650 species and, as can be seen from the map, are almost entirely limited to the Tropics and subtropics in their distribution.

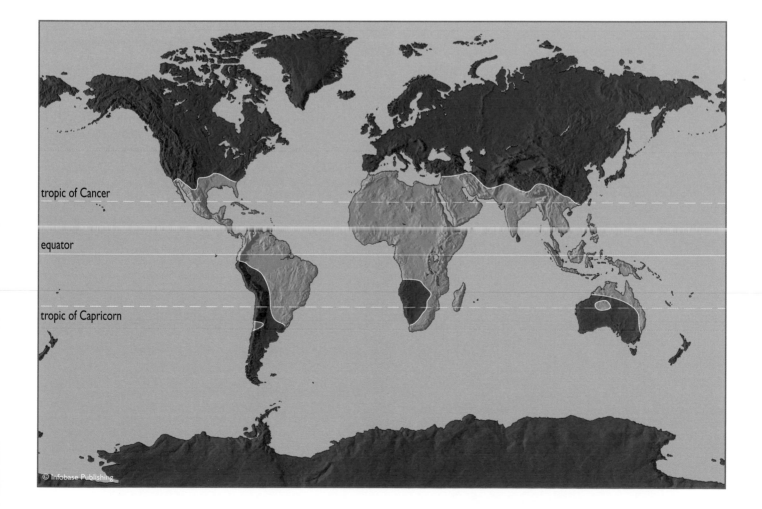

the litter layer on the soil surface as organic matter decomposes quickly and the released nutrients are taken up by the nearest roots. All of these factors mean that there is great advantage in having root systems close to the soil surface. If the soil is waterlogged, then anoxia, or lack of oxygen, becomes a problem for roots that need to respire in order to gain the energy they require for element uptake. Some roots then grow upward until they stand clear of the soil and are able to gain access to atmospheric oxygen for their respiration. These "breathing roots" are called pneumatophores or pneumorrhizae, and are particularly associated with mangrove forests. They may take the form of pegs that rise above the soil or water surface and grow upward to a height of about eight inches (20 cm), or they may loop above the surface only to plunge back downward, leaving an exposed curve of root for air absorption.

■ THE TROPICAL LEAF

The most important function of tree trunks, whether tall or short, is to bear leaves in a position where they receive sufficient light to carry out the food-producing process of photosynthesis. Tall trees have the advantage that they can reach above their neighbors and hold their leaves in unshaded positions, but they face the problem of high light intensity causing overheating and water loss, the difficulties of providing adequate water from the ground far below, and the danger of being struck by lightning or blown over in high winds. Shorter trees, on the other hand, hold their leaves in a more shaded environment, so they need to be able to photosynthesize effectively at lower light levels; in a cooler and more humid atmosphere they are less subject to drought stress.

Some species of tree are adapted to life as emergents and the leaves of these trees need to be able to cope with very different conditions from those of the understory species. This means that their leaves have to operate in very different types of light intensity and quality (see sidebar "Light, Vision, and Photosynthesis," pages 28–29). Some trees grow with parts of their crown held in full sunlight and their lower branches in dense shade; in this case the leaves on different parts of the plant are exposed to very different conditions and need to be adapted accordingly. Plant physiologists have found that leaves vary in their photosynthetic capacities and responses according to the environment in which they have developed. Plants of open, sunny conditions develop "sun leaves," while those that grow under low light intensities bear "shade leaves" (see sidebar).

Since the function of a leaf is light trapping, the most obvious shape for a leaf is circular or ovoid and entire, that is, having a uniform and uninterrupted margin. The trees and other plants of tropical forests are particularly rich in

Sun Leaves and Shade Leaves

The function of leaves is to absorb light, trap some of its energy, convert that energy into a form that can be used for chemical transformations, then use the energy to reduce atmospheric carbon dioxide to simple organic molecules. This is the essence of photosynthesis. Trapping sunlight involves exposing as much surface area as possible to incident light and covering that area as densely as possible with a light-trapping pigment, chlorophyll. Under high light intensity, a leaf can be fairly thick and many layers of chlorophyll-containing cells can be stacked one upon another because the light will penetrate deeply into the tissue. In shade a thick leaf is inefficient because the low light intensity will not permit deep light penetration through the layers of cells. The outcome is that sun leaves tend to be relatively small in area but thick, while shade leaves are larger in area, but thin.

Adaptations to light intensity can be physiological as well as morphological. Plant physiologists examine such adaptations by observing the behavior of leaves, particularly their efficiency at carbon fixation, when exposed to increasing experimental light intensities. All leaves respire, and if the light intensity is low their rate of respiration may exceed their capacity to photosynthesize. A shade leaf has a lower rate of respiration than a sun leaf, so it loses less energy at low light intensity. A shade leaf therefore reaches a break-even point (when photosynthesis equals respiration) at a lower light intensity than the sun leaf. This is called its *light compensation point*. As the light intensity is increased, the shade leaf photosynthesis increases rapidly and exceeds that of the sun leaf under the same conditions, but soon reaches a maximum level of photosynthesis when its capacity to fix light is saturated. The sun leaf responds more slowly under low light intensities, but as the light levels rise its photosynthesis curve expands rapidly and overtakes the shade leaf. It continues to increase in efficiency long after the shade leaf has become light saturated and is eventually considerably more effective than the shade leaf in collecting light and photosynthesizing at high light intensities.

species with simple leaf shapes of this type; indeed, one of the problems of identifying tropical trees from their foliage is that so many species have the same basic shape. Leaves that are subdivided into lobes or into leaflets (called compound leaves) are increasingly abundant in temperate latitudes, but why this should be the case has not yet been explained. There are some tropical forest plants with divided leaves, and some with large holes in the leaves, such as the Swiss cheese plant (*Monstera deliciosa*), which is popular as a house plant. These plants with divided leaves are often more frequent in the lower layers of the forest. It is possible that their form allows better light penetration to the lower leaves on the plant in this dimly lit environment, but this is not a very convincing explanation of such leaf division.

The upper surfaces of many types of tropical leaves are often coated with a thick layer of wax, called the *cuticle*. This feature is certainly not confined to tropical plants and is present among the plants of all the major biomes, including the tundra. In many situations the main function of the cuticle is to prevent any evaporation from the upper epidermal cells and thus conserve water. In the tropical forest environment this is unlikely to be its major function except in those leaves of the upper canopy exposed to intense light and heat. The main value of the cuticle to tropical plants is that the wax rapidly sheds the water that falls upon it. In conditions of heavy rainfall a leaf could become laden with water and could remain wet for much of the time. This adds weight and therefore strain to the tree, and it also leaves the leaf open to colonization by fungi and bacteria that could prove invasive and harmful. It is far better to shed the excess water that constantly falls on the leaf canopy than to retain it on the leaves. One other frequent feature of tropical leaves is an extension, often downward-curved, of the central midrib, forming an elongated tip that narrows to a fine point. This is called a *drip tip* and it serves to channel the water from the leaf's upper surface and assist in the process of water shedding. In one survey of a rain forest in Ghana, ecologists found that over 90 percent of the plant species in the lower layers of the forest were equipped with drip tips.

The trees in the equatorial rain forests are often evergreen, which means that they are always clothed in a leafy canopy. The deciduous habit, in which trees are bare of leaves for a part of the year, becomes more frequent in the higher tropical latitudes where seasonal drought is experienced. The relative advantages of evergreen and deciduous habits have already been discussed (see sidebars "Why Be an Evergreen?" page 57, and "Why Be Deciduous?" page 61). In fact, there is a continuous spectrum of patterns of leaf loss and replacement. At one end of the scale, trees may continuously form new leaves and shed old ones so that leaves are always present on the tree, even though individual leaves may last only a few months before falling. There are also periodic evergreens, which have times of the year when leaf shedding is more apparent. This may occur some time after a burst of new leaves has taken place, in which case the leaf canopy is maintained (as in many members of the Dipterocarpaceae), or leaf fall may take place at about the

time when a new flush of young leaves is bursting from the bud. In the latter case, as seen in species of *Terminalia*, there may be a period of a week or so when the tree is losing its old leaves faster than it is developing new ones and the tree looks bare. But this situation does not last long, and there is no clear season when the tree is bare of leaves. In truly deciduous trees leaf fall takes place well before bud break, leaving the trees bare for several months.

One type of leaf often encountered on forest floors or in the lower canopy is the variegated leaf. These leaves are patchy, with areas of green chlorophyll-containing tissue alternating with pale cream or yellow areas where chlorophyll is absent and only a few carotene pigments remain. Such plants are very familiar as houseplants, for example, the spider plant (*Chlorophytum* species), because their blotched or striped patterns make them attractive to the human eye. From a botanical point of view, however, their patterns are something of a mystery. The function of a leaf is photosynthesis, so why should a leaf, particularly one situated in shady environments, devote only part of its surface area to photosynthetic tissue? There are two possible explanations. Perhaps the striped or spotted patterns make the leaves less visible in the dappled light and shade environment of the forest floor. High light intensity sunflecks, coupled with dark shaded patches of ground, can create patterns of high contrast where a patchy appearance acts as a form of camouflage. Variegated plants can be regarded as the zebras of the forest floor, made inconspicuous by their contrasting dark and light patterns. Camouflage of this type is called crypsis. There is some experimental evidence that confirms this idea, showing that herbivores neglect the variegated species, perhaps failing to notice them. An alternative proposal, which could act in concert with the crypsis idea, is that a variegated leaf looks smaller and less attractive to a grazer than a completely green one, or perhaps it looks infected and diseased, so is best left alone. Whatever the precise mechanism involved, variegated leaves do generally appear less appetizing to the grazing animal.

The leaf, in tropical forests as in all other ecosystems, has the clear and simple function of making food for the plant, but many of the features of leaf shape, structure, and chemistry are associated with the problem of ensuring that this food is retained by the plant and is not appropriated by the wealth of grazing animals avidly seeking sources of energy.

■ DEALING WITH DROUGHT

The equatorial rain forests are wet throughout the year. Dry seasons become increasingly influential in the biology of forest away from the equator. Even close to the equator freak droughts can occur. An example of this phenomenon took place in 1998 in the Lambir Hills National Park of Borneo, when a three-month drought from February to April resulted in less than one-fifth of the normal rainfall arriving at this normally ever-wet rain forest. Even such a short episode of drought resulted in the mortality rate of trees rising from the normal 2.4 percent per year to 7.6 percent per year. The Kalimantan region of Borneo suffered particularly badly, especially when drought was followed by fire (see "Drought and Fire," pages 20–22).

The ecologists who recorded the effects of this drought noted that the trees most affected and most subject to raised mortality were not the rare species but the more abundant and widespread ones. As a result, the forest was more diverse following the drought, in the sense that there was a more even distribution of populations of different species (see "The Meaning of Biodiversity," pages 98–99). This response reinforces the idea that occasional low-level disturbance can increase biodiversity in a forest (see "Patterns of Biodiversity," pages 100–103).

Even when the forest is not exposed to prolonged drought there are parts of the canopy that may experience short-term dry spells. The environment in the upper canopy of the forest is particularly vulnerable because of its exposure to high levels of insolation (sunlight), high temperature,

and the drying action of wind, and the organisms that spend their lives in the upper canopy have to be able to cope with dry episodes. The leaves of emergent trees are more drought resistant than those of the lower canopy, having thicker layers of waxy cuticles on their upper surfaces, and are generally smaller and tougher than those of the lower canopy.

Soils may also become dry quite quickly in times of drought because of the high demand placed on the groundwater resources by the transpiration of the trees. This is exacerbated in the tropical forest by the very high leaf area index (see sidebar, "Leaf Area Index," page 59). Not only do leaves lose water by transpiration, but they also intercept water and prevent it reaching the ground. Although most tropical tree leaves shed water very efficiently (see "The Tropical Leaf," pages 118–119), leaves are so abundant in the forest canopy that even a thin residual layer of water over each leaf results in substantially less water reaching the forest floor than what falls on the upper canopy. Tropical ecologists often remark

---

Epiphytes, such as this bromeliad in the Atlantic rain forests of Brazil, are supported by the trunks and branches of trees and have no contact with the ground.   *(Octavio Campos Salles)*

Epiphytes

Most plants have root systems embedded within the soil that provide physical stability and act as foraging organs that absorb water and mineral elements from the soil. There are some plants, however, that lack any contact with the ground and pass their entire lives in forest canopies. The advantage they gain is that they are held in a position that is better lit than the forest floor, but they do not have to invest in the development of tall woody trunks to get there. The disadvantage is that they have to arrive, germinate, and establish themselves in a fairly precarious position on the upper side of broad boughs or at the junction between branches and the main trunk, where leaf litter and other organic detritus often accumulates in a moist hollow. Once established, their roots need to take a firm grip on the very limited substrate; they do not normally invade the tissues of their host but simply bind to it as a means of support. Epiphytes are green, photosynthetic plants that derive their energy from sunlight and do not resort to parasitism. Mechanical support and security even in high winds and storms are clearly the initial problems that epiphytes need to deal with, although they are then faced with the other requirements normally supplied by soil-based rooting systems, namely water and mineral supplies. Water arrives solely by rainfall, so epiphytes can only survive where rainfall is in abundant supply, which normally means the tropical and temperate rain forests. The roots also absorb some water from the organic substrates that accumulate around epiphyte communities among the tree branches, and some types of root are even capable of taking water directly from the atmosphere when it is saturated. Mineral elements are similarly available largely from the rainfall, where they are generally present in a very dilute form, or from decomposition processes that take place in the small volume of perched soil residing on the branches. The process of reproduction and dispersal is achieved by wind-dispersed spores in the case of mosses, lichens, and ferns, or by seeds in the case of flowering plants. Seeds may also be wind dispersed, which ensures that they are spread widely and relatively randomly through the forest canopy or transported in the guts of animals, including birds, that inhabit the appropriate levels within the canopy.

Although epiphytes demand only physical support from their hosts and are not parasitic in other respects, the weight of a heavy load of epiphytes can place mechanical strains on a tree, sometimes leading to the collapse of branches during high winds and storms. Some trees, therefore, have developed shedding mechanisms, such as the peeling or scaling of bark, thus discarding the additional weight load. Very small epiphytes may even colonize individual leaves and block light penetration. One advantage of the deciduous habit is that leaves affected in this way can be shed and new, more efficient leaves established in their place.

that they can walk through the forest even when precipitation is abundant and do not require waterproof footwear because the soil is relatively dry. Of course, this depends on the drainage properties of the soil, because sandy soils drain more freely than clay soils, and ridge or sloping sites shed water, whereas valley sites accumulate water. But the interception and evaporation of water by canopy leaves certainly adds to drought problems when water becomes scarce.

Drought is a particular problem for those plants that have no rooting system in the soil but attach themselves to the boughs of tall trees (see illustration on page 120). These are called *epiphytes* (see sidebar above).

Epiphytes are particularly exposed to drought conditions, and most have developed mechanisms to cope with this problem. Thick, waxy leaves are very frequent among epiphytes, such as the orchids, and sometimes the leaves are arranged in the form of a cup, so that rainwater is collected in a leafy vessel and retained for times of shortage. The bromeliads (family Bromeliaceae) provide particularly good examples of this type of structure (see illustration), forming waxy chalices of overlapping leaves on the branches of tropical trees. The small pools within these epiphyte cups are of great importance to canopy animals, providing a water resource in a habitat that is otherwise devoid of drinking water. Aquatic insects may even establish themselves in these small water bodies, and the moist habitat of bromeliad leaves provides a valuable sanctuary for many invertebrates that require constantly high humidities. Water storage is one approach to the desiccation problem; the alternative is to become a *xerophyte,* or plants with specialized structural modifications that resist water loss, one of which is succulence. Typical succulent, xerophytic epiphytes are *Epiphyllum* and *Zygocactus,* the latter well known as the houseplant often called "Christmas cactus" because it flowers in midwinter. Both of these plants lack true leaves but have succulent green stems that have a broad and flattened structure and serve the normal photosynthetic function of leaves.

Many bromeliad epiphytes have their waxy leaves arranged in the form of a cup and can collect and store rainwater. They live high in the forest canopy, so they have no root contact with the soil beneath. This can leave them exposed to desiccation if there is a dry period. *(Peter D. Moore)*

Covered by a waxy bloom, they are resistant to water loss and have the appearance of plants one would normally associate with a desert rather than a rain forest. The fact that cacti can become efficient rain forest epiphytes serves to emphasize the desiccation problems faced by those plants adopting this life style. Drought is a very real problem in the high canopy.

Resisting water loss is one approach to overcoming the stress of drought; an alternative is to become desiccated in dry times and resuscitate in wet times. Plants adapted in this way to drought survival are called *poikilohydric,* chief among them being the mosses, liverworts, and lichens. Mosses and liverworts have no waxy cuticle over their leaves and no bark surrounding their stems, so water is quickly lost from their tissues when they are exposed to drought. Plant physiologists have found that many moss species are able to photosynthesize and grow effectively only when the relative humidity is over 95 percent, so they effectively close down as soon as conditions begin to become dry. Their cells desiccate and shrink, and the whole plant may turn brown and shrivel, but when rewetted they quickly spring back into life and begin their cell processes once more. Their powers of

survival are considerable; there are records of mosses being dried out and kept in museum herbaria for over 100 years but being capable of resuscitation when they are soaked in water once more. Mosses and liverworts are thus well adapted to life as an epiphyte in forest canopies, and they are extremely abundant in all forms of tropical forests.

Lichens are also frequent members of the epiphytic community. These organisms, actually a combination of an alga or a cyanobacterium with a fungus (see sidebar [right]), like many mosses, are poikilohydric. They can be desiccated and then rewetted without damage—a considerable asset when living in a periodically dry habitat, such as the forest canopy. The branches of tropical trees are often festooned with lichens, which often adopt hanging growth forms, producing a veil of tissue from the branches. When the photosynthetic component of the lichen is a cyanobacterium, it is possible for the lichen to fix atmospheric nitrogen, adding to the nutrient capital of the epiphyte community.

Many ferns also adopt an epiphytic way of life because of the advantages it brings them in terms of light penetration. Some of these also grow on the forest floor, in clearings or in patches where sunflecks occasionally penetrate, but some ferns are found only as epiphytes. One of the most familiar of these is the stag's-horn fern (*Platycerium* species), which forms two types of frond. One is a long, flat, branching structure that has a shape resembling a stag's antler and is the main photosynthetic frond that hangs from the host tree's branch. These fronds are extremely durable and may last for several years. The second type of frond is in the form of a shield that spreads over the base of the fern, including its rhizome and root structures. This shield is also green and photosynthetic in its early stages but soon dies and forms a papery brown cover that protects the rhizome and leaf buds in the event of drought.

The organic matter collecting on branches around the epiphytes forms a kind of perched soil high in the canopy. An entire ecosystem of soil invertebrates and microbes develops within these soil pockets and a nutrient cycle of decomposition, nutrient release, and absorption by plant roots is established. Rain leaches through these soils, just as it does through the soils of the forest floor, and the lost nutrients trickle down the tree trunks as stem flow (see "Element Cycling in the Tropical Forests," pages 84–89). Nutrient elements, such as phosphorus and nitrogen, are scarce and much in demand among forest plants, so any source becomes a focus of root activity; this is true of the perched soils among epiphyte communities. Studies of nutrient distributions in tropical forests suggest that up to 40 percent of the soil nutrient capital of the forest ecosystem can reside in these epiphyte soil pockets. Some trees are capable of producing roots directly from their stems and branches, called adventitious roots, and in this way the host can tap into the soils that lie on their branches. Even roots from below ground may grow up the trunks of their parent plant or the trunk of a neighbor, guided by the stem flow of nutrients from above, eventually reaching these pockets of soil. Epiphytes and their associated communities thus form a distinct microhabitat within the forest canopy.

Most epiphytes are relatively harmless as far as the host plant is concerned. They do not parasitize the host for water or nutrients in the way that plants like mistletoes do. Their only disadvantage is that of weight and bulk that can cause structural damage. Most epiphytes are relatively small, so bulk becomes a problem only after many years of buildup of epiphyte biomass. There are some epiphytes that increase in size until their presence becomes a serious problem for the host tree. These are the so-called strangler figs (*Ficus* species). Figs have flask-shaped fruits within which are masses of tiny seeds, and when the fruit is eaten by mammals or birds the seeds pass unharmed through the gut and are deposited in feces. Canopy-dwelling birds are particularly effective at dispersing the seeds of strangler figs, dropping them among the epiphyte communities on upper branches or in the junctions between branches and trunks, where organic matter accumulates. Here the fig seed germinates and establishes itself as an epiphyte among the others.

The leaves of figs are mainly thick and leathery, well adapted to life in the canopy and capable of resisting drought. But as the small tree grows, perched high above the forest floor, it develops a rate of transpiration that exceeds the direct supply of water from rainfall. To survive and continue growth, it needs to extend its roots beyond the confines of the epiphyte ecosystem, so the roots begin to grow downward, following the main trunk of the tree and often twining around it for support (see illustration on page 124). A network of roots develops, extending down the trunk and reaching the soils of the forest floor, where they tap into the water resources of the ground. At this stage the epiphytic fig tree is able to grow more rapidly because it has an assured source of ground water, and its leaf canopy expands among the branches of the host. Some species then send adventitious roots directly to the ground from their expanding canopy of branches, as in the case of the Indian banyan tree (*Ficus benghalensis*). In this

Lichens

A lichen consists of an association of a fungus with an alga or with a cyanobacterium. The closeness of that association is quite variable and can simply consist of the two organisms living alongside one another. Such an association would not normally be called a lichen. One definition of a lichen is a stable, persistent, and self-supporting association of a mycobiont (the fungal component) with a photobiont (the photosynthetic partner) in which the mycobiont forms a protective cover for the photobiont. Delicate photosynthetic cells of a micro-alga, or cyanobacterium, are enclosed within a sheath of fungal mycelium. These cells are protected by the surrounding mycelium and although they desiccate quickly under dry conditions, they are able recover remarkably well (poikilohydric). Light penetrates the mycelium, allowing the photobiont to photosynthesize, and the fungus derives its nutrition from this source. Both partners thus benefit from the association, and these benefits allow the combined "organism" to occupy positions that neither component could achieve on their own. They are able to grow on rock surfaces, the surface of stony soils, or on the branches of trees and shrubs as epiphytes, as in tropical forest canopies, where they can cope well with desiccation even in dry and exposed conditions.

The structure of the combined thallus is quite complicated and often takes on a distinctive and recognizable form, which is why lichens can be classified and identified visually. The "lichenized" thallus sometimes produces distinctive chemicals, secondary products of metabolism that protect them from grazing animals by making them unpalatable. Reproduction in the lichens can take the form of small particles containing both components, becoming detached from the main thallus and carried away by wind or water. An alternative process is when the fungal component produces fruiting bodies and releases airborne spores that become dispersed and land on surfaces where they may encounter the appropriatez alga and form a new lichen thallus.

The algal or cyanobacterial component of lichens are well known as independent species that can grow without the aid of the fungus, though usually in a more restricted range of habitats. It was once believed that the fungal component survived only in combination with its symbiont, but recent studies using molecular techniques have shown that the fungi of lichens do grow independently of the photobiont but take on such a very different form that their involvement in the lichen thallus has never been suspected. This poses a real problem for taxonomists involved in classifying organisms and raises the question once more as to whether lichens should be regarded as true species.

Africa and Asia than in South and Central America. In the New World, the most frequent of the stranglers belong to the genus *Clusia* (family Guttifereae, the same plant family as the St. John's wort). These plants vary in the degree of aggressive growth, some being technically hemiepiphytes—that is, epiphytes that develop aerial roots or even roots that extend to the ground but do not establish themselves independently of their hosts.

The epiphytic habit is clearly a very successful one; many different types of plants have taken to the trees in this way. Taxonomists have estimated that there are 28,000 species of plants that can grow epiphytically, belonging to 65 different families. The fact that this way of life is found in such a wide range of families suggests that it has not simply arisen once over the course of evolution but has developed independently on many different occasions. Biologists call this type of multiple, parallel evolution *polyphyletic*, literally meaning having many different sources.

The strangler fig eventually overpowers its host by its rapid growth and eventually overshadows it as it establishes its own canopy and system of prop roots.   *(Peter D. Moore)*

A young strangler fig, which will one day become a banyan tree (*Ficus benghalensis*), has established itself as an epiphyte in the canopy of a host tree. Its roots extend down the host's trunk to make contact with the ground.   *(Peter D. Moore)*

way, the strangler fig becomes an increasingly large structure (see illustration [right]). The competitive ability and growth potential of the fig often exceeds that of the host tree, in which case the host canopy begins to be outshaded by the expanding fig leaf cover, and may eventually die as a result of the intense competition, leaving the fig as an independent plant, the victor in the competitive battle. Strangling figs effectively use the epiphytic habit to give them a start in life, taking advantage of a well-lit location for the early stages of growth and then using that advantage to supplant the host in its location within the forest. Not all epiphytes, therefore, are as harmless as they seem. It is not surprising to find that some tree barks have developed chemical defenses to prevent the invasion and growth of epiphytes. Chemical warfare of this kind is called allelopathy.

The strangler figs occur throughout the tropical rain forests of the world, although they are more abundant in

Many tropical forest epiphytes have developed a form of photosynthesis that is more commonly associated with deserts, known as *Crassulacean acid metabolism* (CAM) (see sidebar below). CAM is a mechanism that helps a plant to cope with drought by enabling it to take up carbon dioxide rapidly and store it in a temporary form. The storage system is also independent of light, so it can be conducted at night. The advantages of the physiological mechanism are mainly associated with the problem of having to keep stomata open for gaseous exchange and consequently losing water. The occurrence of this dry-land type of photosynthesis deep within rain forests emphasizes the importance of drought as regular feature of the upper canopy.

Visual inspection of the epiphyte community does not always allow the observer to differentiate between epiphytes and parasites. Mistletoes (family Loranthaceae), for example, are abundant in tropical forests and are green and photosynthetic, so careful anatomical examination is necessary to distinguish their mode of life from that of the epiphyte. The mistletoes produce adventious roots that do not simply anchor the plant to the host but penetrate the bark of the host tree and enter the vascular system of xylem and phloem.

Crassulacean Acid Metabolism

Crassulacean acid metabolism (CAM) is a biochemical pathway that, as its name reveals, was first found among the plant family Crassulaceae, a family that consists largely of succulent species from arid environments. These plants face problems of desiccation, especially in the daytime if they open their stomata to absorb gaseous carbon dioxide from the atmosphere and consequently lose water from their internal tissues. Instead, they open their pores at night and fix the carbon in a temporary form as organic acids, using the enzyme phosphoenolpyruvate caboxylase (PEP carboxylase). Sealed within the plant tissues, the organic acids then break down, releasing carbon dioxide that can be refixed using the conventional photosynthetic system, involving the enzyme rubisco (ribulose bisphosphate carboxylase/oxidase). The PEP carboxylase enzyme has a higher affinity for carbon dioxide than does rubisco, so the rate of uptake is enhanced by using this enzyme. It is also more energy expensive, so it is appropriate only for plants living in high light intensities. Many different types of epiphytes, ranging from ferns to orchids, have adopted the mechanism, which suggests that its evolution is polyphyletic.

Because they are photosynthetic, they produce some of their own food through their leaves, so strictly they are *hemiparasitic* (partial parasites) rather than totally parasitic. Botanists have conducted a great deal of research to establish precisely what the mistletoe is seeking when it invades its host. Perhaps it requires additional products of photosynthesis, such as sugars, that it can obtain from the host's phloem tissues. Or the mistletoe may need water from its host to allow it to open its stomata and thus maintain its own photosynthesis. Or the parasite may require mineral elements brought up in the water stream from the roots through the host's xylem elements. The answer seems to be all three. By using radioactive carbon atoms fed to the host plant as carbon dioxide, plant physiologists can trace the movement of carbon into host sugars and then into the parasitic mistletoes. Using this method, they have found that on average about 15 percent of the total carbon uptake from a mistletoe plant comes from its host and the remainder is derived from its own photosynthesis. To obtain this sugar the parasite must tap into the phloem tissues in the host wood, because these are the tissues that carry the products of photosynthesis from the leaves to other parts of the plant, including the roots.

But mistletoes also tap into the xylem vessels, which carry water from the roots to the canopy. This is the sole supply of water for the mistletoe that is isolated from the soil, and, unlike an epiphyte, does not have roots to absorb water from its local environment. The mistletoe needs a steady supply of water because, like the canopy of the host plant, it conducts photosynthesis and has to keep its stomata open for gaseous exchange. Plant physiologists have measured transpiration in many different types of mistletoe and have found that these hemiparasites have very high transpiration rates; they are profligate in their use of water. The mistletoe takes all this water from the host, which may well prove a bigger strain on the host's resources than the small demand for food. The growth of a mistletoe plant is therefore closely controlled by the growth of its host; if the parasite grows too quickly it can kill its host by placing too great a demand on the host's root system and the water supply cannot keep pace. At the same time, the leafy shoots of the mistletoe compete with the host's leaf canopy for light, so it must grow fast enough to keep pace with canopy growth. Like all parasites, mistletoes have to moderate their demands and yet grow as fast as circumstances permit.

■ THE FIGHT FOR LIGHT: UPWARDLY MOBILE PLANTS

The highly structured canopies of the forest, the tall supportive trunks of the trees, the complex variety of leaves, and the presence of epiphytic plants that use taller species for support all demonstrate that the demand for light among photosynthetic plants is intense. Finding a place in the sunlight is one of the most critical requirements for a photosynthetic plant, and it must either increase in stature to outgrow its neighbors, adapt physiologically to lower light conditions, or adopt the approach of the epiphyte, which takes advantage of the tall growth of other plants to establish itself in the canopy. There are alternative strategies, one of these being to begin life on the forest floor and climb up through the complexity of the forest layers using other plants for support in the process. This is the mode of life adopted by the climber, or liane.

There are relatively few species of liane in temperate forests, but they abound in the tropical forests and add greatly to the complexity of their structure and their visual impact. They are most abundant in clearings or on forest edges, such as along the banks of rivers. The growth form is found in a wide range of different plant families, so the evolutionary origins are clearly polyphyletic. Among the palms, for example, the rattans are climbing plants, and the longest recorded length for a climber of 732 feet (240 m) belongs to a rattan. The rattan palms are found in both the Old and the New World Tropics, being most abundant in Malaysia, where they are represented by about 480 different species. The arum family (Araceae) also has many members that are tropical forest climbers, and many of these have adhesive roots that stick to the bark of the trees up which they climb. Unlike the mistletoes, these adhesive roots do not penetrate into the living tissues of the host.

Side branches on the liane, and even some types of leaves, may form twining structures that assist in support, called tendrils. Most climbers have evergreen leaves that are simple and elliptical in shape. Although the stems are relatively narrow in comparison to their lengths, the xylem vessels of lianes are among the largest known, reaching cell diameters of over 0.02 inches (0.06 cm) and lengths of up to 23.5 feet (7.7 m). Fast water flow is possible in these large vessels, and plant physiologists have used dyes to measure the rate at which water moves through the liane stem. The fastest rate recorded belongs to an Amazonian climber, *Aristolochia brasiliensis*, in which the flow of dye reached 8.2 feet (2.5 m) per minute.

The main problem faced by a liane is that of initial establishment. They are rooted in the ground, unlike epiphytes, so their seeds need to germinate and establish themselves in the soil, then grow up into the canopy by using more robust-stemmed plants for support. The early days or months in the life of a liane seedling are therefore critical to its survival. Most liane seeds are large, sometimes approaching an ounce (28 g) in weight, and contain substantial reserves of food to allow the young plant to begin its growth in dense shade. Different liane species then behave in different ways. Some may begin life by extending laterally over the forest floor, branching and forming a number of growing points and thus increasing the likelihood that one or more of these

points will find a suitable host and will grow up into the light. The growth of these lateral shoots may be very rapid, up to five inches (13 cm) per day. The alternative is to head straight upward and trust that it will encounter a suitable support before its reserves are all spent. This vertical growth can reach 15 feet (5 m) before a full crown of leaves is formed and the plant develops a self-sustaining photosynthetic system, and the physiology of its growth at this stage is unusual. Most small seedlings respond strongly and positively to light, seeking the best-lit situation for their shoots; they are positively phototropic. The initial growth of many lianes is independent of light but is negatively geotropic, meaning that the shoot grows away from the force of gravity. The reason for this growth pattern is probably the complexity of the light climate on the forest floor (see "Microclimate," pages 25–33). Light may come from many directions, reflected from the trunks of trees and filtered through the leaf canopy, so a seedling responding to light intensity would encounter a confusing mass of signals. Simply growing upward is evidently the best strategy in this situation, and when the shoot encounters a possible support it can wind around it and continue its upward growth.

Stems that twine use relatively small-diameter trunks for their initial support because larger trunks would present too big a problem for their circling motion. When the liane reaches the canopy it is supported by the robust branches of the major trees, and at this stage its connection to the ground may hang loose because the original small stems may well have died away. The forest then has curtains of hanging liane stems attaching their canopy foliage to the forest floor. The climbers rarely use the very tall emergent trees, but concentrate their growth within the B layer of the forest, where light intensity is high but wind and drought stresses are less than those of the A layer (see sidebar "Forest Structure," page 54). On the whole, lianes are more abundant in lowland forest than in montane forest, and may well be most abundant in forests where there has been a degree of disturbance, by either natural disasters or human occupation. Even in forests where lianes are abundant, their biomass rarely exceeds 5 percent of the total above-ground plant biomass. There is some evidence to suggest that the total amount of lianes in the tropical forests of the world is steadily increasing, possibly as a result of climate change or of increasing disturbance. This increase could affect the future structure and composition of the forests.

From the point of view of the tree, the liane is a serious competitor and an even greater liability than epiphytes, so foresters regard them as weeds that need to be eliminated. Not only does the leafy canopy of the liane compete with the tree for light, but also the weight of the biomass creates considerable mechanical strain on its support system. This is especially the case during tropical storms when trees with a heavy load of lianes are much more liable to damage or being uprooted. It is not surprising, therefore, that many trees have evolved mechanisms for discouraging these climbing passengers. Flexible trunks and branches, especially if accompanied by smooth bark, are very difficult structures for a liane to climb. The adhesive roots do not stick, and the tendrils slide off as the sapling bends under the increased weight.

■ PLANTS IN THE SHADE: PARASITES AND SAPROTROPHS

Relatively little light penetrates to the floor in a mature and undisturbed tropical forest. Conditions here are far from ideal for photosynthetic plants, and only a few shade-adapted plants are able to sustain themselves on a permanent basis. Some seedlings of the canopy trees germinate, but few survive because of the problems of maintaining a positive carbon balance under low light intensities (see sidebar "Sun Leaves and Shade Leaves," page 118); many species require the better-lit clearings in the forest for their regeneration. It is not surprising, therefore, that the plants of the forest floor have often evolved alternative strategies for survival and have abandoned the autotrophic (self-feeding) system most characteristic of higher plants for a heterotrophic mode of life in which they depend on other organisms for a source of energy.

Parasitism, in which a plant becomes fully dependent on another plant for its energy, is quite a rare phenomenon. There are relatively few flowering plant families in which this kind of behavior is found; in the Tropics the most frequently met families of parasites are the Balanophoraceae and the Rafflesiaceae. These plants have lost all chlorophyll, so, unlike the mistletoes, they are totally unable to conduct any of their own photosynthesis. As an alternative source of food they invade the roots of other plants and draw on their energy-rich compounds for their metabolism. One of the best known parasites of the tropical forests is the genus *Rafflesia*, whose members are confined to Southeast Asia and are parasitic on the roots of climbers, particularly the genus *Tetrastigma*. Its vegetative parts are normally invisible, consisting of fungus-like threads of tissue invading the root tissues of its host. It only becomes apparent when it flowers, because the flower of *Rafflesia* is the largest known of all flowering plants, often measuring three feet (0.9 m) in diameter and weighing up to 15.5 pounds (7kg). The flower begins to form as a bud within the root of the host, then breaks out and produces a dark red structure above the soil surface that smells of rotten meat and is very attractive to flies, which pollinate it. Male and female flowers form on separate plants (dioecious), so pollination involves transfer by insect over considerable distances in the forest (see illustration below).

Members of the Balanophoraceae also lack chlorophyll and parasitize the roots of tropical trees. Their above-ground parts are often club shaped and bear very tiny flowers that, like *Rafflesia*, are dioecious and pollinated by flies.

One problem that still baffles botanists is how the seeds of these parasites find their hosts. In the case of *Rafflesia*, the seeds are very small, and one possibility is that trampling vertebrates may carry the seeds in soil on their feet, perhaps even damaging the roots of a suitable host and encouraging

Rafflesia, a plant of the jungles of Southeast Asia, produces the largest flower in the world, and one of the worst smelling. It carries the scent of rotting carrion and attracts flies that serve as pollinators. *(Vladimir Pomortsev)*

infection. Most of the research on host invasion by parasitic flowering plants has been carried out on temperate crop plants. Here there is evidence of chemical secretions by the host acting as an attractive agent for the germinating seedling. But this mechanism requires that the seeds be large enough to provide for the seedling while it forages for a host, and it would not explain how small-seeded species such as *Rafflesia* manage to cope.

An alternative way of life for plants on the forest floor is to take on the lifestyle of a fungus and become a saprotroph. This is an organism that depends on the energy contained within dead organic matter for its survival. Saprotrophic flowering plants are mainly restricted to the permanently wet equatorial forest because they cannot cope with drought on the forest floor, but here they are usually more abundant than parasites. They often occur close to the trunks of trees because there is usually an accumulation of dead leaves in such situations, especially in the bays formed by buttress roots. Unlike the parasites, the saprotrophic habit is found in a wide range of plant families, from the orchid family to the gentian family, and some are saprotrophic only in the early part of their lives, later developing chlorophyll and becoming self-supporting.

Flowering plant saprotrophs are invariably accompanied by symbiotic fungi that conduct the decomposition process and liberate the lower-molecular-weight compounds that can be assimilated by the plant. Some saprotrophs may be linked to trees by their associated fungi, because the fungi may also form mycorrhizal association with trees and derive some carbon from them. This complex interaction has been studied in certain saprotrophs of temperate forests but has yet to be studied in the tropical forest ecosystem. Perhaps these forest saprotrophs should be regarded as parasites on fungi rather than part of the decomposer system of the forest floor.

The orchids are among the most remarkable of forest plants for their associations with fungi and their assumption of the saprotrophic habit in their early development. Soon after germination, orchid seedlings become infected with specific fungi and some species remain in this condition, lacking chlorophyll and behaving as saprotrophs for many years before eventually developing into fully autotrophic plants. They usually maintain their fungal partner for the remainder of their days. Botanists believe that this association may have begun as a parasitic attack on the orchid by the fungus; indeed, some of the fungi that enter into association with orchids behave as virulent parasites of trees when on their own, including the honey fungus genus *Armillaria*. But the orchid has evidently reversed the roles and has become parasitic on the fungus, exploiting its decomposing ability to sustain the developing plant. Orchids can be grown in aseptic experimental conditions, where they are kept away from the influence of fungi, but they survive only if provided with a source of sugars and mineral elements that the fungi would normally provide. It is difficult to be sure whether the association of orchid and fungus is a case of parasitism or symbiosis. Perhaps the orchid sustains the fungus when conditions become harsh, such as in drought, ensuring that the fungus gains some benefit from the association. If the orchid eventually becomes autotrophic, then this is clearly beneficial for the fungus, as it can then draw on the products of orchid photosynthesis. There are some orchids, however, that never move beyond the saprotrophic stage. They remain as pale and ghostlike structures on the forest floor, without green pigments and still dependent on what they can derive from their fungal hosts.

■ PLANT DEFENSES

Plants in the tropical forest face two major types of biological pressure. They have to be able to compete for resources with other plants that have a similar way of life, trapping light energy, fixing atmospheric carbon, and taking water and nutrients from the soil. They also have to ward off the numerous animals that feed upon them, the climbers and epiphytes that use them for support, and the parasites and pathogens that use them as a source of energy. Competition with other plants involves adaptation in structure, spatial location, and physiological mechanisms. Deterring predators, parasites, and pathogens requires a wide range of techniques, sometimes involving the use of other organisms.

For invertebrate grazers, structural modification of leaves can prove an adequate defense. Some leaves have teeth, or serrations, along their edges, or even spines on their leaf blades or leaf stalks. Some palms, for example, have strong, downward-directed spines along the petioles (stalks) of their leaves. The presence of such teeth or spines is an effective way of deterring some of the invertebrate and vertebrate predators that feed upon foliage. Caterpillars, for example, often consume leaves by chewing along their margins, usually starting near the leaf blade tip and working down toward the stalk. Sharp, forward-directed teeth on the edge of the leaf, especially if they are strengthened with tough materials, present a formidable obstacle to grazers. The downward-directed spines on a palm petiole, by contrast, are likely to be effective deterrents to invertebrates or even small vertebrate climbing up onto the main leaf blade, which is less well protected. Larger, woody spines are often present on twigs and stems, as in the case of *Acacia* species, where they provide some protection against larger, soft-mouthed vertebrates. Such protection, however, is limited because many vertebrate herbivores seem to be undeterred by the threat and discomfort the spines present.

Hairs, or trichomes, present on many leaves, also make movement and penetration to the leaf surface difficult for raiding insects. They often have additional functions, such as conservation of water or reflection of intense light, but making life difficult for small herbivores is one evident advantage for hairiness in plants. Some hairs are glandular, which means that they have cells associated with them that generate particular secretions. These secretions may be aromatic and smelly, as in the oily products of the leaves of *Pelargonium* species. Although these products may smell quite pleasant to the human nose, they do not encourage insects intent on eating the leaves. Glandular hairs may have hollow, fragile tips that break off when brushed and inject irritants into the skin of the offending animals, such as the stinging hairs of nettles (*Urtica* species). These compounds must be particularly uncomfortable if taken into the mouth.

Not all of the glands on the surface of leaves serve to deter animals; some actually attract them. These specialized glands appear as small spots on the leaf, often secreting sugars or nectar, and are known as extrafloral nectaries. They are particularly attractive to ants, which can be valuable to a plant because they prey upon other invertebrate animals, especially the caterpillars and beetle larvae that present a danger to the plant. The ants may also attack larger animals when they attempt to graze upon their favored plant. This is a form of mutualism, in which the plant is donating some of the energy-rich materials it has gained through photosynthesis to the ant in return for obtaining a degree of protection from its predators. Tropical forests are rich in such examples of mutualistic behavior patterns, which adds to the complexity of species interactions in this complex ecosystem. In the case of ants, many plants supply more than food to encourage their presence; they even provide a home (see sidebar below).

Mycologists, scientists who study fungi, have recently discovered a previously unknown type of mutualism in which the fungi resident in and on leaves can protect a leaf from further infestation by pathogenic microbes. These fungi seem to offer no threat to the plant apart from living within the tissues of the plant and appropriating some of the sugars it produces in its photosynthesis. But leaves containing these friendly fungi, known as endophytes, are much more resistant to harmful microbes, such as those of the *Phytophthora* genus, the fungi that caused the great Irish potato blight of the 19th century. Research so far has focused on the cocoa plant because this is such an important

Ant Plants

Associations between ants and plants are very common in the Tropics, and many plants have a structure that encourages the development of ant colonies within their tissues. In the South American forests, for example, trees of the genus *Cecropia,* which are common in clearings and on forest edges, have hollow trunks and branches that are divided by wooden partitions. Ants of the genus *Aztecia* burrow through the bark and set up colonies within these hollows, and may even practice a simple form of pastoralism within the hollows, tending herds of plant-sucking insects and milking them for their honeydew secretions. During the day, many of the ants roam over the surface of the tree, collecting other insects as food and also inflicting vicious bites on any larger animals that may try feeding on the tree. In this way they act as a deterrent to large grazers, and the benefits gained by the plant compensate for the space and the sap that it donates to the ant colony. There is one other way in which the ants help defend the tree; they bite through the growing points of any climbers and vines that climb up the tree and use it for support. They thus discourage the growth of lianes and keep the tree clear of plants that seek to exploit its mechanical structure. In equatorial Africa a similar association has arisen

between ants and the tree *Barteria fistulosa,* which also has hollow branches. In this case, the ants may even cultivate fungi within the hollow limbs of the tree and consume parts of the fungus. Not only do these ants protect the tree vigorously, but they also clear competitor plants from its vicinity so that it grows more strongly. Monkeys keep well away from these trees because of the painful ant bites they risk when climbing them, and even local people avoid clearing *Barteria* trees. There are records of people being tied to these trees as a very unpleasant form of punishment when they are guilty of social crimes, such as marital infidelity.

Trees are not the only plants to provide ant habitats; some climbers and epiphytes also have special chambers to encourage the presence of ants. The leaves of the epiphyte *Dischidia* are large, waxy, and pitcher shaped, and these become colonized by ants, as do the tubers of another epiphyte, *Hydnophytum,* which are riddled with chambers and interconnecting cavities. The ants fill these leaves and tubers with organic debris, and it is possible that the plant gains some of the nutrients from this detritus in the course of its decomposition. Ants and tropical plants thus have many complex interactions.

commercial crop species, but it is likely that fungi play an important part in the interactions between plants and grazers in the tropical forest.

Some tropical plants discourage grazers by accumulating toxins in their leaves. Many of these toxins have proved extremely useful as drugs, including caffeine, found in a variety of plants including coffee (*Coffea arabica, C. canephora,* and *C. liberica*), cocoa (*Theobroma cacao*), and tea (*Camellia sinensis*). In experiments, ecologists have shown that a solution of 2 percent caffeine, when sprayed onto a slug, is sufficient to kill it, and a cabbage that has been dipped in caffeine solution is highly unpalatable to slugs. Evidently, the development of caffeine production in several tropical plants is a form of chemical defense strategy against potential grazers, although the mechanism by which it poisons slugs and snails is not yet known. Such compounds are known as *secondary metabolites,* meaning that these chemicals play no significant role in the metabolism of the cell, so are present for some other purpose, such as defense. Other species of plant use latex, a white viscous liquid that is both distasteful and unpleasantly sticky, plastering the mouthparts of any insect that attempts to consume the plant. The rubber tree of the Amazon forests (*Hevea brasiliensis*) and many species of figs (*Ficus* species) use latex as a protective substance.

One of the most commonly used toxins in plant defense mechanisms are the alkaloids. These are complex organic molecules consisting of a series of carbon rings with at least one nitrogen atom in their structure, contained within a heterocyclic ring. The nitrogen atom gives a basic reaction to the molecule, hence the name alkaloids. About 20 percent of flowering plants contain protective alkaloids, and in the tropical forests the proportion is much higher, around 40 percent. Most alkaloids are extremely poisonous. It is possible that the first development of these compounds by plants late in the Cretaceous period, some 70 million years ago, may have contributed to the final extinction of the dinosaurs. Alkaloid toxins include many familiar dangerous drugs, such as cocaine, nicotine, ephedrine, solanidine (from nightshades), and conine (from hemlock). The effects of alkaloids on the grazing animal include damage to the liver and lungs in the case of vertebrates. The compounds also affect the nervous system, causing a breakdown in the transmission of nerve signals across synapses, the main impact on insect metabolism. Other toxins include the cardiac glycosides, which affect heart action; this is the main chemical deterrent used by the milkweeds.

The insects, however, are a remarkably adaptable group of organisms. Many insects have developed tolerance mechanisms for these deadly toxins, usually by building enzymes that can break them down, or detoxify them. Even more subtle are the insects that take up the toxins and store them in areas of their body where they will cause no damage. In this way the insect itself becomes poisonous to its predators.

There is little benefit to a plant, however, if a grazer consumes it all and then dies. A deterrent would be much more effective if it caused the animal discomfort at the first mouthful so that it would avoid grazing before too much damage is done to the plant. The simple solution to this problem is the development of distinctive tastes and smells that make the grazer reconsider its intention to eat the plant, and this is the course that many plants have taken. Those insects that sequester toxins are faced with a similar problem of warning potential predators that they are not palatable, and they often resort to bright warning coloration so that the predator can quickly learn which insects to avoid. The course of evolution results in benefits to all those that have effective survival mechanisms, thus it is inevitable that some insects have developed protective coloration without accumulating any of the toxins that would make them poisonous to predators. Such deception is quite common and also very effective.

Plants are less adept at deception than insects, but they face different problems. A grazer can usually take a bite of a plant and then decide that it has made a mistake, resulting in little damage to the object of the attack. But if a predator takes a bite out of an insect it is generally too late for the victim to escape, so a visual warning is the only answer. Plants, therefore, rarely use visual means of deception and normally resort to distasteful compounds in their leaves. There are some exceptions, however, such as the deadnettles (family Lamiaceae), which mimic the appearance of stinging nettles (family Urticaceae).

Plants are also faced with the dilemma that they need to protect their foliage, stems, and roots from predation but may also wish to attract insects to their flowers for pollination and need the assistance of vertebrates for fruit dispersal, in which case those organs need to be at least harmless and usually attractive to the animals that are to be exploited. There is, therefore, a delicate balance between the plants and animals of an ecosystem, an equilibrium between repulsion and attraction. This balance has developed because of a constant fine-tuning in the course of evolution as different species adapt to one another, a process called *coevolution.*

■ CARNIVOROUS PLANTS

The normal flow of energy in an ecosystem is from sunlight to green plant to grazing animal. The grazing animal ingests plant material and extracts not only energy from it but also the chemical elements needed for building its body and for reproduction. Animal bodies are usually richer in proteins than plant bodies and thus have a higher content of nitrogen, which is a vital constituent of amino acids. Herbivores usually need to eat larger volumes of food than carnivores to ensure that they obtain sufficient nitrogen to maintain their protein requirement. Plants, by contrast, obtain their nutrients from the soil (with the exception of carbon, which is obtained from the atmosphere), and elements such as nitrogen, phosphorus, potassium, and calcium are all taken up by the roots. But when the soil is very poor in these elements, as is the case in many tropical forest soils, especially where nitrogen and phosphorus are concerned, the plant has to obtain these vital elements from elsewhere. For some plants this means consuming animals, which are valuable reservoirs of the required elements.

In the tropical forests, there are some plants that resort to this method of enhancing there nutrient supply, most typical of which are the pitcher plants (genus *Nepenthes*), shown in the illustration on the right. They are found from the forests of Madagascar, through out Sri Lanka and Assam, to Queensland, Australia, and New Caledonia. Insects are trapped by means of highly modified leaves, their leaf blades taking the form of a vessel with an open lid at the top. The leaf stalk is long and pendulous, while the vessel at the end remains upright and is partially filled with liquid containing digestive enzymes. The pitchers may be up to 14 inches (35 cm) high and can contain as much as 3.5 pints (2 l) of liquid. The pitchers are often brightly colored (usually red) and scented, and they secrete nectar around the rim by means of extrafloral nectaries. As a result, they are attractive to both creeping and flying insects. The cells around the rim of the pitcher are covered with tiny wax scales, often with several thousand scales on each cell, and these scales are attached to the cell wall by delicate, insecure stalks. When an insect lands on them, they break away and cause the insect to slip and fall into the pitcher, where it is drowned, and the insect is digested by the enzymes present in the liquid. One ecologist counted the number of insect remains present in 10 pitchers and found 1,994 insects belonging to 150 different species, the most abundant being ants. There were also many other small creatures, including diatoms, desmids, worms, crustaceans, spiders, and even tadpoles. Some spiders manage to occupy the lower parts of the pitcher and catch some of the prey insects before they hit the water. This type of prey theft is called *kleptoparasitism.* There are even some mosquito species whose larvae are resistant to the digestive enzymes. They are able to live and thrive within the normally deadly pitcher pools.

Most pitcher plants are thus nonselective in their prey; they accept as food any small creature clumsy enough to fall into their traps. There is one species, however, *Nepenthes albomarginata,* in the tropical forests of Brunei, which has a specialist trapping technique aimed specifically at termites. Just one genus of termite is involved, *Hospitalitermes,* which is unusual in being a foraging species that wanders the forest in columns, seeking fungi and algae as food. This pitcher plant species has a distinctive ring of hairs around the rim of the pitcher, and when a foraging termite discovers this

In pitcher plants the tips of their leaves are modified to form traps for insects. The edges of the pitcher are extremely slippery and visiting insects fall into these vessels, where they drown and are digested and assimilated into the plant's tissues. *(Amanda Rohde)*

it signals to the rest of the column that it has found a food source. The entire column of termites then gathers round the pitcher rim and begins to harvest the hairs, but inevitably the crush results in a cascade of termites into the trap. In one set of observations, ecologists recorded 22 termites a minute falling to their doom. But after about an hour, all the hairs have been harvested and the pitcher is no longer attractive, so the surviving members of the column depart with their hard-won prizes.

There are still many mysteries to unravel and remarkable facts to be discovered about life in the tropical forests, the termite-trapping habits of *N. albomarginata* being first observed not until 2002. Another recent discovery is that a plant found in the tropical forest of South America and Africa, called *Genlisea,* is actually a predatory species. This is a tiny plant with a rosette of green leaves only an inch in diameter. Below the soil surface it has a much more extensive mass of linear rootlike organs, which are in fact modified subterranean leaves that lack chlorophyll and bifurcate toward their tips. These structures are hollow near the tips and have backward-directed hairs on their inner surfaces. These have the appearance of traps, but until 1998 no one had been able to find any creature within them, even though Charles Darwin himself had postulated 125 years earlier that they could be carnivorous. Ecologists then tried an experiment in which they grew *Genlisea* in a dish with large numbers of protozoa. Immediately, the tiny ciliates became attracted to the *Genlisea* "roots" and invaded their tips. Here they wriggled up the tube, unable to turn back because of the hairs of the traps, and the production of digestive enzymes in the upper part of the tube ensured that the valuable nitrates and phosphates from the dying protozoa were made available to the plant. Using prey organisms that had been fed with isotopes of sulfur, the researchers were able to show that there was indeed a transfer of material from protozoa to plant, thus solving an ancient mystery.

■ A TIME TO FLOWER

Temperate plants are exposed to a continuous sequence of environmental signals that informs them about the state of the seasons. Temperature changes, alterations in the relative length of day and night, and patterns of precipitation all give the plant cues for its particular behavior pattern. These signals instruct the plant when to expand its leaf canopy, when to flower, and when to become dormant. The Tropics lack such clear signals. Although there are changes in temperature and daylength, these are very small and subtle. Dry and wet seasons provide clear cues for regions away from the equator, but in the equatorial rain forests even this type of environmental signal is absent. So how does a plant know when to flower?

Perhaps an even more basic question is, does it matter when a plant flowers under such equable climatic conditions? It could be argued that any time is as good as any other, for flowering and fruit production, so establishing some kind of seasonal timing is not really necessary. But the success of sexual reproduction is founded on genetic mixing, and this can only be achieved if gametes are transferred from one individual to another. For this transfer to take place, those individuals need to be in the same reproductive state at the same time, so trees need some kind of cue to synchronize their flowering.

There is one simple way to resolve this problem, and that is flowering all year round. With flowers always being available for pollinators, the plant is assured that another member of the species will be producing gametes at the same time.

But this is a very expensive method for the plant to adopt because it is constantly expending energy on flower production and is failing to focus its energy investment efficiently. Nevertheless, there are some tropical forest trees that adopt this strategy, such as species of *Dillenia, Harungana,* and *Trema.* Continuous flowering is also found in tropical populations of the red mangrove *Rhizophora mangle,* although the flowering becomes increasingly confined to one season farther from the equator, as in the case of southern Florida populations, where seasonal signals are much clearer.

Most of the tropical forest trees adopt a system of synchronized flowering, which may or may not be annual. In some of the dipterocarp trees of Southeast Asia there is a three- to eight-year gap between flowering events, while in some figs from the same regions, such as *Ficus sumatrana,* flowering takes place every three to four months. One of the most common tree genera in the Far East rain forests is Shorea, and there are many species of this tree. The different species have separate flowering times throughout the year, with *S. macroptera* flowering in late March, *S. lepidota* in early April, *S. parviflora* in late April, and *S. leprosula* in May. The obvious advantage of having separate flowering times is that the different species do not have to compete for the attentions of pollinators, so this is an effective way of niche packing (see "Biodiversity and Competition," pages 103–104). Each species has its own slot in the pollination program and more species are thus able to exist side by side, enhancing biodiversity.

Plant physiologists have not yet worked out the biochemistry of flower initiation in plants. Hormones are certainly involved, including the common plant hormone giberellic acid (GA), but there may be other stimulators and inhibitors involved in the process. External stimuli, such as daylength, can be perceived by plants because of specialized pigments, and these often act as a cue for flowering, especially in temperate and high-latitude plants, but if daylength is used by tropical species they must be very sensitive to fine adjustments in timing. Some plants, called day-neutral plants, are not dependent on daylength cues but seem to operate on a timing device. Timing mechanisms in plants are often related to the steady breakdown of a chemical inhibitor in the tissues, and this process may be involved in some tropical trees as wells. It is likely that different species use different cues. Some plants, for example, seem to be dependent on local weather conditions, including some epiphytic orchids that flower following thunderstorms, especially if these have been preceded by a dry period.

In the seasonal forests, which occur away from the equator, the dry season tends to be the period for flowering among many trees, especially the deciduous ones. Flowers develop and open on the twigs and branches bare of leaves, a phenomenon that has many advantages. For wind-pollinated species the movement of air through the canopy is increased during the leafless season, so pollen is transported farther; for insect-pollinated species the flowers are more apparent and accessible in the absence of leaves. This seasonal timing also aids fruit development because the coming of the wet season is favorable to fruit production. Evergreen species in the seasonal forest often flower late in the dry season and produce new leaves and flowers just at the time when the wet season arrives. Plants living closer to the ground, including smaller shrubs and herbs, by contrast, flower in the wet season when the forest canopy is closed.

At present, the timing of flowering in the tropical rain forests is still a matter of active research. The subject is a complex one in which many different factors may be operating.

■ FLOWERS AND THEIR VISITORS

Flowers are specialized organs used by plants to assist in the vital process of genetic exchange during reproduction.

Flowers typically contain both male and female reproductive organs in the form of anthers (containing pollen grains), and carpels (containing egg cells). Genetic mixing and the maintenance of genetic variability within a population are best achieved by cross fertilization between individual plants, which involves the transfer of pollen grains from one flower to another. This transfer is called pollination (see sidebar below).

Although most plants have their male and female organs in the same flower (monoecious), or sometimes as separate flowers on the same plant, there are some plants that have separate male and female plants, each producing single-sex flowers. These plants are called dioecious, literally meaning "having two homes." Only about 4 percent of all known flowering plants have this system, but in the tropical forests it is extremely common. In the forests of Sarawak, for example, 26 percent of the flora is dioecious, while in the evergreen forests of Nigeria the proportion can be as high as 40 percent of the flora. Among the trees and shrubs of the

Pollination

Gymnosperms (a plant group including the conifers) and angiosperms (flowering plants) transfer their male gametes from one plant or flower to another by means of *pollen grains*. The pollen grain is a means of enclosing the delicate male cells or nuclei and protecting them from damage and desiccation on their journey. The pollen grain wall is therefore made of tough material that resists drying yet is light and pliable. It is usually penetrated by one or more pores or slits through which recognition proteins are secreted when the grain arrives at its destination and which provide an opening for the germination and extension of the pollen tube.

The means of conveyance from one flower to another varies considerably among plants, but the main mechanisms for pollen transfer are wind and animals. The problem with wind transfer is that this mechanism is extremely chancy. The statistical likelihood of male pollen being transferred by wind to the stigma of a female reproductive organ belonging to the same species is very low. This is particularly true of tropical forest habitats, where individual trees of the same species are often widely separated. Successful wind transmission can be achieved only by the production of very large quantities of pollen, and this is extremely expensive in terms of the energy invested. A single anther of the wind-pollinated sorrel (*Rumex acetosa*), for example, contains approximately 30,000 pollen grains, so the output from

one plant runs into many millions. Animal pollination is much more reliable because the animal vector, such as an insect, is very likely to visit more flowers on leaving the source of pollen. Because of the predictable behavior patterns of many animals, it is quite likely that the next flower visited will be of the same species. So targeting is much more certain with such animals as insects but is by no means infallible, so many pollen grains are still produced. A single clover (*Trifolium* species) anther contains about 200 pollen grains, but clover flowers are clustered in dense inflorescences, so many anthers are found close together. Attracting an animal involves advertising, so flower structures have to be more elaborate, brightly colored, perfumed, and supplied with energy-rich rewards, such as nectar. Animal pollination is also energy expensive despite its more efficient focusing of resources.

Pollination by animals leaves the plant dependent on another organism for completion of its reproductive cycle, whereas this is not the case for wind pollination. The animal-pollinated plant must therefore time its flowering according to when the vector is active. Wind-pollinated plants, by contrast, are not limited in their flowering and pollen production by the availability of a different organism. In the tropical rain forest there is a high biodiversity of animals that can be exploited as pollinators, but there is nevertheless a great deal of competition for the attentions of an appropriate pollinator.

The flowers of the frangipani tree (*Plumeria rubra*) of Central America are pollinated by hawk moths attracted by the strongly fragrant scent. But the insects are deceived, for there is no nutritious nectar in these attractive flowers. *(Tony Alt)*

forest of Barro Colorado Island in Panama, 21 percent are dioecious. The main biological advantage of the dioecious way of life is that self-pollination is impossible, so it encourages the transfer of pollen between different individuals in a population. In a monecious species, there is always the possibility that pollen will be transferred from the anther to the stigma of a carpel on the same plant. This type of self-pollination greatly reduces the opportunity for genetic mixing and population variation.

Of the two main mechanisms adopted by flowering plants for pollen transfer, animal pollination is by far the more common in the tropical forests. Wind pollination in the tropical forests is rare. The conifers, including *Agathis, Araucaria,* and *Podocarpus,* are all wind pollinated, as are tropical species of oaks (*Quercus* species), which have pendulous catkins like those of their temperate representatives. But these are the exception; most flowering plants exploit various animals for pollination.

The tropical forests are extremely diverse in their fauna, so there is a wide range of animals available for plants to exploit, ranging from insects to bats and birds. There are different strategies available to both plants and animals in the pollination process. A generalist plant will attract and provide rewards for many different kinds of animals and is therefore not reliant on the availability of one particular species. A specialist plant, on the other hand, may become specifically adapted to one pollinator species, a situation that carries certain risks and certain assurances. If the pollinator becomes unavailable for some reason, the plant fails to reproduce, but if the pollinator is present, it is likely to

confine its attention to the one plant species and is therefore more reliable in the transfer of pollen to the right receptor flower. This means that the plant can be more economical in its production of pollen. In the tropical forest, specialization is the more frequently adopted option, and this again adds to the process of niche packing and biodiversity increase.

The conspicuous parts of flowers, usually their petals but sometimes associated sepals or even specialized leaves (bracts), are essentially given over to the task of advertising. The vector must be made aware of the location of the flower, thus visual attraction is an important technique, especially if the animal is day-active when light intensities are relatively high. Size can be important because a large advertisement is more easily noticed. Some of the world's largest flowers are associated with the tropical forests, including *Rafflesia,* the biggest of them all (see "Plants in the Shade—Parasites and Saprotrophs," pages 127–128). *Heliconia* flowers are also large, size being particularly important for this relative of the banana plant because it lives in the understory so needs to make its presence as obvious as possible. An alternative to the production of large flowers is to grow many flowers together in an inflorescence because this can be equally eye-catching. Many tropical trees have adopted this technique of drawing attention to themselves, including the Indian coral tree (*Erythrina variegata*) and the frangipani (*Plumeria rubra*), shown in the illustration above. Both of these trees have become popular for village and town planting in the Tropics because of their abundant blossoms.

The location of the flower or inflorescence must be related to the behavior pattern of the pollinator. Emergent trees and trees of the upper canopy usually bear their flowers on the end of branches, where they can be clearly seen by flying insects and birds. But in the lower canopy, locating flowers at the branch tips may result in their becoming lost in the mass of foliage. One alternative is to produce flowers on the trunks of trees, a habit called cauliflory. Bunches of stalked or unstalked flowers grow directly from buds on the bark of the trunk, and these can be quite conspicuous to an animal climbing the tree or flying through the open spaces between the trees of the lower canopy.

Color is also important for flowers with daylight-active pollinators, the color that is adopted depending on the nature of the pollinator's vision. Animals have different structures to their retinas and hence perceive color in different ways. Entomologists have been able to show experimentally that bees are blind to red, for example. When they are offered varying shades of gray mixed with red, they are unable to distinguish the two. They are, however, able to distinguish yellow and blue very efficiently. Red light has a long wavelength, and bees are evidently insensitive at this end of the spectrum. At the other end of the spectrum, in the short-wavelength region, they are capable of detecting blue, violet, and even very short-wave ultraviolet radiation, which is beyond the human visual range. Thus a flower that is adapted for bee pollination is unlikely to be red in color; blue or violet is the most frequently used color, ranging through to yellow.

Birds are very aware of red colors and are strongly attracted to red and orange flowers, such as the bird of paradise flower, shown in the illustration below. Hummingbirds in the New World Tropics are particularly attracted to red flowers and seek them out, as do their ecological equivalents sunbirds in the Old World Tropics. The inflorescences of the red-flowered bottlebrush tree (*Callistemon ventusa*) are visited by sunbirds and pollinated by them. Birds, like bees, are also sensitive to ultraviolet light, so they are able to detect patterns on petals created by subtle, ultraviolet-reflecting pigments. Lines and patches on petals that are invisible to the human eye are often prominent to birds and bees, directing their attention toward a source of nectar or other reward, or leading them to the correct position for receiving a deposition of pollen ready for transfer.

White flowers reflect light of all wavelengths. White is a good general-purpose color, especially for flowers deep within the shade, where white shows up best against a dark leafy background. The large white flowers of the thorn-apples (*Datura* species) are good examples of shade species. White is the color most often adopted by generalist flowers that accept a wide range of pollinator animals. Beetle and fly pollination is often associated with white flowers. White is also the most sensible color for night-flowering species of plants that depend on the attentions of moths and bats.

Scent is particularly useful as an attractant for insects and mammals, most of which have highly sensitive olfactory systems. Birds have very little sense of smell, so bird-pollinated flowers rarely waste energy on scent production. The sweet scents that humans find attractive are produced by aromatic compounds, especially esters, and these can be very effective in advertising the presence of a flower even when the density of canopy foliage is such that visual attraction fails.

The bird of paradise flower (*Strelitzia* species) has large, bright orange flowers that are visited and pollinated by sunbirds. *(Courtesy Amanda Kohde)*

Cauliflorous flowers are often very highly scented because of their location low in the forest canopy. Flowers that depend on flies and beetles may resort to heavy smells that resemble dung or carrion, often produced by organic compounds such as ketones, and these species tend to be more common close to the forest floor. *Rafflesia* is a good example of a plant that attracts carrion feeders as its pollinators.

Anyone walking in the forest in the late afternoon or evening quickly becomes aware of an intensification of floral scents as darkness descends. This is due to the release of odors by night-flowering plants, which are even more dependent on this mode of advertising than their day-flowering equivalents. Night-flying moths and bats use odor as a means of locating flowers for their feeding activities.

It is clear that animal vectors are able to distinguish between smells and are able to seek out specific flowers, but it is not yet clear just how specific this recognition can be. The different species of *Shorea*, for example, produce scents that cannot be distinguished by the human nose, but it is possible that insects are able to differentiate the species by subtle variations in the chemical composition of their scents. They are often pollinated by small insects, such as thrips and beetles, and they may have specific pollinators, but more research is needed to clarify the degree of specialization that occurs among these pollinator systems.

Animals do not visit flowers simply because they find them attractive; they demand a reward. The usual rewards offered by flowers are nectar and pollen. Pollen, of course, is a particularly precious resource to the plant because the success of reproduction is dependent on its survival and effective transfer. It is also rich in proteins and energy, so the animals that visit flowers are often interested in consuming it rather than transporting it. Those plants that offer pollen as a reward must therefore overproduce it so that some of the material remains uneaten and is carried by the animal to the next flower. Bees, bats, flies, and thrips all consume pollen, and many of these, especially the beetles, are messy, destructive eaters, tearing at the structure of the flower and removing anthers, petals, and sepals as they feed. They become covered with moist tissues, including pollen, and thus prove effective dispersal agents that will head straight for the next flower of the same species.

One advantage of providing pollen as a reward is that the plant does not need to supply nectar as well. Nectar is rich in various sugars, including sucrose, glucose, and fructose, so it has high energy content. This is important to the pollinator because it will probably expend quite a lot of energy in seeking out the flower and feeding from it. Hummingbirds, for example, need a great deal of energy in their food to compensate for that expended in hovering and flying from flower to flower. If a human being were to expend the same amount of energy per unit body weight as a hummingbird, he or she would need the calorific equivalent of 370 pounds

(168 kg) of potatoes or 130 loaves of bread each day. The plant obtains this energy from photosynthesis, but more energy spent on nectar production means that there is less available for growth or seed production. Botanists from Malaysia have measured how much nectar a bee-pollinated flower, *Xanthophyllum discolor*, produces in the course of a day and found that a single flower could generate 0.028 fluid ounces (0.8 ml) of sugar-rich nectar per day. A bat-pollinated flower needs to produce even more to supply the energy requirements of a flying mammal, and the bat-pollinated *Oroxylum indium* produced twice the volume of the bee-pollinated flower, 0.064 fluid ounces (1.8 ml) per day.

The flowers of some tropical trees produce oil rather than nectar, a practice that may be more common than is currently realized. Bees that specialize in oil collection are associated with the oil-producing plants. Oil has many advantages as a food resource for the animal, the chief one being that its energy content per unit volume is considerably greater than that of sugar solution. Scientists who have analyzed the chemistry of the oil produced by *Mouriri parvifolia*, a small tree from the forests of Panama, found 13 different fatty acids as well as amino acids.

Some flowers cheat. They display large, attractive, brightly colored petals, and may even scent the air with gastronomically stimulating odors, but when the animal arrives, there is no reward—no nectar, no oil, no food at all, just a dusting of pollen. The frangipani tree (*Plumeria rubra*) is an example of a plant that resorts to this deception, which is clearly a means of avoiding some of the costs associated with insect pollination. Other flowers deceive their potential pollinators in what may seem a particularly unfair manner; they convince male insects that they are female insects available for copulation. Orchids (see illustration on page 137), particularly those of the genus *Ophrys*, are adept at this form of deceit. Many have developed a form that closely resembles the females of certain types of wasp, having a color, size, shape, and hairiness that entirely fool an approaching male. Even more remarkably, the flower produces a scent that is chemically very similar to that emitted by the female wasp when it is ready to mate. As the male tries to copulate with the flower, pollen is deposited onto its body, which is then transmitted to the next flower that uses the same trick. Evidently, the male wasps do not learn from their mistakes. Different species of *Ophrys* orchids have different pollinator wasps, so each type can develop a flower appropriate to the needs of the pollinator.

Attracting the pollinator is one problem, keeping it there long enough to complete its task is another. Some members of the arum lily family, Araceae, attract their pollinators, usually flies, with heavy odors, and the emission of the organic molecules involved in producing the smells is aided by the flower becoming warm. It is extremely unusual for a plant to generate its own heat, but a central column of the

Orchids, such as this one from Hawaii, produce some of the most complex flowers. Their structure is often adapted for pollination by just one type of vector. *(Commander John Bortniak, NOAA/Department of Commerce)*

exclude ants, while other plants resort to the use of toxic chemicals, such as alkaloids, to repel them.

Flowers have developed a range of different shapes according to the needs and adaptations of their pollinators. Flies have short tongues and are generally associated with open, shallow, dish-shaped flowers, whereas bees and butterflies have long tongues coiled within their heads that can be thrust deep into tubular flowers to obtain nectar. Flowers and their pollinators have evolved together, coevolved, and have thus become intimately linked in their structural adaptations as a consequence of the pollination process. Some hummingbirds, such as the swordbill (*Ensifera ensifera*), have long straight bills that can probe deeply into tubular flowers, such as some species of *Passiflora*. Others, such as the sicklebill (*Eutoxeres aquila*), have curved bills that are better suited to smaller flowers with bending tubes. Many flowers have developed anthers and stigmas that protrude from the main flower tube, thus depositing and collecting pollen from hovering birds or hawk moths, as in the case of *Hibiscus*, which is illustrated below.

When Charles Darwin visited the island of Madagascar on his world tour in the *Beagle*, he found an orchid called *Angraecum sesquipedale*, which had an extremely long spur at the end of which the nectaries were situated. He found no insect during his visit that had a tongue long enough to reach down into the flower and extract this nectar, but his experience with flowers and their pollinators led him to propose that an insect existed on the island that was appropriately

arum flower, the spadix, has cells in which the mitochondria behave in an unusual way. They uncouple the process of respiration from the generation of the energy-carrying molecule ATP and instead produce heat. Insects attracted to the warmth and the volatile compounds emitted by the spadix arrive and crawl down to the base of the flower. Downward-directed hairs trap them here among the ripe carpels, where they deposit any pollen they may have gathered from previous visits to arum lilies. They may remain trapped for several hours, feeding upon nectar, until the protective hairs wither and the flies escape, brushing past the newly ripe anthers and gathering pollen on their way out.

Just as some flowers cheat by promising rewards and then withholding them, some animals cheat by stealing nectar without transmitting pollen. Some beetles, for example, which are unable to penetrate the long tubes of butterfly and hummingbird pollinated flowers, simply bite through the base of the flower and steal the nectar. Ants are also interested in nectar theft, although many flowers have erected physical barriers that make it difficult for raiding ants. Many members of the family Rubiaceae, for example, have rings of hairs around the entrance to their tubes that effectively

The flowers of the hibiscus have a tubular structure so that only long-tongued, hovering hawk moths or a sunbird can penetrate its depth. But the anthers and stigma protrude so that the visitor both receives and donates pollen, thus effecting pollination. *(Peter D. Moore)*

constructed. Such an insect would need a tongue 18 inches (45 cm) long, and the scientists of the time regarded this as a very unlikely adaptation. Sixty-five years later entomologists found a moth, the Madagascan sphinx moth (*Xanthopan morgani*), that fit the description and did indeed pollinate Darwin's orchid.

Perhaps the most specialized of all pollination systems occurs in the figs. The familiar fig "fruit" is actually a false fruit because most of its tissue does not come from the carpel but from a globular swelling that surrounds the true flowers. Masses of tiny flowers develop in the hollow center of the fig and communicate with the outside world by a small entrance hole at its apex. Male and female flowers are separate but both are found within a single fig, and both are extremely simple, reduced in structure to the very basics of anthers and carpels, together with vestigial petals and sepals. Each species of fig is pollinated by its own species of wasp, the female of which enters the developing fig through the aperture, forcing its way through barriers of stiff scales. She carries pollen from the fig she has just left, and this pollinates the female flowers of the newly entered fig, but she also uses her ovipositor to inject eggs into the carpels, so she uses the developing ovaries as a place for the larvae to mature. But not all of the carpels receive fig wasp eggs, so some develop normally and produce seeds. The larvae eventually pupate and then hatch, the males emerging first and fertilizing the females while they are still in their pupal cases. As the fertilized females emerge, the male flowers ripen and shed their pollen on the females, which excavate their way out of the fig aperture and set off in search of a new ripening fig. In this case, the plant is donating part of its reproductive system to the wasps in return for the pollination of some of its flowers. By overproducing flowers it is well able to make this sacrifice and is then assured of a highly specialized and reliable pollinator, which is totally committed to its own species of fig.

Bat pollination occurs in many tropical trees, even those at fairly high altitude in the Andes. Bat pollinated flowers have to be robust to support the weight of this mammal, and the nectar needs to be easily accessible to the bat's tongue. These flowers often open in the evening and emit a strong, musty scent that attracts the bats, but they do not need to be brightly colored, so they are often pale and bell-shaped. In Madagascar, one species of baobab tree, *Adansonia suarezensis*, is totally dependent on bats for its pollination. Bats travel far in their nightly excursions, up to 21 miles (34 km) in a single night, so they are very effective at transporting pollen between populations of trees.

Most anthers ripen and break open as they mature, relying on the physical brushing of the animal vector to collect pollen, but some open explosively, showering the visitor with pollen grains. Some bat-pollinated plants are of this type, such as some species of the *Mucuna* genus, which belongs to the pea family, Fabaceae. Members of the nightshade family, Solanaceae, together with some members of the Euphorbiaceae, also have anthers that open violently, but they are stimulated into their explosions by specific behavior of their pollinators. Bumble bees have the capacity to operate their wing muscles without engaging their wings, thus producing an intense buzzing sound but without taking flight. A trapped bumble be inside a house window often behaves in this way, giving the impression of rage. When a bumble bee buzzes in certain types of flowers, the resonance of the buzz causes the anther to explode, an activity often referred to as "buzz pollination."

Pollination mechanisms are among the most remarkable and best studied examples of coevolution in the whole of nature. The evolution of flowers during the Cretaceous period was closely linked to that of the insects, and the fine-tuning of structure, biochemistry, and behavior involved has led to a very high level of sophistication and specialization in both the insect and the flower. Insect pollination is found in all the world's biomes, from tropical forests to the tundra; bird and mammal pollination is mainly confined to the Tropics and subtropics.

■ FRUIT AND SEED DISPERSAL

Moving pollen from one flower to another is one area where plants successfully exploit the greater mobility of animals in return for certain rewards, usually in the form of food. Moving fruits and seeds to new locations is another problem for plants rendered immobile by their roots, so once again they have developed techniques for using animals.

Getting from one place to another is an important aspect of survival for all organisms. They may be entirely content with their current location at any moment, but the environment is constantly changing; weather, climate, soils, and one's biological neighbors are all subject to change and conditions may not be as comfortable in the future as they are now. There are advantages to be gained by expanding populations as a protection against predators and disease, and by extending geographical range to exploit new areas of opportunity. Dispersal is thus an extremely important aspect of the life cycle of all organisms and it presents particular problems for static plants.

Plants that are evolutionarily primitive, such as mosses, liverworts, and ferns, all disperse by *spores*. These are extremely tiny spherical propagules, only about 0.00004–0.00008 inches (10–20 μ) in diameter. This is just the size of dust particles and smaller than most pollen grains, so they are easily carried by air currents and can move thousands of miles when they are caught up in mobile air masses. The chances of their landing in locations where they can successfully germinate and establish new plants, however, are relatively small, so many of these spores are wasted. Each adult plant produces millions of spores to compensate for these losses so that at least a few will strike luck and continue the species. In the tropical forests many of these lower plants live in high places, dwelling on the branches and trunks of canopy trees as epiphytes (see sidebar "Epiphytes," page 121). Dispersal by airborne spores evidently works well for these plants because the air movements through the canopy carries the spores from one tree to another and they settle upon the moist soils among the establish epiphyte communities on the branches. In such a moist environment with light air movements, the use of wind dispersal by these small plants clearly has advantages. The orchids, many of which are epiphytes in the forest, have dustlike seeds, which are also carried by the air currents below the canopy.

Wind dispersal is also used by many forest trees, including the dipterocarps of Southeast Asia. *Dipterocarpus* has two elongated wings formed by the sepals of the original flower, and when the ripe fruit falls from the tree it spins, carrying it away from the parent plant. In the silk cotton tree (*Ceiba pentandra*), which originated in Peru, the seeds are surrounded by hairs developed from the carpel wall, which are very effective in acting as parachutes and carrying the seeds over long distances. These hairs are used by people as a source of kapok. The seeds of some pea family (Fabaceae) trees, such as *Newtonia* from tropical Africa, have flattened margins that act as wings, while the fruits of *Enkleia,* from Southeast Asia, remain attached to a long stalk with leafy bracts attached, and these act like the blades of a helicopter rotor.

All of these mechanisms assist in moving the falling fruits away from the parent, possibly by several hundred yards, depending on the local conditions, but even this relatively small distance can be of considerable advantage to a plant. In the first place, it takes the seed away from the immediate competition, especially the shade, of its fully grown parent. The second advantage relates to seed predators. The ground of the forest floor contains many animals that feed largely upon plant material falling from the canopy, among which are fruits and seeds, an excellent source of food. Beetles are especially associated with this lifestyle, congregating in large numbers around the bases of fruiting trees. In surveys of seed survival beneath the parental tree, ecologists have found that mortality can be as high as 99.99 percent. The farther one goes from a fruiting tree, the lower the density of these seed predators, so the better the chances of any particular seed escaping their attention and managing to germinate and establish themselves. Ecologists who have studied the statistics of tree establishment have found that for any given tree species there is a particular distance from the parent where germination and growth is most likely. This lies far enough from the tree to reduce the chances of predation, but still lies within reasonable reach for the dispersing seeds. In the case of the toucan-dispersed tree *Virola surinamensis* on the island of Barro Colorado in Panama, for example, the distance from a tree to its nearest mature neighbor averages 110 feet (34 m). It is possible that these combined factors account for one of the observations on some forest trees that has long puzzled botanists. Some trees occur as widely scattered individuals in the forest rather than clumped together in groups. This pattern may well have resulted from each tree having a minimal distance from its neighbors when originally establishing itself.

The dispersal of seeds on the ground can become complicated, however, when animals gather them, transport them, and hide them in stores. This process is called *secondary dispersal*. Many small mammals behave in this way, collecting fallen fruits and seeds and burying them for later consumption; in this way they may double their original dispersal distance from the parent tree. This is of little value to the plant, of course, if the animal returns to the cache and eats the seeds, but sometimes the animal fails to return, either because of premature death or poor memory. The buried seeds may then germinate, but these will establish themselves in a clump rather than as scattered individuals, so they produce a different pattern of tree arrangement in the mature forest. Tapirs in the Amazon forest cause this kind of clumping among some species of palm tree. They eat the fallen fruits, the seed passing unharmed through their guts, so their secondary dispersal results from their movements before voiding the seeds. Tapirs tend to use particular sites as latrines, so these locations become rich in germinating palm seeds and later produce patches of palm trees in the forest. Gorillas and elephants also feed on fallen fruits in the African forests.

Smaller animals such as ants can also play an important part in secondary dispersal of seeds. Some seeds have an oil body attached, called an eliasome, that has no function except to attract the attentions of ants, which carry the seed to their nest, eat the oil from the eliasome, then discard the viable seed. Germinating seeds can often be found along the trails of foraging ants, where workers have stopped for a snack and discarded the seed along the way, but most seedlings are found associated with the middens of waste material near the ants' nest.

Some fruits and seeds are dependent on water for dispersal. The regular flooding of some tropical rivers, such as the Amazon and its tributaries, provides ample opportunity for water travel. Some fish are able to swallow seeds, adding a new dimension to water dispersal. Some of the longest distances traveled by seeds are associated with water, especially for seeds that can be carried by ocean currents. The coconut (*Cocos nucifera*) has very large seeds that can cope with long periods of emersion in the ocean and then germinates on arrival at a suitable beach. Some coastal trees from the tropical forests of the Caribbean, such as species of the genus *Mucuna,* have beanlike seeds that are regularly

The sociable spider monkey from South America lives in groups and feeds mainly on fruits it harvests from the forest canopy. It is an important agent for the spread of some tree species.   *(Lee Pettet)*

carried by the Gulf Stream current across the North Atlantic and land upon the shores of western Ireland, a journey of about 4,500 miles (7,250 km). Unfortunately, the journey is in vain because the climate is unsuitable for its growth when it arrives, but this demonstrates how effective water can be as a means of seed transport.

Animal consumption of fruits is generally more common as a dispersal mechanism than wind dispersal in the tropical forests. In the forests of Ghana in West Africa, for example, about 70 percent of the trees are dispersed by animal consumption of fruits and subsequent gut carriage. Monkeys, such as the spider monkeys (see illustration above), are particularly important in eating fruits and dispersing seeds in this way, as are birds. Toucans are particularly important in the forests of South America, and hornbills occupy a similar ecological niche in Africa and Asia. On the island of Madagascar there are no monkeys, hornbills, toucans, or parrots, but the larger fruits are often carried by lemurs, a group of primates endemic to that island (see illustration at right). Although just eight species of lemur are known to transport seeds by gut transport, these play an important part in dispersal, especially by defecated in clearings and assisting the movement of trees back into open areas for recolonization.

The distance a seed can travel in an animal's gut depends on the behavior and movements of the animal and the time that passes between ingestion and defecation. In toucans, gut passage may be as brief as 10–25 minutes, so most seeds are likely to be deposited quite close to the adult tree. Mammals usually have a longer gut passage time; their dispersal potential depends on their behavior pattern. Howler

monkeys (*Allouatta palliate*) are quite sedentary, so dispersal is usually limited to about 1,000 feet (300 m), wheareas spider monkeys (*Ateles belzebuth*) are more wide-ranging and carry seeds for several miles.

As in the case of pollination, if a plant uses an animal for dispersal it must pay a certain price in offering a reward. The reward in the case of fruits is food. Either the plant must be prepared to sacrifice some of the ingested seeds to the carrier, as in the case of some bird-dispersed seeds, or the fruit itself should be made attractive and edible and the

The crowned lemurs of northern Madagascar live in the relatively high-altitude moist forests. The lemurs are a group of primates entirely confined to the island of Madagascar, off the east coast of Africa.   *(Simone van den Berg)*

seeds protected by hard, digestion-resistant coats. Although the plant must invest energy in providing this food resource, it reaps the considerable reward of being granted mobility.

Fruits need to advertise if they are to attract the appropriate transporter, and they usually do this by being brightly colored. As in the case of flowers, different colors attract different types of animals. Birds are strongly attracted to red colors, as well as to black and purple. Mammals also favor red, but in addition are attracted to yellows and orange. The size of the fruit is also important because animals are limited by the size of their gullets, or in the case of birds, the gape size of their beaks. Particular fruits are thus tailored by evolution to fit certain dispersal agents, and these in turn seek out the fruits they find most appropriate to their tastes and needs. Many seeds germinate more effectively after passing through an animal's gut, sometimes because of the physical scratching and damage to the tough outer coat that the seed receives in its gut passage. There are also chemical changes as digestive enzymes soften the seed coat and dissolve inhibitory compounds that normally delay seed germination. Fruit dispersal, therefore, is another area of tropical forest ecology where the coevolution of plants and animals is actively at work. One outcome of the multitude of interactions is niche specialization and hence the opportunity for increased biodiversity.

■ CONCLUSIONS

Plants and animals are limited in the range of temperature within which they can survive, and their temperature optima and limits are determined in part by the requirements of their enzyme systems. Some organisms control their temperature (homiotherms), while others assume the temperature of their surroundings (poikilotherms). Plants are dependent on their local climate to determine their temperature, and many are particularly sensitive to freezing because it damages their cell membranes and their enzyme systems. The plants of the tropical forests are generally sensitive to chilling injury so are confined to the permanently warm conditions of the Tropics, although some manage to grow at relatively high altitude and have adapted to the problems of low temperature they encounter there.

One feature of plants that is closely associated with their resistance to frost and desiccation is the location of their perennating organs (usually buds). Botanists classify plants according to where these organs are situated, whether well above the ground, close to the ground, at ground level, or below ground. This is called their life-form. The proportions of these different life-forms in any locality are closely related to climate; this analysis is called the biological spectrum of that region. In the case of the tropical forests, the vast bulk

of the flora belongs to the phanerophyte life-form, which holds its buds well above the surface of the ground; in other words, most tropical forest plants are trees.

When the leaves in a tree canopy photosynthesize, they lose water by transpiration because of the open pores on the leaf surface. In order to replenish this water, the tree must elevate large quantities of water from the soil, and this is achieved by the development of a transpiration stream that works because of the high degree of cohesion between water molecules. A column of water evaporating from the top is able to pull water upward because of the tendency of water to hold together rather than break up under the strain. Breaks in the wood vessels, the xylem elements, can occur, however, resulting in cavitation, but these are not normally fatal to the tree because of the abundance of other xylem tubes. Water is usually freely available in the equatorial forest soils, but can become scarce when there is a dry season. This condition is often associated with an increasing abundance of the deciduous habit.

The leaves of tropical trees are generally very similar in their basic pattern, being oval in shape and often having an extended central nerve that ends in a drooping point, or drip tip. The leaves of some plants are variegated, serving as camouflage in the patchy light environment of the forest floor. The physiology of the leaves varies according to their local light conditions. Sun leaves are better able to operate at high light intensities, while shade leaves have lower light compensation points (the light intensity at which overall photosynthesis is matched by respiration) and become light saturated at lower intensities. Shade leaves also have lower rates of respiration in the dark.

Drought can be experienced even in the humid equatorial forests by plants that live in the canopy without any root connection to the forest floor. Epiphytes, which live entirely by being supported on the trunks and branches of host trees, are in this position and need to collect rainwater and store it to avoid desiccation. Many of these have waxy cuticles, or water-collecting forms, that enable them to survive through drought in the forest canopy. Some epiphytes have even developed a photosynthetic mechanism, called Crassulacean acid metabolism, which is normally associated with desert plants and enables the leaves to place a tighter control on their stomatal opening and hence water losses. Mosses and lichens often occupy this position, giving these plants of small stature an opportunity to occupy better-lit conditions than they would find on the forest floor. They survive drought by losing water and entering a dry and dormant state and are said to be poikilohydric, which means that they recover very effectively when rewetted following complete desiccation.

The strangler figs begin life as epiphytes, but then extend their roots down to the ground and eventually develop a larger bulk than their host tree, which eventually becomes completely swamped by the excessive growth of its epiphytic guest. Lianes, or climbers, use the alternative strategy of beginning life on the forest floor and then extend up toward the light, using other shrubs and trees as support. They do not parasitize their hosts, but they may cause damage because of the weight of their biomass on trunks and branches.

In the deep shade of the forest floor, many plants have found an alternative to photosynthesis by parasitizing the roots of other plants. One of the most spectacular examples is *Rafflesia,* which spends most of its life within the root tissues of trees. Saprotrophs are organisms that obtain their energy from dead, decaying organic matter, and some flowering plants have lost the ability to photosynthesize, have no chlorophyll, and feed with the assistance of symbiotic fungi on the dead leaf litter of the forest floor. Many orchids begin life in this way and develop chlorophyll only later in their lives. Some plants are only partially parasitic, including the mistletoes, which have chlorophyll and fix most of the carbon that they need without resorting to theft from the host. They grow in the canopy without root connection to the ground, so their greatest need is water, which they obtain by tapping into the xylem vessels of their hosts.

Perhaps the greatest hazard for a tropical plant is the abundance of grazing animals that look to the plants for their energy supply. Survival often entails the deterrence of these grazers, and this may be achieved by structural adaptation, such as serrated edges to leaves, spines, and stinging hairs. Alternatively, the plant may devote some of its energy to constructing chemical defenses in the form of distasteful or poisonous compounds, such as alkaloids and cardiac glycosides, that prevent the destruction of their leaves by grazers. Some plants have developed a close relationship with ants, providing them with nectar from special glands, or supplying them with hollow stems for residence. In return, the ants protect the plant from attack by grazing animals.

Fast decomposition, together with rapid nutrient uptake from the soil, means that the forest environment is poor in available nutrients, especially nitrogen and phosphorus. Some plants have overcome this scarcity by resorting to the consumption of insects, which are relatively rich sources of these elements. Pitcher plants are the most frequent, trapping insects by constructing vessels from their leaves that have waxy, slippery edges so that visiting insects fall inside, only to be digested by enzyme-rich juices contained within the pitchers.

The flowers of angiosperm plants are organs that assist in genetic exchange during sexual reproduction. To be effective, a species must ensure that the timing of flowering is synchronized between the various individuals of a population. This synchronization can be difficult in the tropical environment, where seasonal signals are weak. Changes in daylength and temperature through the year are small, but seasonal drought occurs in regions away from the equator and often serves as a seasonal cue for flowering. Botanists are still not sure what cues stimulate flowering close to the equator, but perhaps plants are sensitive to even very small alterations in daylength.

The transfer of pollen between flowers can occur by the chance movements of pollen grains in the wind, although pollination by means of animals, insects, birds, and mammals is much more common among tropical plants. The structure of flowers and their colors are determined by the type of animal used in the process of pollination. Hummingbirds, for example, usually prefer red flowers with long tubes, into which they insert their tongues to harvest nectar. Scent is important for attracting insects, but not birds, most of which have a very poor sense of smell. The provision of sugar-rich nectar for pollinating animals is expensive for a plant in terms of energy, but is evidently worthwhile in ensuring the accurate movement of pollen between plants. The specificity of pollinator to particular flowers varies among species but is most highly developed in the fig wasps where plant and wasp species are totally committed to and dependent upon one another.

Plants are by nature immobile; they can extend their distribution range or invade new locations only with the help of physical agencies, such as wind or water, or employ the assistance of mobile animals. Dispersal away from a parent takes the seedling out of its shade and also reduces the likelihood of predation by the beetles and other seed eaters that congregate beneath fruiting trees. Birds and mammals are particularly effective in seed dispersal because of their relatively large guts that can deal with intact seeds and their ability to move considerable distances. Much, however, depends on the time of gut passage and whether the animal is sedentary or wide-ranging in its foraging behavior. Plants sacrifice some energy to provide a food reward for the disperser, and may even lose some of their seeds if they are damaged in the gut, but the association proves very effective in moving a plant species from one location to another.

7

Biology of the Tropical Forests: Animals

On walking through a tropical forest it soon becomes apparent that the vast bulk of the biomass belongs to the plants. But a more detailed survey of the numbers of species of different organisms will show that there are far more animals per unit ground area than there are plants. Most of these animals are very small, but even quite large animals, such as amphibians, reptiles, birds, and mammals, may have more species present than the plants. The reason for this is that almost all the plants present are effectively living in the same way, absorbing light, fixing carbon dioxide from the atmosphere, taking up water and mineral nutrients from the soil, and losing water through their leaf pores. All of the plants, apart from a few exceptional parasites and saprotrophs (see "Plants in the Shade—Parasites and Saprotrophs," pages 127–128), occupy the same trophic level as primary producers (see "Primary Productivity of Tropical Forests," pages 78–82). Animals, by contrast, have a much wider range of feeding opportunities, acting as herbivores, fruit eaters, nectar feeders, detritivores, insectivores, carnivores, parasites, and predators. They can occupy a range of different trophic levels and contribute to food webs of great complexity (see "Food Webs in the Tropical Forests," pages 82–84). Animal life in the tropical forest, therefore, displays a seemingly endless diversity, from the forest floor right up to the highest canopies.

■ INVERTEBRATES OF THE FOREST FLOOR

Although the forest floor may be deeply shaded for much of the day, its dwellers are not short of food. There is a constant rain of dead leaves, twigs, pollen, fruit, dead insects, feces from the canopy dwellers, and just occasionally the corpse of a larger animal. Herbaceous plants are few and

far between, so there is little to graze upon in the mature forest, although openings and clearings offer more opportunities for herbivory. Fungi are often plentiful, however, because they too take advantage of the rain of organic debris as a source of food, and many fungi are nutritious and consumed by vertebrate and invertebrate dwellers on the forest floor. Fungi are important in the initial stages of decomposition, softening wood tissues by the penetration and external digestion of their network of mycelium. Many insects and worms can only begin their feeding when fungi have commenced the process of rot.

Many of the invertebrates that occupy the forest floor inhabit the layers of leaf litter that accumulate on the soil surface, which is constantly humid because of the high rainfall and the lack of light penetration. These animals act as detrivores and include mites and springtails, millipedes, snails, and slugs. Mites are relatives of the spiders (Arachnida) and are tiny ubiquitous creatures, of which there are over 1,700 known species. Their densities in soil can often exceed 10,000 per square foot (100,000 m^{-2}). Many mites feed on decomposing vegetable matter, although there are some that have adopted a parasitic habit, and most beetles carry a load of mites clinging to the joints of their limbs and to their less protected undersides. Some mites are predatory, feeding on the eggs and larvae of flies; the larger ones, which can attain a length of 0.1 inch (0.25 cm), can eat small spiders and millipedes. Some mites inhabit the feathers of birds and the fur of mammals, where they can cause the loss of hair and development of mange.

Springtails (Collembola) are also detritivores, small enough to make a meal of pollen grains and spores that constantly fall from the forest canopy. Many of them also feed on fungi. Springtails are usually less than 0.125 inch (0.3 cm) in length, and their bodies are covered in short, wax-covered hairs and tubercles, which make them almost unwettable, so they are able to run across the surface of patches of water making use of the surface tension. Their name derives from their ability to flick their back legs and leap high into the air, escaping from predators and leaping considerable distances. They are extremely primitive insects, having no wings, but with a very long fossil history, being found in rocks of the Devonian period of 320 million years ago.

Millipedes (Diplopoda) are common detritivores that bore into fallen fruit and rotting wood, passing the material through their guts and depositing finely comminuted materials that are more easily attacked by bacteria. Like many of the larger detritivores, they concentrate particularly on materials that have already been softened by the activities of fungi. Millipedes are divided into rings, each ring bearing chitinous and calcified armor plating above and two pairs of legs beneath, except for the front rings that have only one pair. Tropical millipedes can grow to considerable lengths, the African snake millipede (*Chersastus* species) reaching 11 inches (28 cm). This millipede is bright red in color, with a black head, but most millipedes are entirely black. When they are alarmed, millipedes curl up into a tight spiral, presenting their armored backs to the world. Some species, like the African snake millipede, are able to squirt out a stinking and irritant liquid containing benzoquinone, which deters predators very effectively. The South American flat-backed millipede, which is five inches (13 cm) in length, goes even further in its defense and emits highly poisonous hydrocyanic acid. Millipedes are quite long-lived arthropods, taking two years to reach sexual maturity and then living on for several more years. They play an extremely important role in the decomposition process in tropical forests.

Velvet worms (Onychophora) are found throughout the world's tropical forests and are predatory but will also feed upon the dead bodies of insects. The largest is *Macroperipatus geavi* from Central America and grows to five inches (12.5 cm) in length. They have two large slime glands attached to their bodies, extending over half their length, and they are able to eject slime from these to protect themselves from predators and to capture their prey. The velvet worms are nocturnal predators, lying beneath stones in the day then setting out on hunting expeditions in the night. They seem to detect their prey simply by stepping on it, then grasp it in their strong jaws and cut it into small portions. Like the millipedes, they squirt out secretions when attacked by a predator, but in this case the defensive material is a mass of sticky threads that can be thrown up to 15 inches (38 cm) and entangles the attacking animal. Velvet worms have pairs of legs running down the length of their body, and the velvet appearance results from masses of small papillae over the body segments.

Slugs and snails (Mollusca) are soft-bodied invertebrates that are abundant in warm wet habitats. They form a highly adaptable group, but characteristically they feed by rasping surfaces with their radula, a kind of toothed tongue, which scrapes the epidermis from a plant or the algae from a rock. Most slugs and snails are herbivores or detritivores, although there are some predatory snails, such as *Euglandina* of Florida. This voracious predator discovers its molluscan prey by coming across their slime tracks and following them. It appears to recognize which direction a snail has been heading and it follows the slime track until it catches and consumes its victim. This snail has become infamous because of introductions that have been made on tropical islands, such as *Moorea* in the South Pacific. It was introduced as a means of biological control but has brought many native snails to extinction.

Spiders (Araneida) are among the most efficient of invertebrate predators and are able to deal with many different types of prey, but insects are their staple diet. On and near the forest floor are spiders that roam around to hunt their prey, hang webs from low vegetation, or build tunnels in the ground and leap out at unsuspecting insects. The ogre-faced spider (*Dinopsis longipipes*) is a web-making spider with a difference. It places the web above a regularly used ant trail and hangs from the web by its rear legs. In its front pair of legs it carries a net constructed of silk and uses this to capture passing ants. The so-called bird-eating spiders include a number of particularly large spiders that hunt by stalking and jumping upon their prey. They are active at night and are able to surprise and capture roosting birds, but their prey more often consists of small amphibians and reptiles, together with larger insects, such as grasshoppers, moths, beetles, and other spiders. Among the largest is *Theraphosa leblondi* of northern South America, which has a leg span of over 10 inches (26 cm). Like all spiders, they have a potent venom but are very unlikely to attack a human unless severely provoked, when they can inflict a painful bite but not normally a fatal one. Trapdoor spiders are not confined to the Tropics, but they are present on the tropical forest floors, where they construct tunnels with a hinged flap at their entrance, behind which they hide. They are extremely fast when leaping out upon their prey, which they then drag back into their tunnel. They cover the door and the roof of the tunnel with silk and salivary secretions to make it waterproof, an important consideration in the wet forests. Spiders, like most predatory animals, are not social creatures but tend to live a solitary life.

Although scorpions are more often associated with dry deserts than wet tropical forests, there are some types that prefer the moist Tropics. The Philippine scorpion, for example, is so fond of moisture that it collects it in its claws and throws it over its own body to keep up its humidity. Most scorpions hide in burrows during the day and hunt over the forest floor at night, seeking cockroaches and crickets.

Only one group of insects is able to digest cellulose, the termites (Isoptera). It is not actually the insect that digests the cellulose but associated bacteria, protozoa, and fungi

Insect Architecture

Many termites live in underground colonies or inside rotting wood. Others are much more ambitious in their nest construction and build large above-ground structures that are an important landscape feature in many tropical forests, especially those with seasonal drought. In the savanna woodlands, where drought can be severe and prolonged, some termite species build very large structures. *Macrotermes bellicosus,* for example, builds towers up to 25 feet (7.5 m) in height in the drier forests and woodlands of Africa, and the compass termite (*Amitermes meridionalis*) constructs nests resembling a wall, thin in one plane and long in another, in the forests and savannas of Queensland, Australia. These are often more than 11 feet (3.5 m) in height and are aligned in a north-to-south compass direction so that they receive extra warmth in the early morning and evening but absorb less during the midday period. In the wetter forests, heavy rain may be a greater problem for termites than heat, and the African rain forest species of the genus *Cubitermes* build mushroom-shaped nests, with a cap that sheds excessive rain. The East African species *Cubitermes speciosus* has a series of four or five caps, one on top of another, to ensure that the rain does not penetrate. The South American species of *Constrictotermes* apply many layers of mud to the roofs of their nests and even erect a series of drainage channels arranged in a herringbone pattern to help take water away from the nest.

Termite nests are built to last, and some are known to have been occupied for at least 60 years. They are constructed from mud or soil, mixed with the saliva of the insect and plastered in layers over many years. The main living quarters for the colony are underground, and in the center of the colony lies the egg-laying queen. Beneath her chambers are extensive galleries in which the eggs are stored and tended and where the larvae are subsequently fed. Above the royal cell are the fungus gardens, which are the main site of termite agriculture and provides food for the colony. The bustling activity of a thriving termite colony, together with the growth of the fungi, produces very large quantities of carbon dioxide gas, so the nest must be designed to allow the escape of this waste product and permit the entry of oxygen in its place. A series of chimneys run up the above-ground tower and open into the air above, along with often a main central chimney and a series of side vents leading from it. Air circulates from the atmosphere above, dispersing the gases that accumulate in the colony and dissipating some of the heat generated by the constant metabolism of the insects and their crops.

in the termite's gut. Different types of termite have different combinations of associated gut organisms, but they provide the termite with the ability to derive energy from a compound that is extremely widespread in nature yet relatively inaccessible to animals. Their main food resource is dead wood, which they gather into their nests, preferring wood that has already been partly digested by fungi, which they also consume. This preference has led to one group of termites of the genus *Termitomyces* to develop the remarkable activity of cultivating fungi within their nests. There is no other animal on Earth, apart from humans, that have adopted the practice of agriculture as a means of ensuring a regular food supply. They collect their own feces into "gardens," where they tend the fungal cultures, and some termite species are extremely careful about the types of fungi that they cultivate. *Odontotermes redemanni,* for example, is so concerned to keep its fungal culture pure that it applies a selective fungicide, caprylic acid, to its cultures to protect them from invasion by other fungal species. The termite queen secretes this chemical from her anus, and the worker termites collect it and apply it to their fungal gardens to pre-vent the spores of invasive fungi from germinating. The efficacy of the chemical can be demonstrated experimentally because if termites are excluded from their garden, other species of fungi soon invade.

Termites are social animals that resemble ants in appearance and in their nest-making habits, but they are actually closer related to the cockroaches. Termites are most abundant in the tropical forests with seasonal drought and in surrounding savanna woodlands and grasslands. They survive well under such conditions because they build large structures in which they can effectively control their microclimate and ensure that the humidity is maintained (see sidebar).

Life in the termite colony is highly organized, with a series of social castes. At the top of the social order are the queen and king, and sometimes there are several pairs within one nest. These are the only members of the colony that reproduce, the queen having the task of constantly producing more eggs. The queen usually has an enlarged abdomen for this purpose, and thus may be considerably larger than the worker termites. In the African species *Macrotermes,* the queen can grow to a length of seven inches (17.5 cm) and is capable of producing an egg every two seconds. The king, by contrast, is a rather puny creature no larger than a worker termite. The workers and soldiers spend their entire lives underground in the dark, so they are wingless and blind; their functions, as their names suggest, are to tend the needs of the colony and defend against predators, respectively. The soldiers are equipped with armored heads and strong jaws, some having the ability to squirt irritant fluids at invading organisms.

Some larvae develop into reproductive termites that have wings and, on maturity, are released from the nest in a frenzied swarm of flying termites. Such swarms usually attract the attention of their many insectivorous predators, and very few of the male and female termites succeed in mating. Females attract males by emitting a chemical signal, or pheromone, and the male follows the female, landing and running over the ground until they find a suitable location for setting up a nest. These two become the king and queen of the new colony. The queen lays eggs, which initially all hatch into workers so that the construction of the new nest superstructure can begin. The queen will then spend the rest of her life in pampered luxury laying her 30,000 eggs per day, while the king continues in attendance and fertilizes the eggs.

Ant colonies are also found on the forest floor, including some that do not make a permanent nest for colony residence but spend their lives on the move. These are the army ants, of which there are about 240 different species. Their lives involve periods of rest and temporary settlement followed by times of active migration from one area of the forest to another. In their stationary phase, the ants construct bivouacs beneath a log or in a tree hole, sometimes even using the living bodies of the ants themselves, linked tightly by their strongly hooked claws, to form a roof shelter for the colony. The queen lays eggs and the workers feed the growing larvae from the local supplies of food, mainly captured insects. By common consent, the colony then sets off on the move, usually starting at dawn and forming a living column of moving ants. Scouts run out ahead and lay down odor trails that the main body of the colony follows. Any prey they encounter is soon overpowered by a seething mass of ants and is carried back into the main column for communal consumption. Temporary bivouacs are set up for overnight rests, but eventually the ants discover a suitable location to settle for awhile, and it is here that the queen's abdomen begins to swell and a period of egg laying begins before the next call to move on.

Many other groups of invertebrates are found on the forest floor, including ants, planarian worms, annelid worms, fly larvae, cockroaches, crickets, and beetles. Of these, the beetles are among the most prolific, both in numbers and in diversity (see illustration above). All the various beetles have their own food preferences, either acting as detritivores and

Beetles are among the most diverse of all insect groups and have many different ways of life. Many are herbivorous, and some have elaborate horns and projections, such as the one pictured, which are used in battles for mates. *(Paul Gertler, U.S. Fish and Wildlife Service)*

consuming the masses of falling organic matter, or preying upon the detritivores and occupying higher trophic levels in the forest floor food web. They vary greatly in size and in structure. Some have hornlike extensions that can give them a very aggressive appearance, as in the beetle shown here, but these elaborate projections are used mainly in battles between males for the attentions of a female. Beetles are a major source of food for many insectivorous birds and are the pollination agents for many types of flowers (see "Flowers and Their Visitors," pages 133–138).

■ CANOPY INVERTEBRATES

The forest canopy is a complex environment with many layers of leaves, intertwining branches loaded with epiphytes, cascades of lianes with hanging stems and clambering shoots, masses of brightly colored flowers emitting powerful scents, and clusters of succulent fruits nestling among the branches. The penetrating light is bright and the radiant heat rapidly raises the temperature of surfaces exposed to it, while the shade is cool and moist, its temperature varying little between day and night. In this three-dimensional structure there is an almost unlimited range of microhabitats, microclimates, and food resources for invertebrates to use, and it is not surprising that the forest canopy is the site of immense biodiversity.

Canopy invertebrates can be divided into herbivores and carnivores, although this is a great simplification of the range of feeding options available. The herbivores include leaf feeders, wood borers, bark strippers, pollen and nectar eaters, and the consumers of fruits, the *frugivores.*

Carnivores have such a range of prey available that there are great opportunities for generalist feeders and for specialists that concentrate their attention on just a single type of prey. Almost all groups of terrestrial invertebrates can be found somewhere in the canopy, though some are perhaps less well represented than one might expect. Aphids, for example, the sap-sucking greenfly, and blackfly, so common in temperate gardens, might be expected to abound in the paradise of plant tissues in the canopy. But, in fact, aphids are relatively scarce in tropical forest, perhaps because of their specificity in the plant hosts they feed upon, coupled with their poor ability to seek out their target plants. In the tropical forest there are often large distances between individual plants of the same species, so aphids find it very difficult to locate new sources of food.

Some canopy insects change their food sources in the course of their life history, as is the case with butterflies. The larval stages of butterflies, the caterpillars, usually graze upon foliage, often specializing in one particular food plant. Unlike aphids, however, the adult butterflies are strong in flight and can detect their preferred larval food plant by olfactory and visual means, so they can find the right place to lay their eggs.

One group of butterflies, the heliconids, has developed a highly intricate set of relationships with the passion flowers (*Passiflora* species) of the forests of the New World and the East Asian regions. These butterflies of the genus *Heliconius* have larvae that feed mainly on passion vines in the forest, but different caterpillars feed on different parts of their foliage. There is intense competition between the caterpillars of different species, some resorting to cannibalism to ensure that they survive in the struggle for food. One butterfly, *Heliconius nattereri,* lays its eggs on the tips of the vine tendrils, and the caterpillar lives out its larval existence and even pupates on the tendril, avoiding the leaf blade, where it is more likely to meet up with aggressive competitors. Passion flowers have evolved a number of mechanisms for defense against the grazing larvae (see "Plant Defenses," pages 128–130). Some plants have extrafloral nectaries that they use to attract ants (see sidebar "Ant Plants," page 129), which rigorously defend the plant against invading caterpillars. Others, such as *Passiflora adenopoda,* have hooked hairs on their leaves, which can penetrate the skins of the heliconid caterpillars and impale them or cause them to hemorrhage to death. There are also passion flowers that accumulate a cyanogenic glycoside that breaks down to produce the highly toxic hydrogen cyanide when consumed. But the heliconid larvae have adapted to this problem and even turned it to their own advantage by accumulating the cyanogenic glycosides in their own tissues and thus becoming poisonous to their predators. They retain these toxins through the pupal stage and into the adult butterfly, so use the plant defenses to construct their own protection plan.

To be effective the insects must advertise the presence of the toxin; they achieve this by adopting bright warning colors so that any insectivore with predatory intentions can quickly learn to avoid them.

Insect evolution has moved one step further than this, however, because other species have taken advantage of negative signals to predators and have copied them, regardless of whether they contain any toxic chemicals. This type of deception is called *mimicry* (see sidebar on page 148).

Butterflies are among the most conspicuous of tropical forest insects because of their size, their bright colors (see illustration below), and their constant flight between the flowers where they obtain nectar. Some of the largest butterflies are found in the forests of Southeast Asia, including the Rayah Brooke's birdwing (*Ornithoptera brookiana*), a massive black and yellow butterfly with a wingspan in the female of 7.5 inches (19 cm). The male is slightly smaller. Some moths are even larger, such as *Aurivillius triramis* from West Africa, a yellow and brown species with a wingspan of 10 inches (25 cm). This species has large round spots on its wings made up of concentric circles with a black center. "Eye spots" of this type are common among butterflies and moths and can deter a predator when the insect suddenly flicks its wings open to reveal a pair of sinister-looking eyes. The shock tactic gives the insect a vital moment of hesitation in which it can escape.

The colors of butterflies and moths range from green and red through to electric blue, and the purpose of the colors includes predator deterrence, camouflage in the patchy light, and the attraction of mates. The colors of the wings are partly produced by pigmentation like flowers, but also by iridescence. The wings are covered by tiny scales, each of which is highly sculptured with papillae and grooves. The

Butterflies are among the most conspicuous of insects in the tropical forests. Many are strongly territorial, males driving interlopers from their patch in the light. *(U.S. Fish and Wildlife Service)*

Mimicry

Some insects, such as the heliconid butterflies, are toxic to predators because of the poisonous chemicals they have taken up from their food plants and stored in their tissues. Being consumed by a predator and then poisoning it would be little comfort to the animal that has been eaten and would be of no value in natural selection, which depends on survival to produce more offspring. Warning coloration gives the predator a signal that it would do well to avoid this prey organism. Once such a warning signal has evolved, there are advantages for other insects to take on the same colors and share in the advantages of deterrence. The combination of black and yellow, for example, which is found in wasps and some bees, is a common indication of an insect being unpalatable. It makes sense for other unpalatable species to adopt the same signal, so that predators can learn more effectively and make fewer mistakes at trying to eat these insects. Among both wasps and butterflies, different unpalatable species take on similar coloration. In heliconid butterflies, for examples, the female *Helliconius nattereri*, which contains toxins obtained by its larva from the passion flower food plant, is closely mimicked by another unpalatable butterfly, *Dismorphia astyocha*, from a completely different family. Both have black, yellow, and orange wings and are very similar in size and shape. This type of mimicry is called Müllerian mimicry, having been first described by the German zoologist Fritz Müller (1821–97).

Mimicry has also developed among organisms that lack any chemical and unpalatable protection, however. In bees and wasps there are mimics among the hoverflies that are not able to sting and do not warrant warning coloration. Among the heliconids, the male *Helliconius nattereri* has a different wing pattern from the female, carrying a strong pattern of black and yellow on its wings. An unrelated butterfly, *Aeria olena*, is patterned in precisely the same way and the two are visually almost identical. But the mimic does not contain any toxins, so it is using deception in its adoption of warning coloration. Mimicry involving deception in this way is called Batesian mimicry and was first recorded by a British naturalist, Henry Walter Bates (1825–92), who spent much of his life exploring and collecting insects in the Amazon Basin of South America.

scales combine to create a shiny, shot-silk effect, in which light is broken as by a prism into its component colors, giving the metallic sheen to the wings and accounting for the changing color when viewed from different angles.

Just as birds take on bright colors to attract mates and display, so do butterflies. Males are often more brightly colored than females, although females are often a little larger. Perhaps their responsibility for egg laying means that they need a larger body size, which in turn requires larger wings for air transport. Butterflies can often be seen in pairs, fluttering around in the forest sunflecks and apparently engaged in some kind of aerial dance. This behavior is often not a mating ritual but an exhibition of aggression. Many butterflies are strictly territorial and males defend their territories against other males that attempt to invade, so the outcome is a show of strength in which they flutter around one another until one gives up the challenge and retreats.

Butterflies can also be cryptically colored, meaning that their coloring serves to make them difficult to detect in their forest environment. The green charaxes butterfly (*Charaxes eupale*) of the West African rain forests, for example, has wings that are pale greenish-yellow with dark green wing tips above and entirely dark green below. When resting with its wings closed, or when basking with its wings open, this butterfly is extremely difficult to detect until it moves. Like many butterflies, the green charaxes also has some odd feeding habits. Besides nectar and rotting fruit, which attract most butterfly species, this insect is also partial to animal carcasses and even fecal material on the forest floor. These materials are good sources of nitrogen-containing proteins, which are less available in most plant tissues. The butterflies may also be seeking certain elements that are scarce, such as sodium. Soil, feces, and even animal sweat are good sources of sodium, which male butterflies in particular need to construct sperm sacs that they deliver to the female. The muddy banks of tropical rivers are often rich in swarms of butterflies licking at the salty sediments (see "Element Cycling in the Tropical Forests," pages 84–89).

One normally thinks of moths as night-flying equivalents of butterflies, and on the whole this is true. Most moths spend the day in hiding and are especially in need of cryptic coloration to avoid the attentions of day-hunting predators. Many moths, therefore, have brown and gray upper wings that fold over the lower wings and make them inconspicuous when sitting on bark. But some moths have adopted a day-flying habit, such as *Alcides zodiaca*, a member of a group called swallowtail moths (Uraniidae), from the rain forests of northern Australia and New Guinea. This group looks more like

The mud of river banks is a rich source of chemical elements, including sodium, and butterflies often gather to harvest this element that they need in their reproductive processes.   *(Vladimir Gurov)*

a butterfly than a moth, having large wings with yellow and pink patches edged with black. Bright colors are of no value to a night-flying moth, but for the day active species color is important for both sexual signaling and for camouflage.

All butterflies and moths have a long proboscis, or "tongue," which they store coiled beneath their heads, extending it into the tubes of flowers to collect nectar. The proboscis has a channel running along its entire length, so that the insect can suck up the fluid by using it as a drinking straw. The length of the proboscis varies according to the needs of the insect, which is dependent on the length of the tubes or the spurs of the flowers on which it feeds. In the Madagascan hawk moth (*Xanthopan morgani*), the proboscis needs to extend to 18 inches (45 cm) to reach to the base of the orchid *Angraecum sesquipedale* that it feeds from.

The larvae of moths and butterflies are almost all herbivorous, and the vast majority feed directly on leaves by cutting along the edges with their strong jaws. Many caterpillars are green to make them inconspicuous to their predators, and some, especially larvae of the hawk moths, have adopted defensive techniques such as flicking brightly colored eye

spots when disturbed in hope of deterring a predator. An additional option for escaping predatory attention is living within the plant leaf tissues, and some moth larvae burrow beneath the surface layers of the leaf and eat their way through the internal tissues, forming burrows. This technique is known as leaf mining. Other caterpillars are hairy, the number of hairs being as great as two million per caterpillar (see illustration on page 150). These hairs often have a gland at their base producing irritant chemicals that make them unpalatable to predators. The larva of the Venezuelan emperor moth (*Lonomia achelous*) has toxins that cause serious and potentially fatal hemorrhage in its victims. In some species the hairs are barbed and cause extreme distress to an animal when taken into the mouth or gut. Caterpillars with chemical defenses and protective hairs have no need of camouflage and are often brightly colored as a warning signal to predators, which quickly learn to avoid them.

Carnivory is very rare among caterpillars, but it is found among the pug moths (*Eupithecia* species) of the Hawaiian Islands. The larvae resemble twigs in appearance and add to this effect by keeping still on a branch, holding it by claspers at the end of their abdomens and holding the front part of the body erect. When a hapless plant-sucking bug comes wandering by, they make a lightning strike with their fore parts and devour their prey.

Most caterpillars are solitary, or occur in small groups where the adult female has laid her eggs. The so-called army-

Caterpillars are a major source of food for insectivorous birds in the forest. Some are green and difficult to find in the foliage, but others are conspicuous and arm themselves with irritant bristles to deter their predators. *(Nicola Stratford)*

worms (family Noctuidae), by contrast, are communal and formidable in their ravages of vegetation. The African armyworm (*Spodoptera exempta*) moves in very large swarms and is a serious pest of crop plants.

Among the colonial insects of the forest canopy, wasps and bees are particularly prominent. These insects construct nests in hollow trees, building cells for their eggs and larvae by using wax, partly produced by the insect and partly derived from plant resins. Many bees and wasps are armed with a sting and readily use this in the defense of the nest, but there are also stingless species, such as the stingless bees (tribe Meliponini). These insects have very powerful jaws, however, and are able to deliver a painful bite into which they inject irritant saliva. The stingless bees are great architects and build elaborate nest structures from a material called batumen, which is a mixture of wax, resin, vegetable material, and mud. Often they begin with a hole in a tree, sometimes derived from an abandoned ant or termite nest, and extend outward to form a dome on which they construct a tubular entrance corridor.

Stinging bees inject toxin through the barbed sting at the rear of their abdomen. The location of the barb means that the bee leaves the sting behind when it tries to release itself, and the damage resulting to the bee's body is invariably fatal. For a bee, an attack is essentially a suicide mission. Behavioral ecologists from the time of Darwin onward have been intrigued by this pattern of suicidal behavior on the part of the individual to protect the community, as it seems to go against the self-interest demanded by the theory of natural selection. But genetic studies have now produced an explanation for this apparently altruistic behavior (see sidebar [right]).

Bees, Behavior, and Genetics

Darwin's theory of natural selection is a means of explaining how the process of evolution is driven. The theory is based on the fact that individuals of a species are in competition with one another and only those that adapt best and are the fittest survive to breed and pass their genes to the next generation. In this way, the genetic quality, the fitness, of a species is both preserved and enhanced. Accordingly, each individual must have behavior patterns based on self-interest, a concern that their genes must survive whatever the cost to other members of the species; there is no room for altruism toward one's fellows. Yet bees seem to be an exception to this principle because they are prepared to sacrifice their lives for the sake of the colony. When they sting an enemy in defense of the nest they invariably die as a consequence, so is this a denial of Darwin's concept requiring an emphasis on self-protection? Geneticists have come up with an explanation for the self-sacrificial behavior of bees that is in accord with Darwinian theory. Most worker bees are female and are diploid—that is, they have two sets of chromosomes, one derived from their mother and one from their father, as in most other animals. But males are different. They are haploid, having just one set of chromosomes, because they develop from unfertilized eggs. When the male produces sperm, all of the gametes have an identical genetic constitution. The male mates with a potential queen that then lays large numbers of eggs, and all of the diploid daughters of the mating share more genes than would be the case for any other animal. They all have an identical paternal input (50 percent of their total genes), together with a common share of half their mother's genes (25 percent of the total). So the genes that the workers all have in common amount to 75 percent of their total genetic constitution, whereas in sisters in most animal families there would be only 50 percent of genes in common. Behavioral ecologists argue that because the worker bees share so many genes, their death in the defense of the colony represents very little danger of loss of their own genetic constitution. Their behavior is not altruism but a form a self-interest, because in saving the colony they are automatically saving 75 percent of their own genetic heritage. The theory of natural selection, evolutionists contend, is not weakened by the apparent self-sacrifice of bees, but is confirmed by it.

Bees are vegetarian and play an important role as pollinators in the forest, while wasps are more omnivorous, and some species carnivorous. Both bees and wasps have solitary species as well as colonial ones. Ants, on the other hand, are invariably colonial, but like wasps, may be herbivorous or omnivorous, depending on the species.

One of the most characteristic ants of the forest canopy is the leaf-cutting ant of the Central and South American forests. The colonial nest is usually on or below the ground, but the worker ants spend much of their time in the canopy, harvesting plant tissues. The workers vary in size, the large workers being responsible for cutting discs of leaf tissue from a food plant and transporting it from the canopy back to the nest. But while it is carrying the leaf portion, which is often considerably larger than the ant itself, the large worker is very vulnerable to attack by enemies. The enemy may be a predator, but more often it is a parasitic fly (Phoridae) that lands on a defenseless ant and lays its egg on the back of the ant's neck. When this egg hatches, the larva bores into the body of the ant and ultimately results in its death. While carrying the leaf disc, the ant is unable to ward off parasitic flies. It overcomes this problem by carrying a smaller worker as a passenger on the disc, and this individual is responsible for defending the carrier from attack.

Like termites (see "Invertebrates of the Forest Floor," pages 143–146), leaf-cutter ants build fungus gardens in their nests, inoculating their compost heaps of leaf fragments with a fungus first brought to the nest by the founding queen. The worker ants take care of the garden, provide fecal material to fertilize it, and masticate the leaf tissues so that they are more easily infected by fungi. They then feed upon the fungal mycelium, which they harvest in a sustainable manner. Most fungi are aerobic, so the buried gardens need to have an adequate supply of air, and the ants achieve this by constructing a series of ventilation shafts from the atmosphere down to the garden that ensure a constant circulation of air. A colony can contain over two million ants, so the food demand is high and the supply of leaf tissue to the fungus garden must be maintained. Filling this need can result in quite serious defoliation of the canopy, not only in the forest itself but also in surrounding croplands, where the leaf-cutting ants are regarded as a serious pest.

Camouflage among canopy insects is certainly not restricted to the moths and butterflies but is widespread, both as an aid to predation avoidance and, in some cases, a mechanism that fools the prey and assists the predator. In an environment filled with leaves and twigs, it is not surprising that leaf and stick mimicry are the most common forms of camouflage. Some of the most convincing mimics are the katydids (family Tettigoniidae), close relatives of the grasshoppers and crickets (order Orthoptera). Some of the katydids, such as *Brochopeplus exaltatus* from the rain forests of Sri Lanka, are bright green with leaflike forewings, equipped with venation patterns that closely resemble those of their background foliage. Others, like *Sasima spinosa* from New Guinea, have similar venation patterns but are brown and resemble dead leaves. Acanthodis, a katydid genus from Malaysia, is a bark mimic and is brown and flattened, making it almost invisible against a tree bark background. The fulguroids (family Fulguroidea) are leaf hoppers, belonging among the plant bugs (Homoptera), and they are even more remarkable in their appearance. They are brown and barklike in their basic coloration, but the heads are swollen and extended to a grotesque degree, having eye and mouth markings that give it a sinister and reptilian appearance. Potential predators are undoubtedly dismayed by its alarming appearance. Stick insects (family Phasmidae) are also bugs but their protection is based entirely on their resemblance to small bare twigs, the hope being that they will be overlooked rather than appear shocking to their predators. There are some beetles, such as species of *Gymnopholus* from the forests of New Guinea, that bear a growth of algae, liverworts, and lichens on its back, providing it with a perfect camouflage among bark also covered with epiphytes.

All of these insects employ camouflage to avoid predation; there are also insect predators that use cryptic form and coloration so that they can more effectively approach and capture their prey. The assassin bugs of Australia (family Reduviidae), for example, have a flattened, barklike abdomen, and capture flies and bees that land close to them. The mantids (order Mantodea) often take on the appearance of leaves, twigs, or even flowers, where they lie in wait for their prey (see illustration on page 152). Their camouflage also assists them when they stalk their prey with stiff and jerky movements. The South American casque-headed mantis (*Choeradodis rhombicollis*) is very specific in its mimicry, being an almost perfect reproduction of the orchid (*Epidendrum ciliare*), with winglike extensions of its thorax, which is so realistic it has markings resembling fungal infections. The mantids are famous for their cannibalism, and in many species the female kills and eats the male during or immediately after copulation. Behavioral scientists have recently discovered that removal of the head of a male mantis, which is usually the part the female devours first, causes a nervous reflex in the rest of the male's body that actually stimulates continued copulatory movements, so cannibalism seems to improve the male's performance.

Invertebrates are thus abundant within the forest canopy and occupy a very wide range of ecological niches. Because of their small body size, their densities greatly exceed those of the larger vertebrate animals. With their abundance they provide a valuable food resource for larger predators. But even within the invertebrates quite complex food webs can develop as they feed upon and parasitize one another.

The praying mantis is an important invertebrate predator of the forest canopy. Its twiglike appearance provides camouflage as it stalks its insect prey and protects it against its own predators. *(George Whitaer)*

■ FOREST FISH

Fish may not seem an obvious component of the tropical forests, being restricted to aquatic environments, but there are aquatic habitats within the forest where fish are found. Most lowland forests have rivers and streams running through them, many of them large with numerous tributaries, so the forest and the waterways are never far from one another (see illustration on page 153). Changes in the courses of rivers often leave meanders and bends isolated from the main river flow, forming oxbow lakes and slow-moving or stagnant pools. These bodies of water are very important for many forest animals and are the homes of many species of fish. Many of these fish form an important food resource for the resident human populations. The Amazon Basin is notorious for frequent flooding (see "Tropical floods," pages 16–19), which creates a temporary habitat for large numbers of fish. These fish come to interact closely with the plants and animals of the normally terrestrial forest areas. Coastal forests, especially the mangrove forests (see "Mangrove Forests," pages 71–73), are rich in fish species. Fish must be regarded as animals that reside close to and sometimes within the tropical forests and form an important part of many forest ecosystems.

It is within the South American forests that fish play the most important roles in the function of the forest, for only in the jungles of the Amazon Basin have fish developed the habit of eating fruits and seeds. Many different species of fish have adopted the seed-eating habit, one of the most important being the tamaqui (*Colossoma macropomum*). Not only is this a large and abundant fish in the Amazon Basin but it is also one of the most widely exploited fish for human consumption in the region. It has a fruity flavor, which is not surprising given its diet. It belongs to the family Characidae, which it shares with the more famous piranhas, and is quite large, growing to more than three feet (1 m) in length and over 45 pounds (20 kg) in weight. The floodplains within the Amazon forest are very extensive, so when the water rises and the rivers overflow their banks, the tamaqui, along with many other fish, are able to invade the forest and forage for fallen fruits and seeds. During its immature stages the fish feed upon zooplankton in the river, but once mature, after about three or four years, they can cope with a fruit diet. They are not fastidious about their food, as they are prepared to eat the fruits and seeds of many different plant families, including palms (Palmae), figs (Moraceae), members of the bean family (Fabaceae), and the Euphorbiaceae, including the rubber tree (*Hevea brasiliensis*). Sometimes the main source of food is the fleshy fruit, as in the case of *Cecropia* (Moraceae), and sometimes it is the hard seed itself, as in the case of the rubber tree. The size of fruits and seeds taken by tamaqui depends on the size of the fish; a large specimen can swallow seeds of two inches (5 cm) in diameter.

The tamaqui has very large nares (nostrils), which suggests that it has a highly developed sense of smell, so it is probable that the fish seeks out its food by olfactory means. The tropical trees that the fish exploits are rich in chemicals, such as resins, alkaloids, and other organic compounds, so the fish may well hunt out target trees first and then seek the fruits. They have large stomachs and can hold about one-tenth of their body weight in seeds within the distended stomach. Tamaqui also have extremely large, molarlike teeth, which is extremely unusual in a fish. With these teeth they can grind their food and digest the contents. There is always a chance that some seeds survive gut passage, and for these the fish provide a highly efficient means of dispersal.

Some piranhas are also vegetarian, such as the pacu (*Piaractus mesoptanicus*), and they have longer jaws than their carnivorous relatives and are slower swimmers. The flesh-eating piranhas (*Serrasalmus* species) have short but powerful jaws. They are quite small, rarely exceeding two feet (0.60 m) in length. When they attack their prey they are only able to bite off about one cubic inch (16 cm³), around a thimbleful, at time, but a shoal consisting of hundreds of these fish can strip an animal the size of a cow in a matter of minutes. The flooding of the forest, therefore, brings both

herbivorous and carnivorous fish into the food web of this normally terrestrial environment.

There are some fish that do not require floods to prey upon the animals of the forests. The archer fish (order Perciformes) consist of several different species found in India, Southeast Asia, and northern Australia. These fish are able to prey upon terrestrial insects by spitting at them from below the water. The roof of the mouth of the archer fish has a narrow groove leading forward, and its tongue is narrow at the front and thick and muscular at its base. When the fish presses its tongue to the roof of its mouth it creates a narrow tube and can force a thin jet of water along this tube, aimed at an insect perched on a leaf or branch above the water surface. Once hit by the spurt of water, the insect falls and is rapidly consumed by the waiting fish. Some archer fish have a remarkable range and can hit an insect from five feet (1.5 m) away, though aiming from below water can be difficult. When viewing an aquatic object from above the water, its precise location can be difficult to determine because light is bent when it passes through the air–water boundary, a process called refraction. Light is similarly bent, but in the opposite direction, when an underwater organism looks up into the aerial world, so aiming at an object above the water is quite a problem for an archer fish. Thus they position themselves directly below their prey before spitting, because there is no refraction when light strikes the water at right angles.

Although fish are aquatic organisms and may not seem a part of the terrestrial forest ecosystem, they do interact with the land plants and animals and thereby enter the food web of the forest.

Wetlands penetrate deeply into many tropical forests. The idea of the jungle being a tangled mass of impenetrable vegetation probably arose because early explorers often traveled along waterways where the edges were very dense.   *(James P. McVey, NOAA Sea Grant Program)*

■ AMPHIBIANS AND REPTILES OF THE FOREST

Some animals live in two worlds, spending part of their lives in the water and the remainder on land: These are the amphibians and include newts, frogs, toads, and salamanders. They use water mainly for breeding because they lay eggs that are unprotected by shells, which would rapidly desiccate if exposed to dry air. The eggs are normally produced in very large numbers, sometimes up to 20,000 in a clutch, and the young larvae (tadpoles) suffer very high levels of mortality from predation, cannibalism, and the instability of the environment if they use small, temporary water bodies. The amphibians are thus generally r-selected in their reproductive strategy (see sidebar "r- and K-selection," page 96). Tadpoles take in oxygen in a dissolved form from water through their gills, but the adult amphibian has lungs and breathes air. The adult is also able to respire through the moist skin.

Tree frogs are among the most abundant amphibians of the tropical forest. The high humidity enables them to live far from water, often high in the canopy, where they prey upon insects. This species, *Phyllomedusa tomopterna*, lives in the rain forests of the Amazon. *(Claus Meyer/Minden Pictures)*

Many amphibians use the margins of large tropical rivers and their associated wetlands for breeding; others are content with small streams, pools on the forest floor, muddy wallows created by the activities of large mammals, such as wild pigs, and water-filled hollow tree stumps in which to lay their spawn. Although the eggs and larval stages are the most sensitive to drying, even the adults of many amphibians prefer moist atmospheres in which their skin is permanently wet. Toads are an exception to this generalization, as they are able to survive in desert environments. Water evaporates easily from an amphibian's skin, unlike that of a reptile, so high atmospheric humidity is ideal. This condition accounts for the rich diversity of amphibians in tropical rain forests. On Costa Rica alone there are 119 species of frogs, tree frogs (see illustration above), and toads (jointly called anurans).

Most anurans lay their spawn and then leave them to the chances of nature to survive or perish. Some, however, take parenthood seriously and care for the eggs and develop-

ing larvae. Among these is the dart-poison frog *Denrobates pumilio*. The male and female frogs combine in pairs, and as the female lays the spawn, the male sheds sperm over it to fertilize the eggs. The male then guards the eggs for about 12 days until they hatch, when the female takes over and transports each tadpole to the water-filled goblet of an epiphytic bromeliad plant (see sidebar "Epiphytes," page 121). Each tadpole thus has its own individual pool in which to develop, which greatly reduces the risk of mass predation or catastrophe. To ensure that each tadpole has a good chance of survival, the female revisits its young regularly and lays a single unfertilized egg to supply it with food.

In Darwin's frog (*Rhinoderma darwini*) the male takes protection to an extreme level. It holds the fertilized eggs in its mouth and in the vocal sac in the throat to prevent desiccation or predation. Only when the tadpoles hatch, after 10–20 days, does the frog release them into the risky wet world. Although its vocal sac is distended by the brood of eggs, this does not deter the male frogs from calling. Some of the poison-arrow frogs carry the tadpoles around on their backs. Among the leaf frogs (subfamily Phyllomedusinae), the female constantly empties her bladder over the developing eggs, which are laid on the underside of canopy leaves. The Old World tree frogs include some species that are known as "foam nesters." The females produce large volumes of fluid and thrash it with their hind feet until it forms a foam cover over the eggs. It then solidifies into a protective mass with the texture of meringue. Generally speaking, the more parental care and protection afforded to a frog's brood, the fewer eggs it produces. Thus some frogs tend toward a K-selection strategy rather than r-selection in their reproduction.

In the moist atmosphere of the forest, amphibians are able to live far from standing water, and many only visit pools for a few days during their breeding season. The tree frogs are arboreal, possessing suction disks on their feet that enable them to cling onto foliage and stems in the upper canopy. The disks are made of cartilage but their shape can be changed by muscular action. The secretion of mucus adds to their adhesive properties. Most of these tree frogs are green, providing them with camouflage against predators. For example, the gaudy leaf frog (*Agalychinis callidryas*) is bright green with protuberant red eyes. This frog is so thoroughly arboreal that even mating and egg laying takes place in the canopy. The female lays eggs on the underside of a leaf, where the male fertilizes them, and when the tadpoles hatch they fall into the water down below.

Some frogs, such as the dart-poison frog, secrete highly poisonous toxins as a means of protection against predation. The South American Indians use the poison of the dart-poison frog, called batrachotoxin, to coat their arrows. Even a wound in a hunted mammal will be sufficient to kill it. The Indians extract the batrachotoxin from the frog by holding the frog over a fire, and the poison drips out of the skin. One frog produces enough poison to coat the tips of 50 arrows. When injected into a mammal, batrachotoxin has a strong impact on the heart and nervous system, causing paralysis and death. Frog toxins are some of the most highly poisonous substances found in nature. In the case of the Koikoi poison-arrow frog (*Phyllobates bicolor*), for example, 0.00000004 ounce (0.00001 g) is sufficient to kill an average-sized human. This frog is bright red in color, so it makes no attempt to camouflage itself, but instead displays warning coloration to potential predators. Toads are also well known for their toxins, and there are reports that dogs have died after picking up the giant cane toad (*Bufo marinus*) in their mouths. Toad toxins, mixed with other plant and fish toxins, are used in voodoo ceremonies, inducing a deathlike trance in people.

Newts and salamanders are, perhaps surprisingly, more characteristic of temperate rather than tropical habitats. The lungless salamanders (family Plethodontidae) are quite well represented in the New World Tropics, and these differ from the frogs in having internal fertilization of their eggs. The fertilized eggs are still laid in aquatic situations because they lack any protective shells. Many are arboreal, like tree frogs, but they lack the suction pads on their feet. The fingers and toes are very short and the webbing is retained, unlike most salamanders, where the webbing is lost as the animal develops. The short, webbed feet are very effective in providing adhesion to leaves and branches.

Reptiles, including tortoises, turtles, lizards, and snakes, lay eggs with a hard shell around them, so they are far less dependent on aquatic habitats for their reproduction. Some snakes give birth to live young (viviparous), thus avoiding many of the problems of egg predation. The skin of reptiles is also very different from that of amphibians, consisting of waterproof scales that prevent water loss and allow the animal to spend time in hotter and drier environments.

Lizards, which include geckos, iguanas, and chameleons, are found both on the forest floor and in the canopy. The geckos have feet that resemble those of the tree frogs, with toes that have considerable adhesive properties, enabling them to walk upside down on leaf and branch surfaces. Gecko toes are variable, some being thin and clawlike, while others are flat and covered with fine ridges that give them grip. Chameleons have long toes ending in a claw and they grasp firmly onto twigs and branches. Snakes have prehensile bodies with between 200 and 400 vertebrae and are well adapted for canopy life because they can coil around branches and cling securely with their tail end while exploring new areas with the front part of their bodies.

As with frogs, there are many different behavior patterns among reptiles that are associated with care for the eggs and young. Most lizards merely lay eggs on the ground, often excavating a hole, or finding a suitable rock or log beneath which they can lay their brood. The Brazilian skink

(*Mabuya heathi*) is one of the exceptional lizards that retains its fertilized eggs in the oviduct and nourishes the developing embryos through a placenta very similar to that found in mammals. Some snakes, particularly pythons, lay eggs and then care for the clutch while it is developing. Among the pythons (family Pythonidae), the females usually coil around the clutch of eggs once they have completed laying them and remain with them until the eggs hatch. Reptiles are cold-blooded, so they are not able to incubate eggs in quite the way that birds do, but they can protect them from attack, and they can leave the clutch to bask in the sun, returning to transfer heat to the eggs. Scientists have closely observed egg care in the Indian python (*Python molurus*) and have found that when the egg temperature falls below about 86°F (30°C), the female sets up a series of muscle contractions in its body at a rate of one every two seconds. This practice generates heat and can raise the temperature of the eggs by 9–12°F (5–7°C). The Indian python may remain with the eggs for as long as three months until they hatch and it does not leave to feed during this entire time.

Most tropical forest reptiles are cryptically colored either to escape predators, such as certain birds and mammals, or to hide themselves among foliage to ambush their prey. The vine snakes, such as *Theltornis kirtlandii* of West Africa and *Dryophis prasinus* of India and Southeast Asia, are particularly good examples of snakes adapted for camouflaged ambush. They can grow to several feet in length, but are pencil thin and can twine their bright green bodies through foliage to become almost invisible. They eat birds, lizards, frogs, and insects that come close. This type of ambush hunting requires very fast reaction, and few reptiles are faster than the chameleon. These animals are cryptically colored lizards that move in a slow, jerky manner and lie in wait for their insect prey, approaching it with great stealth. When they finally shoot out their long, sticky tongue to capture the insect, the entire operation lasts a mere $1/25$ of a second, so few insects can react fast enough to escape.

Chameleons take cryptic coloration to a new level by being able to alter their shade according to the background (see illustration on page 156). They are usually green, often with blotches of paler and darker colors over their bodies. An intricate network of pigment cells, called chromatophores, cover the chameleon's body, and these can expand and contract under the control of the nervous system, having the effect of lightening or darkening the skin color. The reptile may change color on acccount of its background, or it may do so because of changes in light intensity, becoming lighter in bright sunlight to avoid becoming overheated. Color changes can occur in a matter of seconds, so the animal is capable of rapid alteration of shade as sunflecks come and go.

Snakes are important predators in the tropical forests, taking all manner of prey from insects to large mammals. Visual hunting of prey is limited in many species because

The chamaeleon is a canopy reptile that feeds mainly on insects, which it catches with its long and sticky tongue. Its skin contains pigment patches that can alter its color according to its background. (*Angela Foto*)

of poor sight, but they make up for it in their other senses. Some snakes, including pythons and boas, can detect infrared radiation, so can locate warm-blooded prey even in the dark. The sensory organs are situated in small pits on the lips of the reptile. These are the most sensitive heat detectors known in nature, picking up heat differences as small as 0.002°F (0.001°C). Other snakes hunt by chemical odors that they pick up by means of their tongues and some highly sensitive organs in the roof of the mouth. They collect chemical signals from the atmosphere by waving their tongues in the air or from the ground by touching their tongues on the soil, then flicking it back into their mouths, where the chemical analysis is carried out. Using their tongues in this way, they can locate both prey and predators or even potential mates.

Many tropical forest snakes, including the pythons and boas, kill their prey by crushing them, the size of the prey depending on the size of the snake. The emerald tree boa (*Corallus caninus*) of Amazonia concentrates on small rodents, birds, and lizards, while some pythons (family Pythonidae) are large enough to tackle pigs, deer, or even crocodiles and leopards. There are accounts of pythons taking children or small adult humans, but these are rare. The boas (family Boidae) are very similar to the pythons, but are mainly found in the New World. The best known is the boa constrictor (*Boa constrictor*), which is widespread in the Neotropics and can grow to a length of 18.5 feet (5.6 m). But the biggest of the boas is the anaconda (*Eunectes murinus*), which can reach 38 feet (11.4 m) and weigh over 240 pounds (110 kg). They spend much of their time on river banks in Amazonia, lying in wait for their prey, which consists mainly of mammals such as capybara, but can also include crocodiles and caimans.

Snakes often take prey that is considerably larger than the size of their heads. They do not have the required dentition to carve it into manageable portions, so prey must be swallowed whole. The upper and lower jaw are not tightly attached but are held together only by muscles and ligaments, so that a snake can dislocate its jaws to swallow large prey. The backward-pointed teeth in the jaws help the snake to shuffle the prey gradually down its throat, and hold the animal tightly so that it cannot escape even if it struggles. The body structure of the snake is sufficiently flexible to allow the gradual passage of a large swelling back through the gut without injury to intestine or skin. Since snakes are cold-blooded, they require less energy and therefore less food than warm-blooded animals such as birds and mammals, and they do not need to feed very often. A large python may eat only once a year, and rarely more than once a month.

Some snakes are specialized in their prey requirements, concentrating on fish or even snails. Asian snail-eating snakes have lower jaws that are modified to hook snails out of their shells. Egg-feeding snakes may swallow the egg whole and then constrict their muscles to break the shell, or they may have modified vertebrae in the neck that cut the egg open as it is swallowed.

Killing or at least disabling prey before swallowing it avoids the problem of dealing with a struggling mouthful of fur and muscle. Constriction is not the only weapon available to snakes; some use venom. The poison is stored in a gland in the cheek of the snake, just below the eye, and it connects to the hollow fangs at the tip of the upper jaw by ducts. Snake venoms typically contain protein neurotoxins that attack the nervous system of the recipient, together with enzymes that destroy the blood-clotting mechanism and cause internal hemorrhage. These enzymes also cause the breakdown of muscle tissues and a drop in blood pressure. Finally, the impairment of circulation and the products of tissue breakdown combine to cause kidney and heart failure. The most important groups of venomous snakes in the tropical forests are the front-fanged snakes (family Elaphidae), which include the cobras, coral snakes, and mambas, and the viper family (Viperidae), which includes the tropical pit vipers as well as many temperate venomous genera (see illustration below).

The bushmaster (*Lachessis muta*) is one of the biggest of the New World venomous snakes, growing to a length of 12 feet (3.6 m). Like the pythons and boas, it hunts mainly warm-blooded prey and is equipped with heat sensors in its upper lips. Its fangs and gape are large, making it a particularly dangerous species to humans; its venom is able to kill a man in a matter of hours. Most venomous snakes bite their prey and then release it, following the dying animal until it collapses, but the bushmaster holds on to its prey, maximizing the injection of venom. More dangerous to humans in the New World tropical forests, however, is the fer-de-lance (*Bothrops atrox*), an extremely irritable and aggressive snake that strikes on being disturbed. It is a large reptile, up to 10 feet (3 m) long, and is probably responsible for more human deaths in Central and South America than any other forest species.

There are no vegetarian snakes; all are predators occupying the higher trophic levels in the tropical forest food web. Some, such as mature anacondas, are top predators and are not subject to further predation, but young snakes and even some adults are regarded as potential food by certain mammals and birds, such as the harpy eagle (*Harpia harpyja*).

■ TROPICAL FOREST BIRDS—THE FRUIT AND NECTAR FEEDERS

In general, animals are much richer in proteins than plants. Animals that feed entirely upon plants, therefore, need to take in larger quantities of food than predatory animals, which have a richer diet. Plant tissues are also generally poorer in their energy content per unit weight than the equivalent animal tissues, thus herbivores constantly need to eat. Leaf tissues are often protected by the presence of distasteful or poisonous chemicals, which serve to deter herbivores. There are certain parts of plants, however, in which both proteins and energy are concentrated, and these parts are usually associated with reproduction. Flowers offer food rewards, usually in the form of nectar, to visitors that may transport their pollen (see "Flowers and Their Visitors," pages 133–138). Seeds are supplied with stores of energy that can be used by the developing embryo in germination, so they provide herbivores with

The Gaboon viper of tropical Africa grows to over six feet long (2 m) and lives mainly on the forest floor. It injects venom into its prey, which includes mammals, amphibians, and birds, and then waits for it to die before swallowing it whole. *(Kevin Tate)*

a rich diet. And fruits sometimes have enveloping layers of energy-rich, digestible materials that encourage animals to consume them in the hope that some seeds will pass through the gut and be transported to new areas. Thus many of the animals that eat plant material concentrate their attentions on flowers, fruits, and seeds.

Plant-feeding birds in the tropical forests are almost entirely nectar, fruit, or seed eaters, but there are some exceptions, the best example being the hoatzin of South America. The hoatzin (*Opisthocomus hoatzin*) is about the size of a scrawny chicken—two feet (0.6 m) long, but only 1.8 pounds (0.8 kg) in weight. It has a ragged crest of head feathers that give it a particularly bizarre appearance. It is closely related to the cuckoos and roadrunners, which in many ways it resembles. The hoatzin lives in groups, usually close to riversides or pools in the forest, where it has a diet consisting of over 80 percent leaves. A leaf-eating animal is called a folivore. Like many herbivorous animals, it needs a specialized gut to deal with the large quantities of leaf tissue it consumes, both to digest material and to rid it of the toxic secondary compounds found in many leaf tissues. In the guts of all birds the esophagus leads into a sacklike crop, where food is initially taken and stored. In the hoatzin this is an enlarged organ where bacteria commence the breakdown of plant material, assist in digestion by fermentation, and detoxify some of the unpleasant compounds found in some elements of the diet. Behind the crop is a two-compartment stomach, consisting of the proventriculus, where enzymic digestion commences, and a gizzard, which is muscular and often contains grit ingested by the bird to aid in the process of grinding and comminuting the food. This is the organ that compensates for a bird's lack of teeth and is particularly important for birds that eat tough seeds. The length of the remaining part of the intestine is dependent on diet, and the hoatzin, with its intake of plant tissues, needs a long intestine to allow adequate absorption of nutrients from the relatively indigestible food.

Hoatzins breed in groups, building rather crude and fragile nests of twigs in tree branches. At each nest, besides the breeding pair there may be up to five additional non-breeding birds that act as helpers, especially in feeding the young hoatzins. The young birds are remarkably primitive and reptilian in appearance because they bear claws on their wings that help them climb and scramble among the branches of trees. In this respect they resemble the fossil bird *Archaeopteryx*. These claws, however, are lost as the bird matures. Nonetheless, even the adult seems more at home clambering among the riverside branches than when flying.

Nectar feeding provides energy-rich food but requires high-energy output to harvest the rewards. In the New World Tropics, the hummingbirds (family Trochilidae) are the most abundant group of nectar feeders, and they adopt a hovering technique to tap nectar through their long bills (see "Flowers and Their Visitors," pages 133–138). They are all tiny, weighing between 0.1 and 0.3 ounces (2.5 to 8 g), and have a very high power-to-weight ratio to sustain their energetic flight. The hovering is achieved by using a figure-of-eight motion, sweeping the wings backward and forward and thus holding the head perfectly still. They can also fly very fast in a straight line, attaining speeds of 60 miles per hour (96 km/h). Many species of hummingbird are migratory, spending their winter in Central and South America and migrating into North America for the summer.

Hummingbirds are not found in the Old World; their ecological niche is filled by the sunbirds (family Nectariniidae). Like the hummingbirds, they are small, ranging from 3.75 to eight inches (10–21 cm), and have long, often down-curved bills that are used for extracting nectar from flowers. Like hummingbirds, they are often brightly colored, with metallic sheens produced by iridescent feathers, but they prefer to perch on a flower when extracting its nectar rather than hovering by it. Africa is the richest region for sunbirds, with 66 species. One of these species, the scarlet-tufted malachite sunbird (*Netarinia johnstoni*), has been recorded at an altitude of 12,000 feet (3,650 m) in the Ruwenzori Mountains of East Africa. Life in cold conditions is difficult for a small, energy-demanding bird, and they may have to reduce their body temperature in order to survive, leading to a dormant state similar to hibernation. The New World hummingbirds migrate to avoid unfavorable climatic conditions. Many hummingbirds spend the summer in high latitudes but then fly to the tropical forests for the winter. One of these is the calliope hummingbird (*Stellula calliope*), shown in the illustration below. It is North America's smallest bird and yet

Hummingbirds, such as this calliope hummingbird, spend the summer in the temperate regions of the New World but return to the Tropics for the winter. The tropical forests are thus important for the conservation of many temperate bird species.   *(Timothy Wood)*

travels south to the forests of Central and northern South America for the winter.

Among the fruit-eating birds of the tropical forests, the toucans (family Ramphastidae), hornbills (family Bucerotidae), and parrots (family Psittacidae) are perhaps the most spectacular. The colorful plumage and bills of these birds might be considered a disadvantage in the jungle, making them conspicuous to predators, but the canopy of the forest is such a mixture of colors—greens, yellows, and the red of flowers and fruits—that brightly colored birds can often be well camouflaged and difficult to spot. The most distinctive feature of most fruit- and seed-eating birds is the bill, which is adapted in various species to deal with a range of different materials. The bill of the toucan is a massive structure; ecologists have long debated why it should be so big. Breaking open hard fruits and crushing seeds requires a tough beak with strong musculature, but in the toucan and the hornbill the beak is also very long. Perhaps the length is valuable because it enables the bird to perch on thin and delicate twigs in the canopy and yet stretch out to obtain fruits that would otherwise be out of reach. There may also be an element of sexual selection in the evolution of the toucan's bill, females being impressed by such a large and brightly colored appendage. But this explanation does not account for the fact that the female also has a similar bill.

The hornbills, which are the Old World equivalent of the toucans, also have excessively long and heavy bills. The male usually has a larger bill than the female, especially in its upper horny extension, called a *casque* (see illustration above), so sexual selection may well be at work here. These birds are bigger than the toucans, the largest being the great hornbill (*Buceros bicornis*) of Southeast Asia, which is five feet (1.5 m) long. Hornbills have unusual nesting habits, first choosing a hole in a tree and then plastering the entrance while the female is still inside, leaving only a small hole through which the male can pass food. The female is responsible for the sealing of the hole and uses her own fecal material for the job, though in some species the male assists, bringing pellets of clay. The female remains there and undergoes a molt until the young have reached such a size that the male can no longer cope with the food supply, usually after two or three weeks. The female then breaks out and assists in the feeding while they replaster the hole to protect the growing young.

Toucans eat many different kinds of fruits, particularly from *Ficus* and *Cecropia* trees. They pick off the fruit with the tip of the bill and then throw their heads back to swallow it whole. Toucans will also eat insects, but do not have such a varied diet as the hornbills, which are perfectly ready to accept eggs, nestlings, and tree frogs. The hornbills that live below the main forest canopy, such as the white-crested hornbill (*Tropicranus albocristatus*), eat mainly insects and

Hornbills are remarkable because of the extreme size of their beaks, which are adapted to deal with large fruits and seeds. They nest in holes in trees, where the female is plastered in for the duration of the incubation period. *(Goh Kheng Liang)*

are even adept at catching insects in flight. Both toucans and hornbills use their bills in an extremely dexterous manner and are able to perform acts of considerable skill with twigs and other implements. Such ability can be useful when dealing with poisonous snakes and scorpions, which can be handled with the tip of the bill, avoiding the danger of being struck.

Parrots are found throughout the tropical forests of the world, but are most abundant in South America and Australia. They all have massive but relatively short and hooked bills, and many are very brightly colored. Some, like the dusky-orange lory (*Pseudeos fuscata*) of New Guinea, feed upon flowers, taking pollen and nectar by cutting open the flower base, but most parrots are fruit and seed eaters. Although the bills are the most distinctive feature of parrots, their feet are also unusual, having two toes pointing forward and two pointing back, rather like woodpeckers. This gives them great manipulative ability and they can handle objects such as seeds very effectively with their feet. The feet are also powerful and are useful for gripping branches when climbing, as is the bill, which parrots use a great deal when clambering in the canopy.

Most parrots are gregarious and noisy, so they are easily detected in the forest canopy. The macaws of South America are particularly spectacular in their coloring and form impressive flocks along the riverbanks of the Amazon Basin. They are particularly fond of palm nuts, crushing these strong fruits with their bills (see illustration on page 160) and shifting them into position with muscular tongues. Some of the seeds eaten by macaws contain extremely potent toxins. One way in which they deal with this problem is to eat clay minerals from the banks of rivers, which have the effect of detoxifying the poisons in their gizzards.

Macaws are brightly colored parrots found in South America. Their strong bills enable them to crack hard nuts, some of which contain toxic chemicals, but they overcome this by eating clay minerals from river mud that counteracts the toxicity. *(Kevin Tate)*

have taken both color and form to an extraordinary degree. Sexual selection has led to the development of long plumes from the head and tail that greatly impede flight, as is also the case for peacocks, but these features are evidently of value in winning the attentions of females. In Papua New Guinea there were virtually no predators for birds of paradise in the forest, with the possible exception of a species of tree python. Without predators, there was no evolutionary check on the development of exotic plumage, and this is precisely what has taken place. Humans have long valued the decorative symbolism of bird feathers, and the people of New Guinea have harvested birds of paradise for ritual dress. The arrival of European civilization brought with it cats and rats, in addition to an excessive demand for decorative feathers in the millinery trade, so most species of bird of paradise have suffered serious decline.

Macaws are among the biggest parrots; at the other end of the size scale are the sparrow-sized parakeets belonging to the genera *Bolborhynchus* and *Forpus,* found in the montane forests of the Andes. The smallest of all parrots are the pygmy parrots (*Micropsitta* species), found in New Guinea, and are only four inches (10 cm) long. These little parrots are the only birds known to feed upon fungi, scraping it from decaying wood with their short bills. It is likely that they eat termites at the same time.

Although many of the parrots have spectacular plumage, they are not alone. The male peacock (*Pavo cristatus*), which lives in the dry tropical forests of India, has a remarkable tail that it erects during mating displays (see illustration [right]). Perhaps the most impressive of all tropical forest birds are the birds of paradise (family Paradisaeidae). These birds have been a source of awe for tropical explorers since their fist discovery by those who returned from Ferdinand Magellan's (1480–1521) circumnavigation of the world in 1522. They brought back some skins of birds of paradise that they had collected from Papua New Guinea, causing a great stir in the Spanish court. Many felt that the remarkable plumes associated with these birds could not possibly be natural, and there were undoubtedly many fake plumes on the market at that time. More confusion arose because the skins returned to Europe had no legs, as these were removed when the birds were killed. This led to the supposition that the birds of paradise spend their entire life in the air, mating in flight, with the male carrying eggs on its back. Although the birds of paradise are a remarkable group of birds, they are not quite this remarkable.

Bright colors are not necessarily a problem in the tropical forests because they can merge with the light and shade of the canopy trees. The birds of paradise, however,

One of the most spectacular birds in the world, the peacock, which is closely related to the pheasant, originates from India, where it lives in open deciduous monsoon forest and scrublands. The display of the male bird takes place in open clearings. *(Birmingham Zoo Press Room)*

The birds of paradise are mainly fruit eaters, but they have very strong claws with which they can scratch at bark and dead wood to supplement their diet with insects. They build a nest of leaves high in the canopy and the female is responsible for all the incubation. Females are relatively dull birds compared with the males, which bear a remarkable variety of plumage bearing red, yellow, and bright blue colors. These colors are used to best effect during displays, when the male also performs acrobatics to win the favors of a female. The male of the Prince Rudoph's blue bird of paradise (*Paradisaea rudolphi*) even hangs upside down from a branch to display its wings and tail. It then performs a series of regular swinging motions to reveal the deep blue of the tail, using the elongated plumes to maintain its balance.

There are more than 40 species of bird of paradise in New Guinea, and although they differ considerably in their plumage, they also interbreed. Clearly, the females are attracted to male displays in general rather than those confined to their own species, which suggests that the group is still in the process of active evolution and speciation. Species replace one another along elevation gradients in the montane forests of the region, so it is likely that geographical separation is the key to the genetic isolation that permits the development of new species.

The New World equivalent to the birds of paradise is the quetzal of Central America. The quetzal (*Pharomachrus mocinno*) is a magnificent bird with an emerald green head and back, and green tail feathers that may reach a length of two feet (60 cm). The underparts are bright red, and the wings are black and white. Like the birds of paradise, the feathers of the quetzal have long been valued by local people for decoration and the bird has been revered as an incarnation of the god of the air. The ancient Aztecs and Mayas used quetzal tail feathers in their ceremonies, and sculptures carved by these people often display their god Quetzalcoatl in the form of a serpent with quetzal feathers. The bird is still used as the national symbol of Guatemala and the name is used for their currency.

Trogons (family Trogonidae), the family of birds to which the quetzal belongs, is distributed throughout the tropical forests of the world. Many are brightly colored and they generally occupy the canopy and lower layers of the forest, feeding mainly upon fruits, but also taking insects and tree frogs. Like the parrots and woodpeckers, they have an unusual arrangement of toes, with two pointing forward and two back, which enables them to clamber though the network of branches in which they live. They breed in holes in trees, enlarging these but not building an elaborate lining. The gartered trogon (*Trogon violaceus*) uses the nests of wasps, first consuming the resident wasps and grubs. When the males incubate eggs, their long tails can prove an encumbrance, thus they bend their tails over their heads so that the tip pokes out of the entrance hole.

Many fruits fall to the forest floor, where there are many insects that readily consume them, but some birds have also adopted this niche, foraging around for the freshly fallen produce of the canopy. The curassows (family Cracidae) are large birds, up to the size of a turkey, that spend most of their time on the forest floor or in the lower parts of the canopy. Although they are generally dull colored, some have brightly colored faces and helmetlike structures that can give them a bizarre appearance. The male curassows have a distinctive song, ranging from booming noises to high whistles, produced with the aid of an elongated, looped windpipe resembling the structure of a bagpipe. Females lack this modification of the trachea. Curassows are able to fly but are rather heavy, like a grouse. Nevertheless, they are able to fly rapidly even in dense forest, navigating their way expertly between trees. Often, however, they prefer to run along the ground and generally seek out habitats with dense ground cover, where they can hide from predators. They nest close to the ground but not on it, building a nest of dead sticks low in the scrub canopy. They lay two or three eggs, and once hatched the young develop quickly, being able to fly within a few days. This is often the case with birds that nest close to or on the ground; the young are said to be *nidifugous*, meaning that they leave the nest very early.

The tinamous (family Tinamidae) of Central and South America are similar to the curassows in their way of life, spending much time on the ground, but they are weak flyers and are less able to cope on the wing in dense forest. They are closely related to the ostriches and emus (the ratites) and ornithologists regard them as very primitive birds from an evolutionary point of view. Like the ostrich, they have three strong, forward-directed toes, while the hind toe is small or absent, an adaptation for rapid running along the ground. They nest on the ground in a communal fashion, several females often using one nest and laying up to 12 eggs. They are often polygamous and some species are polyandrous (where a female may have more than one male partner). The young birds are strongly nidifugous, often leaving the nest on their first day after hatching.

The bowerbirds (family Ptilonorhynchidae) of New Guinea and northern Australia are closely related to the birds of paradise. Although some species are brightly colored, especially the males, many are much duller than their very splendid relatives, and they spend more time on the forest floor, especially in the breeding season. The males of the various species have a variety of methods for attracting the female, though most resort to some form of construction on the forest floor. The male of the Archbold's bowerbird (*Archboldia papuensis*) of New Guinea is 15 inches (38 cm) long, one of the largest of the bowerbirds. This species clears an area of forest floor and then decorates it with leaves, snail shells, and other objects to impress the females. The vogelkop gardener bowerbird (*Amblyornis inornatus*), also from New Guinea, is smaller; it goes even further in its construction activities, building a mossy stage ringed with sticks and covered by a roof. Again, the construction is decorated and the male sings to bring females to the bower for mating. Once mating has taken place, the male loses all interest in the female, which has to construct a nest, incubate, and bring up the young on her own.

Like most ground-dwelling birds of the forest, the bowerbirds feed mainly on fruits, together with some insects and amphibians. Such food is constantly available on the forest floor, which is why the male bowerbirds have the time available to build their elaborate constructions.

The bowerbirds are among the most unusual in their mating behavior, as are the cotingas (family Cotingidae) of the New World Tropics and subtropics, which are also well known for their elaborate mating displays. The orange cock-of-the-rock (*Rupicola rupicola*) of the Brazilian rain forests has the most display, which is conducted in the rain forest understory, usually close to cliff and rock outcrops where they eventually build their nests. Several male birds gather in an area of ground that they have cleared in preparation for their communal display, which is called a *lek* (see sidebar [right]).

During the lek, a collection of 50 or more males perch around the clearing, often in close proximity to one another. The appearance of a female inspires a communal outbreak of displays. The males being impressive birds, with bright orange plumage contrasting with black wings and tail, surround the female, turning to display their profiles, especially the fanlike crest developed on the front of the male's head. At times the males freeze, apparently in a trance, but then leap into the air, attempting to win attention by shock tactics. The females often pretend to be unimpressed and may go away without coming to a decision, but they usually return and eventually copulate with the selected male. They sometimes return for further matings, but always with the same male. The females build cup-shaped nests on rock faces within the forest and often nest socially, only a few feet away from one another. As is so often the case, they are responsible for all of the work associated with incubation and bringing up the young.

The forest floor is thus a place of considerable activity for birds. Foragers include the flightless cassowaries of New Guinea (family Casuariidae), which can weigh up to 100 pounds (45 kg) and have a vicious kick to defend themselves against predators. They also include the pheasants and junglefowl, including the red junglefowl (*Gallus gallus*), which is the ancestor of the domestic chicken. This is one of the few tropical forest animals that has been brought into domestication. All of these birds take advantage of the very considerable harvest of fruits and seeds that fall to the forest floor.

Among temperate birds, the timing of breeding is generally determined by changing daylength, but, as in the case

The Lek

One of the major forces that drives natural selection is the choice of a mate. Physical hardships, obtaining sufficient food, and competition between individuals are all factors that sort out the fit from the unfit; only the best adapted and strongest animals are able to survive and available to pass on their genes to the next generation. In order to do so, however, they still need to attract a mate. For this purpose, impressive appearance and elaborate behavior have been developed by many animals, especially birds. The gender most deeply involved with spectacular plumage and courtship is usually the male, as in birds of paradise and bowerbirds. The males often court individually, frequently using calls or songs to alert females of their presence. But there is nothing quite as good as a marketplace for examining all the goods available, so it is not surprising that mating systems based on marketplace choice have developed in the bird world. When males are gathered together in a group, females are able to compare all of their various attributes and make a more informed choice about their future partner. Such an arrangement is called a lek.

When the males gather together they are also able to exercise their machismo by sparring and displaying among themselves, thus establishing a pecking order or, rather, a mating order, for only the most dominant males will succeed in mating with the females that attend the lek. Subordinate males attend the lek and may spend years competing in vain for a chance to mate; eventually they may reach a dominant position. Many fail to do so and complete their lives without establishing themselves in the mating game. Behavioral ecologists argue about whether the lek is really based on female choice or on male assertion of dominance. Whichever is true, the lek is a behavioral mechanism for ensuring that only the very best genes are passed on to the next generation. It is quite a widespread activity among birds and even some mammals, but is particularly associated with various species of grouse, some shorebirds, and especially the tropical cock-of-the-rock.

of plant flowering (see "A Time to Flower," pages 132–133), the length of day and night vary little in the Tropics. The presence of a wet and dry season is often a valuable cue that induces breeding in birds, as does the availability of appropriate food. In the case of the omnivorous bulbul

Andropadus latirostris from Gabon, for example, the female increases in weight during a period of high fruit and insect availability and then begins to build a nest and lay eggs. In years when food availability is low, the bird may not breed at all. So some birds are evidently opportunistic breeders. They may follow the regular cycles of forest productivity under normal conditions, but then react to interruptions in productivity by breeding restraint.

■ TROPICAL FOREST BIRDS—THE INSECT EATERS

Insects are one of the most abundant sources of food for forest birds, so it is not surprising to find that insect feeders, insectivores, are well represented in the forest fauna. *Insectivorous* birds have adopted a variety of techniques for catching their prey, some taking flying insects on the wing, others extracting insects from bark and wood, some scarching for leaf-feeding insects, and others picking up the numerous insects of the forest floor. The collection of insects is thus an excellent illustration of niche partitioning within a food resource, which ensures that a maximum number of species can be supported, each specializing in a particular mode of feeding.

Among the birds that capture their prey on the wing, most operate by day because they use visual means of locating and hunting flying insects. The flycatchers consist of two large groups. The Old World flycatchers (family Muscicapidae) are replaced in the New World by the tyrant flycatchers (family Tyrannidae). Both have a similar mode of hunting—sitting on a perch and waiting for an insect to fly past, when they flit into the air and snatch their prey in flight before returning to their perch. Some flycatchers are brilliantly colored, such as the black-naped monarch flycatcher (*Hypothymis azurea*) of Southeast Asia, which is bright blue, and the vermilion flycatcher (*Pyrocephalus rubinus*) of South America, which is red. The motmots (family Momotidae) of Central America have a similar way of hunting insects and may sit motionless in the canopy for long periods. Despite their bright blue, green, and yellow plumage they can be difficult to locate. Their tails are remarkable, consisting of two central feathers, the tips of which are shaped like tennis rackets.

Although the daytime is more appropriate for hunting flying insects than the night, there are some insectivorous birds that hunt by night and compete with bats for the night-flying insects, including moths and mosquitoes. The nightjars (family Caprimulgidae), potoos (family Nyctibiidae), and frogmouths (family Podargidae) are closely related groups of birds, all having large gapes and flying with open mouths to collect the aerial harvest of insects. They all have large eyes, resembling those of owls, which assist in visual hunting in twilight conditions. Because they are night active, they need to sleep during the day, and they remain motionless in the canopy, protected from predators by their cryptic coloration and by sitting vertically alongside tree trunks (potoos) or horizontally along branches (nightjars), merging with the cover of bark and lichens.

Many birds feed upon insects that they pick from the surfaces of leaves and twigs, and these foliage gleaners are often very specialized in their feeding habits. Among the tanagers of the Caribbean island of Trinidad, for example, the speckled tanager (*Tangara guttata*) takes insects from the undersides of leaves, the turquoise tanager (*T. mexicana*) harvests its prey from fine twigs and leaf stalks, and the bay-headed tanager (*T. gyrola*) hunts for insects along the main trunk and branches. The birds thus partition the insect resource among three species.

The antbirds (family Formicariidae) of South America are varied in their appearance and hunting techniques. Some feed upon ants and termites in the lower canopy of the forest or on the forest floor, whereas others take advantage of the raids by army ants (see "Invertebrates of the Forest Floor," pages 143–146) and catch the insects fleeing from the advancing army. This is a dangerous game to play, as the ants are quite capable of overwhelming and killing a bird, so the birds perch just out of reach of the ants to catch their prey, or ensure that their tails and wings are held high if they land on the ground. They keep ahead of the advancing column of ants, where panicking insects and spiders are flushed from their hiding places.

Woodpeckers are also found in tropical forests and have a similar lifestyle to those of the temperate regions. They concentrate their activity on dead and rotten wood where insects, such as beetle larvae, are present, and extract these by hammering the wood with their stout bills, flexible skull structure, and strong neck muscles, then inserting their long, penetrating tongues into the damaged wood. The toes of woodpeckers are specifically adapted to supporting the bird vertically on a tree trunk, having two toes pointing forward and two backward, unlike the usual system of three forward and one back. This adaptation, together with the stiff tail feathers, enables the bird to cling tightly to the tree bark while hammering with its head.

■ FLESH-EATING BIRDS OF THE FOREST

Some birds occupy the positions of top predators in the forest. Eagles, hawks, and owls feed largely upon other birds and forest mammals, and the vultures scavenge dead flesh, taking their place among the detritivores that mop up the energy remaining in dead animal tissues.

Many falcons (family Falconidae) are long-winged, fast-flying birds of prey that capture their prey, mainly other birds, by sheer speed. These birds are better suited to open conditions than to the dense and complex canopy of a tropical forest, although there are some falcons that have become adapted to the forest environment. The barred forest falcon (*Micrastur ruficollis*) of Central America, for example, has relatively short, broad wings and a long tail, which provide the bird with a high degree of agility in flying between tree trunks at high speed and taking its prey by surprise. It hunts by sight when there is daylight; it also has acute hearing and can detect its prey by sound when it hunts in low light conditions.

Hawks (family Accipitridae) are generally better adapted to forest conditions than falcons, having the short wings and long tails that give them formidable speed among trees and foliage. The African long-tailed hawk (*Urotriorchis macrourus*), for example, has a particularly long and flexible tail, giving it the ability to pounce with unerring accuracy on small mammals, such as squirrels, which it seizes with its powerful talons.

The largest birds of prey are the eagles, but their size makes them relatively inefficient hunters within the forest canopy and understory. They are best adapted to the upper canopy layers and the emergent trees, where they can detect their prey by soaring over the treetops. The harpy eagle (*Harpia harpyja*) of the Amazon forests is the largest of the eagles, being three feet (0.9 m) in length and weighing over 20 pounds (9 kg). It is a fearsome bird with extremely strong feet that can be as large as a human hand, armed with lethal claws that can kill a monkey on impact. It flies at speeds of up to 50 miles per hour (80 km/h) through the upper canopy, surprising and terrifying its prey. Monkeys are the main prey, as these are often exposed in the emergent trees and upper canopy, but it will also take birds, such as macaws, and snakes. Monkey-eating eagles are found in all the major tropical forest regions of the world. In Africa, the crowned eagle (*Stephanoaetus coronatus*) soars above the forest canopy, but drags its prey down to the forest floor to consume it. The Philippine monkey-eating eagle (*Pithecophaga jeffreyi*) is a rare species that has suffered population collapse as its forest habitat has been cleared. Apart from preying upon forest canopy mammals, it will also resort to stealing dogs from nearby villages.

At night the owls take over as the main aerial predators. Shelley's eagle owl (*Bubo shelleyi*) of the African jungles is two feet (0.6 m) in length and hunts mammals in the canopy under low light conditions. Most birds have eyes on the sides of their heads, which gives them good peripheral vision and allows them to detect approaching predators, but is not very efficient for binocular vision directly

The king vulture occupies the dense tropical forests of the New World. Its strong bill is adapted for tearing the flesh of dead animals, and the lack of feathers on its head and neck prevents problems of blood contamination in a position where preening would be difficult.   *(Han Hegerman)*

in front. Owls, by contrast, have flat faces and eyes that can coordinate to provide binocular vision, making them proficient hunters. The eyes are also large and have sensitive retinas for operation under low light conditions. Hearing is also very sensitive among owls, so they can locate moving prey by the sounds that they make.

Vultures are soaring birds that hunt by sight and feed mainly on dead carcasses of mammals lying on the ground. In a forest environment, this is not a very efficient way of foraging because of the dense canopy foliage, but there are some species that are found in this habitat. The king vulture (*Sarcorhamphus papa*) (see illustration) and the turkey vulture (*Cathartes aura*) are extremely unusual among birds in having an acute sense of smell. They can detect a decaying corpse even though it is out of sight below the canopy and can home in on it by scent alone. Tropical ecologists have demonstrated that turkey vultures can even find the carcass of a chicken buried in the forest soil.

Tropical forest birds have thus adapted to almost all of the different feeding patterns available within the ecosystem, from herbivore to predator and detritivores. Birds fill many different kinds of niches within the forest food web.

■ MAMMALS OF THE FOREST FLOOR

Like birds, mammals have come to occupy a wide range of roles in the tropical forest ecosystem. They have diversified in the range of food that they use and in the locations that they occupy in the stratified layers of the forest, from the forest floor to the highest levels of the canopy.

Herbivores are present on the forest floor, living mainly by gathering fallen fruits and seeds or by digging in the ground for roots and tubers. The peccaries (family Tayassuidae) are a group of mammals closely resembling pigs that are efficient rooters and grubbers in the forest soils. The collared peccary (*Pecari tajacu*) is a widespread species throughout Central and South America that weigh about 45 pounds (21 kg) and travel in bands of up to 20 individuals. The collared peccary prefers scrub habitats in canyons, but its close relative, the white-lipped peccary (*Tayassu pecari*), inhabits dense jungle, where it digs in the soil for roots. In Africa, wild pigs, such as the giant forest hog (*Hylochoerus meinertzhageni*), occupies the same ecological niche as the New World peccaries. It is a much larger animal, however,

weighing up to 600 pounds (275 kg), but it is also a sociable beast, living in herds of up to 20 individuals.

Tapirs (family Tapiridae) consist of just four species worldwide, three of which live in the New World. The fourth species, the Malay tapir (*Tapirus indicus*), is the largest, weighing up to 800 pounds (360 kg), and lives in the swamp forests of Southeast Asia. Unlike the peccaries and pigs, it is a solitary animal that wanders through the forest, feeding on plant materials and avoiding predation by tigers. The South American species of tapir are smaller than their Asian counterpart but are also important herbivores of the tropical forests. They are fond of fruits and play an important part in dispersing some tree species, particularly palms.

Other mammals, such as armadillos (see illustration below), also root in forest soils for their food but concentrate on the invertebrates they find there. The giant armadillo (*Priodontes giganteus*) and the giant anteater (*Myrmecophaga tridactyla*) are both inhabitants of the Amazon forests and are extremely efficient diggers. Both have powerful forelimbs equipped with strong claws and are able to rip open termite nests to consume their inhabitants. The giant anteater is so strong and well armed that it is prepared to confront a jaguar when the need arises. It has no teeth, but its long, pointed snout contains a sticky tongue that assists it in extracting termites from their nest. Besides these large insectivorous mammals there are also many smaller animals roaming the forest floor, including the moon rat (*Echinosorex gynmurus*) of Southeast

The armadillo feeds upon invertebrates on the forest floor. Its scaly surface protects it from predators.　*(Terry J. Alcorn)*

Ecological Equivalents

Many groups of organisms have evolved in isolation, confined to particular continental landmasses or islands, and have developed independently of the inhabitants of other regions with similar climates. This is particularly true of regions that have long been separated from one another as a result of continental drift (see "Tectonic History of the Tropics," pages 38–41). Under these conditions, unrelated animals have often developed similar structures, physiology, and behavior patterns as they have taken up similar ecological roles in their particular regions. The armadillos of the New World, for example, are not found in the Old World, but their ecological niche is filled by the pangolins, which are very similar in general appearance and have similar patterns of feeding and behavior. The forests of South America contain a tiny group of deer, called brocket deer (genus *Mazama*), which are not found in the Old World Tropics but are replaced by another small group of deer, the duikers (genus *Cephalophus*), all less than 31 inches (80 cm) in height. The capybara, a forest rodent of South America, is replaced by the pygmy hippopotamus (*Choeropsis liberiensis*) in West African forests, where it has a semi-aquatic lifestyle similar to that of the capypara. Although the toucans of South America are not found in the Old World, the hornbills occupy a very similar ecological niche there, having massive bills for feeding on fruits and dealing with large seeds.

The existence of ecological equivalents in different parts of the world demonstrates the process of convergent evolution. Totally unrelated groups of organisms, when placed under similar environmental pressures and exposed to similar feeding opportunities, tend to develop similar body structures and behavior patterns. Very different animals can thus come to resemble one another physically as they adapt to equivalent ecological niches in different parts of the world.

Asia, which is regarded as one of the most evil smelling of all mammals, and the tenrec (*Tenrec ecuadatus*) of the Malagasy Republic. These two mammals also have elongated snouts that they use to hunt for insects in the forest litter.

Tropical forests provide many excellent examples of ecological equivalents (see sidebar above), where unrelated animals adopt a very similar role on different continents. The armadillos of the New World are replaced by pangolins (family Manidae), sometimes called scaly anteaters, in the Old World. Like armadillos, these are armored with hard plates to protect them from predators, have sturdy forelimbs with strong claws, and can dig in the hard soils for their food. Unlike armadillos, however, many species are able to climb trees. Their favorite food consists of ants and termites, and a single animal may collect up to seven ounces (200 gm) of these creatures in one night of hunting. Like anteaters, pangolins have no teeth but collect their invertebrate prey using a long sticky tongue. The stomach is lined with horny toothlike structures, which serve to break up the ants and termites and aid digestion. Some pangolins have been observed swallowing small stones, which may also aid in the grinding of food in the stomach.

In a mature forest there is relatively little leaf material available for grazing animals confined to the forest floor, but clearings and areas of forest regenerating from catastrophes such as wind damage or flood may contain dense understory growth, and some grazing and browsing mammals can survive here. The okapi (*Okapia johnstoni*) is a West African herbivore closely related to the giraffe but looks more like a zebra with its black-and-white striped hindquarters. It has a very long tongue for tearing off the leaves from low branches.

The gorilla (*Gorilla gorilla*) is the largest of the great apes (see illustration below) and is found in two main populations in equatorial Africa—in the lowlands of the west

Gorillas are among the most charismatic yet endangered mammals. Hunting them for bush meat and as trophies, together with disturbance by ecotourists, is placing this peaceable animal under threat.   *(Courtesy Kevin Tate)*

The mandrill is a tropical African monkey that feeds mainly on fruits, seeds, fungi, and roots. Only the male has the elaborate face pattern shown here. *(Birmingham Zoo Press Room)*

dispersal for the plants. The mandrill and chimpanzee are confined to the African tropical forests; in the New World the coatis (*Nasua* species) has taken up the role of forest floor insectivore and fruit gatherer (see illustration below). They have long, mobile snouts and sharp claws for digging.

Carnivores do not have the same problem as vegetarians because animal flesh is relatively rich in protein. On the forest floor, many mammals are insectivorous and some seek larger prey. Cats are among the most efficient of predatory mammals, and the tropical forests are rich in feline predators. In the South American forests, the largest cat is the jaguar (*Panthera onca*), which can weigh up to 300 pounds (135 kg) and grow to a length of nine feet (2.7 m) (see illustration on page 168). It feeds mainly on peccaries and capybaras, large wetland rodents of the Amazon forests, but is also able to take tapirs, sloths, monkeys, and fish. The jaguar often buries its prey and comes back to feed on several occasions, even when the flesh is in an advanced state of decay. Jaguars are accomplished swimmers and are adept at fishing. They sometimes lie on a low branch over the water where falling fruits attracts fish, and they are able to scoop fish from the water with their claws. There are reports that jaguars land the tips of their tails in the water to attract fish, but this may not be a deliberate act of luring their prey. The ocelot (*Panthera pardalis*) is smaller than the jaguar and has a beautifully marked coat, making it a prime target for hunters. It feeds mainly on small mammals, especially rodents, and birds. Tigers (*Panthera tigris*) are found in a number of habitats in Asia, including the tropical forests, and especially in the mangrove swamps of the Sundarbans. They are capable of killing mammals as large as buffalo, but in the forest they mainly hunt wild pigs. They are so strong that they

and in the mountains of the east. An adult male gorilla can reach a height of 6.5 feet (2 m) and weigh up to 660 pounds (300 kg). They live in troops of between 15 and 40 and live quiet and leisurely lives, consuming herbage and fruits in the forest understory. Gorillas can climb and may sleep in low trees, where they are less likely to be surprised by a nocturnal attack by a leopard, their only predator. They are very selective in their vegetarian diet, picking only the most nutritious shoots, usually the plant tissues with the highest proportion of nitrogen to carbon. Plant tissues are relatively poor in proteins that account for the bulk of the nitrogen content, and a herbivore often needs to eat large quantities of vegetable matter to provide for its protein requirements.

Some other primates, including chimpanzees (*Pan troglodytes*) and mandrills (*Papio sphinx*) (see illustration above), spend considerable amounts of time on the forest floor. Both of these primates are omnivorous animals, rooting around for grubs, fungi, and plant roots. They also gather and eat fallen fruits, carrying the seeds in their guts and discharging them far from their original source, thus providing very effective

Coatis are sociable mammals that live mainly on the forest floor, where they eat insects and fallen fruit. They are found in South and Central America. *(Bill Lucey)*

can drag a pig carcass of 200 pounds (90 kg) for more than half a mile (1 km). In the wet forests of the coastal regions they may even tackle crocodiles, but are also able to subsist on lizards, frogs, and birds.

Smaller cats are also present in the tropical forests, such as the golden cat (*Felis aurata*), the two spotted palm civet (*Nandinia binotata*) and the servaline genet (*Genetta servalina*), all of equatorial Africa. Civets and genets belong to the family Viverridae and are mainly solitary and nocturnal hunters, hiding during the day in holes in the ground or in hollow trees. They climb well, so may hunt in the canopy, where they prey upon frogs, snakes, and birds. Civets are well known for the musky secretions they produce from glands around their genitals and that have been used by humans in perfumes. The cats use the scent to mark out their territory.

Cats are not the only mammalian predators that operate close to the ground. The bush dog (*Speothos venaticus*) of the South American forests hunts in packs and preys upon larger rodents. It is a true dog of the family Canidae and has a stocky build with short legs; its height at the shoulder is

The jaguar is the top predator of the tropical forests of the New World. *(Claudia Meythaler)*

only about 12 inches (30 cm). But its diminutive size and compact shape make it ideal as a hunter in dense scrub and along river banks. They swim well and sometimes supplement their diet with fish. Males mark their territory by raising their hind leg and urinating, just like domestic dogs. Grisons are predators that belong to the stoat and weasel family Mustelidae. They are found in Central and South America and have partially webbed feet, which allow them to swim and to hunt frogs. They live in social groups in the forest and also extend into tropical grasslands.

The sloth bear (*Melursus ursinus*) is found in the rain forests of southern India and Sri Lanka and, like many bears, is omnivorous, taking carrion, birds, eggs, or reptiles, but is particularly fond of termites. It has strong forepaws armed with claws that are ideal for attacking termite nests but that also allow it to climb trees. Unlike anteaters, the sloth bear does not use its tongue to extract termites from the nest but pouts its lips into a tubular form and sucks up its prey. Two of its incisor teeth on both the upper and lower jaw are missing, which helps in creating a suction tube. When the bear is active in this way, the noise it makes is both loud and distinctive and some observers claim that it can be heard for a distance of 600 feet (180 m).

The forest floor is thus a place of considerable activity where hunter and hunted mammals live in a delicate balance. Some are confined to the ground level, but others are able to climb trees and enter another realm of the forest ecosystem.

■ CANOPY MAMMALS

Many mammals are able to exploit the conditions of the forest floor, including its rich resources of soil animals, roots, and fallen fruit. They are also able to climb trees to take advantage of the additional supplies of food, together with the relative safety from ground-dwelling predators found in the lower canopy. There are other mammals, however, that spend most of their lives in the canopy, rarely if ever descending to the forest floor. Such animals need to be light and agile to cope with an aerial life, sustained only by the flimsy branches and twigs of the taller trees, and many have highly adapted hands, paws, limbs, and tails that enable them to cling tightly to their precarious supporting structures. These animals can exploit the rich resources of flowers and fruits in the upper canopy or feed on the abundance of invertebrate life, rather than wait for food to fall to the ground.

Many species of primate, including lemurs, monkeys, and apes, are well adapted to life in the canopy. Among the most agile of these are the gibbons of Southeast Asia (genus *Hylobates*). Gibbons have extremely long limbs, with arms almost twice the length of the torso and legs 30 percent longer than the trunk. The arms are used particu-

The orangutan, seen here in Chiang Mai, Thailand, is a large primate of the lowland rain forests of Southeast Asia. Its survival is threatened by logging and habitat destruction. *(Patrick Roherty)*

larly for locomotion, enabling them to swing from branch to branch, gripping the branch with extremely long and slender hands. They rarely use their feet when swinging and leaping, thus the excessive length of their arms. A gibbon on the ground is a most unwieldy creature, with arms so long that they have to be carried above the animal's head to avoid dragging them along the ground. Although the animal is relatively small, weighing only 11–18 pounds (5–8 kg), the spread of its arms from hand tip to hand tip is seven feet (2.1 m). The gibbon's thumb is longer and more deeply cleft than a human thumb, giving it a stronger grip, and its wrist has a ball-and-socket joint, rather than the hinge joint of a human, which again provides greater flexibility when swinging. The shoulder muscles need to be especially strong to prevent dislocation of the joint as it supports the whole weight of the moving gibbon.

The orangutan (*Pongo pygmaeus*) is an exception to the general rule that canopy animals need to be small (see illustration above). The males may grow to a height of five feet (1.5 m) and weigh up to 220 pounds (100 kg), but the eight feet (2.4 m) total spread of their arms is only slightly greater than that of the gibbon. It swings with its arms like a gibbon, but is also more inclined to use its feet, gripping branches and walking along them with its prehensile toes. Orangutans are found in the forests of Borneo and Sumatra, especially in the wet bog forests of the coastal regions. They are totally vegetarian, feeding on the abundance of fruit in the forest.

Both gibbons and orangutans are agile and fast-moving members of the canopy fauna; other mammals are content to cling to branches and conduct their lives at a more leisurely pace. The tree sloths (family Bradypodidae) have a well-deserved reputation for their lack of alacrity (see illustration on page 170). An alarmed mother sloth, hurrying toward its endangered baby, has been known to achieve a speed of 15 feet (4.6 m) per hour, but this rapid pace is exceptional. Their sedentary nature, sleeping for 18 hours a day, can even result in a coating of algae developing over their fur. Their lack of movement may well enable them to avoid the attentions of predators, such as harpy eagles and jaguars. Like all animals, sloths are warm-blooded, but their lack of activity makes it difficult for them to maintain their body temperature, so this varies with the environmental conditions and makes them very sensitive to cold. All of the tree sloths have strong, three-inch (7.5 cm) claws used for gripping on to branches and trunks, which the sloth can climb using a hand-over-hand technique. Some species have three toes, while others have only two. When resting, which occupies much of their time, sloths often hang upside down from branches. The tree sloths spend most of their time in the canopy but will descend approximately once a week to defecate, digging a small latrine with

A three-toed sloth hangs upside down in the rain forest of Costa Rica. *(Keith Sirois)*

their stumpy tails. Their food consists mainly of foliage, some species being very specialized in the type of vegetation than they find acceptable.

Sloths, gibbons, and orangutans all have very short tails, a feature that is perhaps surprising because many canopy mammals rely on their tails as a fifth limb for gripping and balancing (see sidebar below).

The squirrel monkey (*Saimiri sciureus*) (see illustration on page 171) is a sociable species that prefers clearings in the forest, or riverside forests and mangroves. Groups of up to 30 individuals live in polygamous harmony and spend their time gathering fruits and catching insects in the canopy.

Agility and a firm grip are obvious prerequisites for an animal inhabiting the high layers of the forest canopy, and most of the canopy mammals possess these attributes. Jumping from tree to tree is often necessary, both to reach new food resources and to escape treetop predators, and several groups of mammals have developed the ability to glide across openings in the canopy to reach new resources and flee from enemies. These range from the tiny pigmy gliders, only three inches (7.5 cm) in length, to the giant flying squirrel, which has a body length of 23 inches (58 cm) and a tail that projects an additional 25 inches (64 cm). These large flying squirrels can glide up to 1,500 feet (450 m) through the forest canopy.

Some mammals are capable of true flight, with forelimbs that have developed into true wings. The bats play an important role in the forest canopy fauna. There are

Tails

Limbs are vital appendages for mammals living in the tropical forest canopy, and the possession of an extra limb in the form of a tail can provide many advantages. Most monkeys, arboreal anteaters, kinkajous, and cats all have long, prehensile tails. The tail is an elongation of the spine beyond the trunk of the body and consists of a series of vertebrae, just like the backbone. But the vertebrae of the tail are much more flexible in their junctions, so that the organ is able to bend much more freely than the backbone contained within the torso. The tail is well supplied with muscles, especially at the base, where it joins the body. In monkeys and many other canopy mammals, a tail can be wrapped around a branch and provide an additional grip or support, or it can be held as a balancing organ, much as a tightrope walker might hold a pole. The flexible tail of the spider monkey (*Ateles geoffroyi*) has bare skin on its underside, providing additional grip, and is patterned with grooves rather like the palm of a hand. The tail of the

spider monkey is so strong that it can support the entire weight of the animal, leaving its arms and hands to gather the fruits that it eats. Tails can also act as parachutes and rudders for mammals that leap from tree to tree, as in the case of the black and white colobus monkey (*Colobus polykomos*), which has a richly hairy tail that helps direct its leaps as it sails through the air. Some flying squirrels, which actually glide from tree to tree rather than fly, also have large, bushy tails that are used as rudders. One such squirrel is the Southeast Asian giant flying squirrel (*Petaurista petaurista*), which is the size of a cat and is the largest of the flying squirrels. Other flying squirrel species have a membrane that connects their hind limbs to the tail and acts as a parachute. In the pygmy glider (*Acrobates pulchellus*) of New Guinea, the tail has laterally projecting stiff hairs that give it the appearance of a feather. This plays an important part in directing the course of a glide by adjusting the angle of the hairs.

There are also predatory mammals that inhabit the trees, one of the largest being the clouded leopard (*Neofelis nebulosa*) of Southeast Asia (see illustration on page 172). Despite its considerable size, attaining a length of up to six feet (2 m), it is a highly adept climber and is able to catch and kill monkeys in the canopy. The clouded leopard is able to run down the trunk of a tree headfirst, and is even known to swing from a branch by one paw before dropping on its prey.

Mammals have diversified within the tropical forests to take advantage of all the different levels of its complex structure and all the various food resources available. Mammals are found at ground level, digging beneath the soil, within the lower strata of the forest, in the canopies of the highest trees, and even in the air above and between the trees. Mammals, along with all the other major groups of the Earth's animals, have evolved to fill every possible niche within the forest and contribute to its overwhelming biodiversity.

Squirrel monkeys live in large groups, mainly in forest clearings and at the edges of forests along river banks. Insects form a major part of their diet. *(Kevin Tate)*

Fruit bats, as their name implies, are fruit-feeding mammals of the upper canopy of rain forests and play an important part in seed dispersal. *(Echo Lake Aquarium and Science Center)*

many different kinds of bats and they differ considerably in their diet. Among the largest and most apparent of the bats are the fruit bats (family Pteropidae), which can easily be observed hanging upside down in the canopy as they roost during daylight hours (see illustration [right]). The largest of the fruit bats is *Pteropus gigantus,* which can weight up to 2.2 pounds (1 kg) and has a wingspan of up to 69 inches (175 cm). At dusk the bats awake and set off in search of fruit. The flying foxes (genus *Rousettus*), so named because their heads have a foxlike appearance, have been recorded flying 155 miles (250 km) to seek their food. The large fruit bats and flying foxes have well-developed eyes, unlike the small, insect-eating bats of temperate regions. Those tropical bats rely largely on sight when flying and foraging, and their retinas are extremely sensitive in low light conditions but lack color vision. The membrane of the bat's wing connects to the hind limbs, thus providing a large surface area for flight, but they have no significant tail to act as a rudder.

The clouded leopard is a tree-dwelling cat that lives in the forests of Southeast Asia and preys upon monkeys, squirrels, and birds. *(Michael Lynch)*

CONCLUSIONS

Although the biomass of animal life in the tropical forests is considerably smaller than that of the plant life, its biodiversity is much higher. All of the major groups of animals are present within the forest, occupying the different levels of the complex forest structure and using all the numerous food resources available.

Invertebrates, including mites and springtails, abound on the forest floor, taking on the role of detritivores, extracting energy from decaying plant tissues that constantly fall from the canopy as litter. Millipedes, worms, slugs, and snails also consume plant litter and reduce it to a comminuted mass as it passes through their guts, thus becoming more accessible to the bacteria and fungi that complete the decomposition process. Invertebrate predators, including spiders and scorpions, prey upon the detritivores, adding complexity to the food webs of the forest floor. Termites are important invertebrates in the forest because they are responsible for wood decomposition, liberating the chemical components of this durable material back into the environment for reuse. They are not able to digest the wood themselves but instead use the services of fungi to break down complex chemicals, such as lignin and cellulose, making its energy available to their colonies. The architecture of termite mounds adds diversity to the landscape of the forest floor.

Invertebrates also abound in the forest canopy, with many butterfly and moth caterpillars feeding directly on the foliage. Many plants produce chemical deterrents to prevent such grazing, but certain caterpillars have adapted to these challenges by developing enzyme systems to deal with toxins, or even storing them within their tissues so that they can in turn deter their own predators. Much of the forest diversity is due to this specialization of grazers for certain food plants, effectively partitioning the resources among different species. The complexity of species is increased by the development of warning coloration among the protected insects and of mimicry adopted by other species, some toxic (Müllerian mimicry) and some nontoxic (Batesian mimicry). Other insects adopt cryptic colorations and patterns that serve as camouflage against the diverse background of bark, lichens, and leaves.

Colonial insects, especially bees and wasps, are present in the canopy, and these display remarkable self-sacrifice in defense of the colony. Such behavior is not true altruism, however, because their fellow members of the colony are closely related, so lives may be lost but genes are preserved. Leaf-cutting ants are another group of colonial insects found in the canopy. Although they gather their vegetable food there, they nest on the ground and, like termites, cultivate fungal gardens in which the leaves are decomposed.

Some forests, such as those of the lowland Amazon, are frequently flooded, so fish there play an important part in the forest food web. Many are seed eaters and can be important in the dispersal of seeds.

Amphibians, especially frogs, are also abundant members of the forests and occupy a very wide range of microhabitats. Each has specific requirements for breeding, making it another example of the partitioning of resources among species. Some frogs spend their entire life in the canopy, even breeding there, and some protect themselves from predation by accumulating toxic compounds in their bodies. The dart-poison frog of the Amazon is one of the best known of these. Skinks, salamanders, and geckos are amphibians that are also found in the forest canopy, where they avoid falling by use of a variety of techniques, including claws and suckers. Snakes are predatory reptiles and often adopt camouflage to wait in the canopy and ambush their prey. The chameleon is able to change its pigmentation to match its background as it awaits the arrival of an unsuspecting insect. Snakes need to disable their prey before swallowing them. They may use constriction and suffocation, or they may inject toxins through their hollow fangs to suppress the struggles of their prey.

Relatively few birds of the tropical forest canopies feed on leaves, but one that does is the hoatzin of South America. Many more birds feed on nectar, such as the hummingbirds and sunbirds, or on fruits and seeds, such as toucans, hornbills, and many parrots. Many canopy birds are brightly colored, including some parrots such as macaws, and the Southeast Asian birds of paradise. Although the coloration is usually associated with mate selection, it is possible to be brightly colored in the rain forest canopy without necessarily being conspicuous to predators. Colored flowers and the patterns of light and shade can make even the brightest plumage difficult to detect. Birds of the forest floor, such as curassows and tinamous, are usually more drab in their plumage, and they feed on fallen fruit or the invertebrates that live in the soil.

Mating activities among tropical forest birds are often complex, one of the most elaborate ones being the lek, in which a gathering of males competes for the attentions of the visiting females. Only the most dominant of males thus succeed in propagating their genes. Bowerbirds are less social in their mating habits, the individual males creating aesthetically attractive constructions to win the favors of females.

Insect-eating birds of the forest adopt a variety of techniques for catching their prey. Some, such as the flycatchers, sit on a perch and make excursions into the air to capture a flying insect, while others forage among the leaves and twigs of the canopy to obtain their less mobile prey. The nightjars take over from the day birds as twilight falls, and harvest insects from the air by flying with open mouths, trawling for any small organism in the atmosphere. One of the most sophisticated insect-hunting techniques is employed by the ant bird, which uses the advancing column of army ants in the jungle as a means of flushing alarmed insects from their cover.

Some birds occupy the position of top predator in the upper layers of the forest, such as the harpy eagle and the Philippine monkey-eating eagle. Eagle owls take over in the night as canopy predators. There are few vultures in the tropical forests because most of these use sight to locate corpses, and the forest canopy interferes with this. But some are equipped with an acute sense of smell, enabling them to locate hidden corpses beneath the canopy.

Mammals occupy all of the different layers of the tropical forest. The ground level supports peccaries, hogs, tapirs, and armadillos. Small deer and the okapi are also found at ground level, and each of these animals has equivalent forms, often from very different animal groups, in the forests of other continents. These are known as ecological equivalents. The armadillo, for example, is replaced in Africa by the pangolins, which have a very similar appearance and way of life. Such equivalence is a consequence of convergent evolution.

Some primates, including the gorilla, are essentially ground-dwelling animals, although they will climb into the canopy of the understory. Some large mammalian predators, including large cats such as the jaguar, are essentially ground based, though they, too, will take to the canopy when they need to do so. Smaller cat species, together with bush dogs, hunt on the ground or in the lower canopy. Agility in climbing among forest mammals is thus a valuable asset because it enables an animal to avoid many predators.

Canopy mammals include the monkeys, the gibbon being one of the most highly adapted species for canopy life, with its extremely long arms and very strong hands for swinging from branch to branch. Although orangutans are larger and less agile, they also spend much time in the forest canopy. Neither of these mammals has a long tail, as many of the monkeys do; this additional limb assists in balance and in holding onto branches when the animal is moving through the canopy. Some monkeys, such as spider monkeys, have tails that are strong enough to support the entire weight of the animal, leaving the hands free to gather food. Flying squirrels use their tails as rudders to direct their glides through gaps in the forest.

Bats are mammals capable of true flight, and these play an important role as fruit and insect feeders within the forest. They are often social; fruit bats can form large colonies and fly long distances to obtain their food.

Animal life, therefore, has diversified greatly in the tropical forests, making them the most diverse habitats on Earth. The degree of specialization found among the animals here is unrivaled in any other of the world's biomes, the food resources being tapped in every conceivable manner. The very high primary productivity, the complexity of the forest architecture, and the long history of evolutionary development have all contributed to the biodiversity of the tropical forests and have led to the development of an intricate web of relationships between plants and animals. Ecologists still argue about whether such intricacy and complexity has led to greater stability in the forest, or whether this highly diverse biome is more fragile as a result of the varied interrelationships found between its many members.

8

Ancient History of Tropical Forests

The early 21st century is a time of rapid global climate change, so it should come as no surprise to know that the Earth's climate has changed very considerably during the 4.6 billion years of its existence. The biology of the Earth has responded to these climatic changes, with new forms arising to take advantage of the changing conditions, and the distribution patterns of living things constantly altering in response to the dynamics of the climate. The study of the ecology of past communities is termed *paleoecology*.

■ CLIMATIC AND BIOLOGICAL HISTORY OF THE EARTH

The early Earth was lifeless. Traces of structures in rocks that may well be the *fossil* remains of primitive living organisms are present after about 3.8 billion years ago, and life was certainly present from 2.5 billion years ago, but until 600 million years ago, all living things were microscopic and aquatic. The conquest of the land by primitive plants did not take place until the Silurian period, 440 million years ago, and more complex terrestrial ecosystems involving invertebrate animals were in place during the following Devonian period, from about 410 million years ago. Plants at this time were low in stature, so the architectural complexity and structure of such ecosystems was simple. By the Carboniferous period, from 360 million years ago, however, plants related to modern horsetails and ferns had evolved into treelike forms and assembled into swamp forests in the hot and wet regions of the world. These wetlands can be regarded as the first tropical forests (see sidebar "The First Tropical Rain Forest," page 176) and were a source of large geological deposits of peat that eventually became consolidated as *coal*.

During this long period of Earth's history, over four billion years, the climate was not static. Geologists have found rocks that indicate the presence of ice. When ice moves over the ground it crushes and grinds rocks into fine clays, and

these become mixed with larger rock detritus that falls onto the surface of glaciers, resulting in a mixture of materials sometimes called boulder clay, or till. Over the course of millions of years, this material becomes effectively fossilized, being crushed and compacted as other rocks are laid down on top of it, and such ancient tills are given the name tillite. Geologists have discovered tillites at various stages in Earth's history, suggesting that there have been specific times when this planet has experienced cold conditions and has developed ice caps. The following table shows the occasions when these ice ages have occurred on Earth.

Looking at the timing of ice ages, particularly that of the more recent ones, which are better recorded in the geological strata and better dated, there is a distinct periodicity. There is a suggestion of a cycle in which cold periods alternate with warmer ones. The Earth seems to experience ice ages roughly every 150 million years, but as yet, no one has come up with a fully acceptable explanation of the mechanism that causes this long-term regularity. There are many theories that have been put forward to explain the Earth's occasional development of glaciers and ice caps, and these ideas are extremely varied in nature. The Earth's crust consists of a series of plates that are constantly on the move, sometimes separating the landmasses and sometimes leading to collisions. It is possible that the configuration and spacing of the plates is involved in determining ice ages. Perhaps there have to be large landmasses in the right position (which is not necessarily at the poles) for ice to begin forming. The appropriate location would have to be cold (therefore close to the poles) but also where precipitation is high, so that abundant snowfall can form ice (thus some distance away from the poles, which are virtually deserts). Ocean currents also play an important part, as they help to distribute heat energy (ultimately derived from the Sun) over the surface of the planet, and any disruption to the circulation of these currents can lead to an ice age. The right positioning of continents could do precisely this, modifying the direction in which warm waters flow. When mountain ranges arise, they lead to changes in precipitation patterns

The Occurrence of Ice Ages in the History of the Earth

ERA	GEOLOGICAL PERIOD	APPROXIMATE AGE
Proterozoic (= Precambrian)		2,300 million years
		900 million years
		750 million years
		600 million years
Paleozoic	Ordovician	450 million years
	Carboniferous/Permian	300 million years
Mesozoic	Jurassic	150 million years
Cenozoic	Tertiary/Quaternary	10 million–10,000 years ago

Here the term *ice age* is used to indicate times when ice sheets have been present on Earth.

and create high landforms where glaciers can form. There is always the possibility that astronomical factors are involved, such as the distance between the Earth and the Sun, or the angle of the Earth on its axis. Further geological research may help geologists to sort out exactly what causes the Earth to enter an ice age at certain points in its history.

More is known about the current ice age than any other, simply because the records have been better preserved than for any previous cold episode. The ice age that the Earth is currently experiencing has lasted around 10 million years, beginning in the Miocene epoch, becoming more evident in the Pliocene, and then really making its mark on the Earth during the last 1.75 million years of the Pleistocene. There is no reason to believe that the Earth has fully emerged from the recent ice age, despite current fears of global warming. Ice sheets and tundra are still present on Earth, and there remains the possibility that they may one day expand once again; indeed, the past pattern suggests that this may well eventually come to pass. During the past 10,000 years (a mere blink of the eye in geological terms), conditions have been warm and ice caps and glaciers have retreated far from their former positions. There have been many warm interruptions to the general cold of the last two million years, and there is no reason to believe that the present warm period is anything more that another such interruption. The cyclic pattern of recent history, in which cold and warm episodes have alternated, suggests that the cold will return.

Geologists divide the history of the Earth into convenient sections, trying to use natural breaks wherever possible. These are points at which the fossil content of rocks changes relatively abruptly, often indicating that there is a sudden change in the climate or other in conditions on Earth. The separation of the Cretaceous from the Tertiary periods some 65 million years ago is a good example of a natural break, because the dinosaur fossils disappear and other groups of organisms exhibit considerable change at about that point in time. This boundary is now believed to have been caused by a catastrophic cosmic event—the collision of a massive meteor, or bolide, from space onto the surface of the Earth. The kind of abrupt and cataclysmic environmental change that resulted from this catastrophe, however, is relatively rare, and the divisions of geological time are usually far less abrupt and less easy to identify. This is certainly the case with the progression of the current ice age, involving the formation of permanent ice caps and the development of glaciers in various parts of the world. Although ice began to appear back in the Miocene, the first evidence for extensive glaciation occurs in the Late Pliocene, especially after about 2.4 million years ago, and cold conditions became even more acute around 1.75 million years ago, which is generally regarded as the beginning of the Quaternary period and the Pleistocene epoch within that period. But there is still much argument about this, and some geologists feel that the change at 2.4 million years would be a better marker for the beginning of the Pleistocene. This argument illustrates that the last few million years of Earth's history have seen gradual and progressive cooling overall and that any subdivision is somewhat artificial and arbitrary.

The development of ice sheets on the Earth is indicative of general global climatic cooling, so all the biomes are likely to have been affected in some way. Although tropical forests are farthest from the polar ice caps, global cooling is nevertheless likely to have affected this biome.

■ GEOLOGICAL HISTORY OF TROPICAL FORESTS

Ice ages are relatively rare and widely spaced in the history of the Earth. The first of these cold episodes to occur at a time when a kind of tropical forest existed on Earth was in late Carboniferous times. At this time, some 300 million years ago, many of the present-day continents were all joined together in massive supercontinents. North America and Europe lay together, forming the supercontinent of Laurasia. They had recently become separated from another larger landmass to their south. South America, Africa, and Antarctica, India, and Australia were all joined and formed this southern landmass, called Gondwana, which lay over the South Pole. Perhaps this coincidence of a huge landmass over a polar position accounts for the development of an ice age. An ice sheet formed over what is now Antarctica and spread out over South America, Africa, India, and Australia, all of which were much farther south than they are at present. What are now the continents of North America and Europe lay over the equator; their climate, therefore, was a hot tropical one, and the extensive swamp forests that formed at this time led to the formation of the great deposits of coal buried within these continents. This was the first tropical rain forest (see sidebar below). The fossil trunks of the component trees show no signs of annual growth rings, suggesting that the climate lacked strong seasonal variation, just like the present equatorial regions. There is no evidence of monsoon climates with

dry and wet seasons at that time in the interior regions of the continent, where such conditions might be expected.

During the Devonian period that immediately preceded the Carboniferous period, the Earth experienced "greenhouse" conditions. The level of carbon dioxide in the atmosphere was about 18 times that of the present day. Since carbon dioxide allows short-wave radiation from the Sun to pass through to the Earth's surface but absorbs the long-wave, infrared radiation generated by land surfaces (see "The Forest as a Carbon Sink," pages 209–211), the outcome is that temperature rises and the Earth becomes a hothouse. It was precisely these conditions that led to the development of the Carboniferous rain forests in Laurasia. But the accumulation of large quantities of peat beneath the swamp forests, combined with the generation of masses of calcium carbonate in the shallow, warm oceans, acted as a sink for atmospheric carbon dioxide and led to a sudden drop in its concentration, down to a level similar to the present day. It may well be the case that this removal of carbon from the atmosphere contributed to the climatic changes resulting in the late Carboniferous/Permian ice age.

The First Tropical Rain Forest

The hot, rainy conditions of the equatorial regions during Carboniferous times (360–300 million years ago) led to the development of the Earth's first tropical rain forest. The first trees evolved toward the end of the Devonian period, and they had diversified considerably by the Early Carboniferous, sometimes called the Mississippian epoch. None of the trees found at that time are still found on Earth, but the general structure of the forest was surprisingly similar to that of modern tropical forests. The canopy was dominated by tall *lycopsid* trees, up to 130 feet (40 m) in height, consisting of genera such as *Lepidodendron, Sigillaria,* and *Cordaites,* all closely related to modern horsetails (such as *Equisetum*) and club mosses (such as *Lycopodium*). Elevated branching systems effectively increased the volume of the inhabitable ecosystem and its spatial complexity, so just as in modern tropical forests, there arose a great potential for the expansion of biodiversity. A branched tree has an enormous surface area that can be occupied by epiphytic plants and canopy animals. A moderately sized beech tree may have an above-ground surface area of 117,000 square feet (11,000 m²), all of which can be colonized by small organisms, creating a very complex ecosystem. The trees of the Carboniferous swamps generally had shallow rooting systems, just like those of

modern swamps, presumably to ensure that they were occasionally aerated and able to respire in the waterlogged conditions. Ferns were present in these forests, indicating the existence of a layered structure. Tree ferns and seed ferns (Pteridosperms) grew to a height of about 20–30 feet (8–10 m), so they would have formed a canopy stratum below the tall trees, and small ferns, mosses, and liverworts would have covered the surface of the ground.

As in modern wet tropical forests, amphibians abounded, along with insects related to dragonflies and mayflies. The largest of the amphibians, *Eryops,* was up to six feet (2 m) in length and had a large head with powerful jaws. It was a predator, probably mainly of fish, which occupied an ecological niche similar to that of modern crocodiles. There were no mammals or birds, of course, and egg-laying reptiles first evolved in drier conditions, where egg shells avoided the need for aquatic breeding practices. So there were no large aerial predators apart from insects like giant dragonflies. The food webs of the Carboniferous forests, therefore, were largely fueled by detritus such as spores falling from the canopy and the fungi that invaded dead plant tissues. Much of the insect life thus consisted of detritivores, and predatory animals fed on them.

The Permian period, starting around 300 million years ago, was a time of mass extinctions, which may have resulted from the great climatic changes of that time. The southern landmass of Gondwana had moved north and collided with Laurasia, creating a single landmass of Pangea and causing the uplift of the Appalachian Mountains in the process. The tropical forests gave way to vegetation that reflected the cooler, drier conditions. The Gondwana regions became occupied by woody vegetation dominated by a seed-forming shrub called *Glossopteris,* fossils of which are found throughout modern Antarctica and at the tips of the southern continents. This plant grew as close as 5° latitude to the South Pole of that time and shows evidence for seasonal growth in its wood, which is not surprising because such southerly latitudes would have been in complete Antarctic darkness for much of the year. Farther north, the equatorial forests contracted as the climate became cooler in late Carboniferous times, becoming sandwiched between extending deserts from the south and the north. Coniferous trees, which first evolved in the Late Carboniferous (the Pennsylvanian epoch), then became dominant in the world's flora.

Closely related to the conifers are the cycads, which also rose to importance in the tropical regions in Permian times. These are palmlike plants that grow slowly, can live for over a thousand years, and, like conifers, produce pollen grains as a means of sexual reproduction (see sidebar "Pollen Analysis," page 179). The conifer pollen, containing the male gametes, was transferred from male to female reproductive structures by wind, but because pollen grains are rich sources of nutrients, insects found them to be a valuable food resource. The insects acted as detritivores, consuming the large quantities that fell to the ground, having failed to reach the female receptor. But it would not have taken these organisms long to discover that they could attack this resource right at its origin, in the male cone. Among cycads, insects were certainly attacking the reproductive structures, the seeds, and the leaves of the plants. The plants responded by erecting chemical defense systems in the seeds, accumulating poisons to protect their tissues. But they did not protect their pollen grains in this way, the likely reason being that the insects consuming the pollen were also effective in conveying some of the pollen sticking to their surfaces from one plant to another. Insect pollination thus evolved out of wind pollination and a whole new set of plant–insect partnerships and, hence, biodiversity came into existence (see sidebar "Pollination," page 133).

The vegetation of conifers, cycads, and maidenhair trees (*Ginkgo* species) reflects the drier conditions that prevailed generally throughout Permian times. The evolution of giant reptiles, such as the dinosaurs, in the following Jurassic period suggests that the vegetation of those times was less complex in structure than that of the Carboniferous, although there was undoubtedly great local variation and the coastal edges probably carried luxuriant vegetation similar to modern mangroves. Large animals, such as the gigantic *Diplodocus,* with a length of 60 feet (27 m) and weighing 18 tons (18.2 tonnes), could not have survived in dense forest, so conditions must have been relatively open through Jurassic and Cretaceous times. Indeed, the presence of large herbivores undoubtedly led to constant disturbance of the vegetation and the maintenance of extensive open areas of landscape. During the Cretaceous period, approximately 145 to 65 million years ago, flowering plants (angiosperms) and mammals evolved and diversified, and the ecosystems of the world began to move toward a more recognizable, modern form. By the later stages of the Cretaceous, the flowering plants had developed into a considerable range of shapes and size, including tall trees. They were then able to compete with the conifers and gradually replaced them as the dominant life form in the tropical forests of those times. The replacement was almost completely successful, as far as the flowering plants were concerned. Only in northern Queensland in Australia and in New Guinea do certain conifers still survive as major components of the rain forests. The giant *Agathis* trees of these regions are all that is left of the conifer-dominated tropical forests of the early Cretaceous.

Great changes took place in the history of the Earth around 65 million years ago, many of which were strongly affected by the collision of the Earth with a bolide that struck our planet close to modern Mexico. Massive atmospheric and climatic changes resulted, and the Earth that emerged into the Cenozoic era was a very different place. The dinosaurs had already been on the decline and now they were gone. The mammals that had played a minor role in the planet's ecology up to this point were now in a position to take advantage of the vacant niches available as a result of the loss of the big reptiles. And the angiosperms were set to take over the dominance of world vegetation.

Forests very similar to the modern tropical rain forests began to appear in many parts of the world during the early Cenozoic era, often referred to as the Tertiary. Several fossil rain forests have been found that are far from the equatorial regions, allowing for the fact that continental drift has transported landmasses over considerable distances in the intervening 60 million years. Geological deposits in Alaska at a present latitude of 60°N show the presence of "tropical" rain forests in the early Tertiary. Many of the tree genera present, such as *Macaranga* and *Parashorea,* still exist and are currently found in the rain forests of Malesia. Out of a total of 35 genera founds as fossils in the Alaskan deposits, 32 are currently found in the Southeast Asian forests. The vegetation was largely evergreen, and the presence of drip tips on the leaves testifies to the similarity of conditions to those of the present equatorial rain forests. These forests were lost and replaced by deciduous forests as the climate of Alaska cooled over the following 20 million years.

At about the same time, the estuary of what is now called the Thames River, below London, England, at latitude 51°N, also bore rain forest vegetation. Just as in the Alaskan forest, the London rain forest contained many tree genera that are now characteristic of Southeast Asia. There were extensive mangrove swamps and an abundance of a palm belonging to the genus *Nypa,* now found in Malesia. Animal remains consist largely of mollusk shells and the teeth of sharks but also indicate warm conditions. On the basis of the fossil evidence, paleontologists have estimated the climate of London at that time to have been frost-free and with an average temperature of 77°F (25°C). There is some evidence for growth rings in the fossil wood, so it is possible that the climate was seasonal, perhaps with a dry season alternating with a wet season.

These examples of warm rain forests in the high latitudes show that the global temperature of the Early Tertiary was considerably greater than that at present. Earth temperature declined through the Tertiary and it is likely that the "tropical" forests became increasingly restricted to the equatorial latitudes, although there is very limited geological evidence of the timing and course of these changes. By about two million years ago, at the commencement of the Pleistocene epoch, the boundaries of the world's biomes, including those of the tropical forests, probably resembled those of the present day.

■ CLIMATIC CHANGES DURING THE LAST TWO MILLION YEARS

The climate of the Earth has fluctuated very considerably during the last two million years. There are many reasons for this variation, the most significant underlying cause being an astronomical one, relating to the orbit of the Earth around the Sun and the way in which the axis of the Earth changes in time. The shape of the Earth's orbit around the Sun is not constant. At times it is roughly circular and at other times it is more elliptical; during the latter times the Earth spends part of the year farther from the Sun, resulting in a higher degree of seasonal variation in temperature. The time taken to complete a full cycle from circular to elliptical and back to circular is 96,000 years. The Earth is not fully upright on its axis but tilts at an angle of approximately 21.5° to the vertical. But this position varies, and in the course of 21,000 years it tilts yet farther to 24.5° and then returns closer to the upright position once more. When the angle of tilt is greatest, there is a stronger climatic difference between the equatorial and polar regions of the planet. Finally, there is an additional wobble of the Earth on its axis that follows a 10,500-year cycle, adding another variable to the global climatic pattern. When all three variables are added together, an overall pattern of change in the Earth's climate can be predicted, and this corresponds quite closely to the alterations in climate that geologists have been able to reconstruct from their studies of rocks and sediments from the past two million years.

There are additional factors that can influence global climate, such as the frequency of volcanic eruptions that throw large quantities of dust into the atmosphere and shield the Earth's surface from some of the incoming radiation. The composition of the atmosphere changes over the course of time, and studies of the gas bubbles trapped in the ice of the world's great ice sheets in Greenland and Antarctica have revealed that warmer times in recent Earth history have coincided with larger quantities of greenhouse gases, such as carbon dioxide and methane, in the atmosphere. It is difficult to determine, however, whether these gases cause warmer conditions or whether their production is enhanced during warmer climates. Cause and effect can often be hard to separate when dealing with correlations of this kind. The ocean currents that redistribute the energy received by the Earth from the Sun are also a factor. When strong currents take warm waters from the equatorial regions to the high-latitude oceans, the polar regions do not become starved of energy and develop ice caps. There have certainly been times in the past two million years when changes in global climate have been strongly affected by alterations in the strength and course of ocean currents. The same may apply to mountain-building processes, which can alter the flow of the Earth's atmosphere and result in heat redistribution being interrupted.

Many factors can thus affect global climate, and geologists are able to confirm that the world's climate has indeed shown many, sometimes abrupt, changes in relatively recent times. The evidence for these changes comes from a variety of sources. Evidence for the presence of glaciers in the past is left in the form of landscapes carved out by great depths and weights of moving ice and the deposition of detritus carried over large distances by the ice. Sea levels change during times of climatic variation because water becomes locked up in the ice and removed from the oceans at the same time as low temperature reduced ocean volume. The presence of glaciers on a land surface also depresses the Earth's crust, so that the land becomes lower in relation to sea level, further complicating the geological picture.

Important evidence of climatic change also comes from fossil plants and animals, both *macrofossils* (large enough to be visible) and *microfossils* (requiring microscopy for observation). These may be trapped in ice and preserved, or buried beneath glacial debris, but fossil materials tend to be more abundant in times of warmth, when vegetation and animal life are more abundant and are more likely to become preserved in lake sediments and peat bogs. One of the most important sources of fossil evidence for climatic

Pollen Analysis

Pollen grains are produced by seed plants (gymnosperms and angiosperms) in considerable numbers as a means of genetic transfer during sexual reproduction (see sidebar "Pollination," page 133). Plants using the wind as a means of pollen dispersal, which includes most gymnosperms (for example, conifers) together with many trees, grasses, sedges, and plantains, produce larger quantities of pollen than insect-pollinated plants because dispersal by wind is a very chancy mechanism compared to insect transfer. The tough outer coat of pollen grains survives well in wet conditions, even after the cell contents have decayed. Bacterial activity is partially inhibited in constantly wet conditions because oxygen is not easily available, so the pollen coats accumulate as microfossils in lake sediments and in peat. In these materials they become stratified over the course of time as new sediments accumulate above them, so they reflect the changing vegetation of the past.

The relatively inert chemical structure of pollen grains enables scientists to extract them from sediments by dissolving away the surrounding matrix. The pollen grain coats are so inert that it is possible to boil them for a limited period in hydrofluoric acid, which is strong enough to dissolve glass and removes sand and silt from the sample, leaving the pollen unharmed. The structure and delicate sculpturing of the pollen coats allow analysts to identify them, sometimes even to the taxonomic level of species,

but more often to generic levels, such as oak, or pine. The abundance of the fossil pollen in most sediments is useful because the different types can be counted and their proportions in the total assemblage calculated. In this way it is possible to follow the changes in various plants over the course of time, reflecting changes in vegetation. Vegetation change is often related to environmental change, such as climatic variations or, in recent times, changes associated with human land management. The interpretation of pollen assemblages is fraught with difficulties. Different plants produce different amounts of pollen, so some species are underrepresented and others overrepresented. Some pollen grains disperse more effectively than others, and some are more delicate, so they are damaged or decomposed more easily. Then there is the problem of how far away the source plant is from the lake or peat bog where the pollen is finally preserved. Wetland plants are more likely to grow nearby, so will be better represented than distant dry-land plants. All of these factors need to be taken into consideration when interpreting a fossil pollen assemblage in terms of the vegetation that gave rise to it. Work in the Tropics has been particularly frustrating because of the high diversity of plants present and the fact that most of the trees are insect pollinated. Consequently they produce small quantities of pollen that is not very effectively transferred into lake sediments or peat deposits.

changes come from pollen grains, which are abundant in the atmosphere, relatively easily identified, and become well preserved in lake and peat sediments (see sidebar).

Chemical evidence, such as differing proportions of isotopes of certain elements, can also be important in the reconstruction of past climatic fluctuations. Oxygen, for example, has both a light and a heavy form. Most oxygen on Earth occurs in a light form with an atomic weight of 16. But 0.2 percent of the world's oxygen is a heavy isotope with an atomic weight of 18. Both isotopes are present in water, H_2O, but water with the heavy isotope does not evaporate as easily as that with the lighter isotope. Ice sheets and glaciers grow because water, usually in the form of snow, falls on their surfaces faster than it melts, and the water that feeds the ice comes from evaporation, so tends to be poor in ^{18}O. When the Earth is cold, ice enriched with ^{16}O accumulates and the residual ocean waters are enriched with ^{18}O. Under warm Earth conditions, by contrast, less ice accumulates but contains more of the less volatile ^{18}O, so the oceans are less

enriched with the heavy isotope. By coring into ice sheets, or into the sediments of the deep oceans, geologists can reconstruct past Earth temperature in some detail by documenting the relative proportions of the two oxygen isotopes. This method has proved extremely valuable in reconstructing the climatic history of the past two million years in great detail.

When all of the different sources of evidence are put together, a reasonably clear picture of the Earth's recent climate history emerges. It displays a strongly fluctuating climate, varying between warm and cool conditions in a cyclic manner. The cycle began over two million years ago but has become more intense in the last one million years, with a greater amplitude of temperature variation between the warm and cold stages. These stages correspond with the glacial (cold) and interglacial (warm) episodes that have been recognized from geological and paleontological studies. The wavelength (distance apart) of the cold glacial events during the past million years has been approximately 100,000 years, which corresponds relatively closely to the predictions derived from the model based on astronomical cycles. The difference in average temperature between the warm and cold periods can be estimated and is likely to have been about 9–14°F (5–8°C). The oceans vary much less in their temperature during the cycle because they both gain and lose heat more slowly than the atmosphere.

The pattern of temperature change is not a simple cycle of alternating warm and cold events, however. Warm episodes (interglacials) are generally much shorter in duration than the cold episodes (glacials). The peak of warmth usually occurs within the first 10,000 years of the interglacial and is followed by a stepwise cooling of global temperature that may take about 80,000 years before it reaches its coldest trough, which is when glacial ice would be at its most extensive around the world. The reversal to warm conditions then takes place quite rapidly, according to the isotope records in ice and ocean sediments; this pattern is repeated in each glacial–interglacial cycle. Biogeographers and climatologists have become very interested in just how fast the climate can change, because this information will be of great importance in anticipating future changes. The last glacial reached its maximum extent about 20,000 years ago and warming began in earnest 14,000 years ago. There was a rapid increase in temperature, but this was interrupted in some parts of the world, particularly in the North Atlantic region, by a temporary return to cold conditions. By 10,000 years ago even this part of the world was affected by global warming, and evidence from the fossils of beetles in northern Europe suggest that the temperature may have risen by 12.5°F (7°C) in just 50 years. Rapid changes in global climate may take place, therefore, even in the absence of human intervention. Such changes had considerable influence on the distributions of different organisms over the face of the Earth.

Climatic changes of such magnitude and speed were associated with alterations in the quantity of ice on the surface of the planet, and this in turn affected sea levels around the world. When the oceans were cold, they contracted and their volume was reduced. But even more important was the extraction of water by evaporation that became locked up in the ice sheets and glaciers. The reduction in ocean volume as a result of these two effects caused *eustatic* changes in the relative level of land and sea. Other changes in the relative sea level, however, resulted from the weight of ice on the Earth's crust, bending it and lowering it in relation to the oceans. Where the burden of ice was at its greatest, in the neighborhood of the major ice sheets of North America, Greenland, Eurasia, and Antarctica, these *isostatic* changes in relative sea level created some complex patterns of shoreline changes.

When conditions were at their coldest, the volume of ice was so great that the average lowering of sea level was about 340 feet (100 m). This meant that there were very large areas of land available for plants and animals to colonize that are now submerged beneath the waves. In the region between Alaska and eastern Russia, called Beringia, for example, there was an extensive bridge of land linking the two continents during the height of the last glaciation. Exposed lands between islands and continents created new opportunities for animals and plants to spread into new areas, increasing their biogeographical range and exploiting the new opportunities for population expansion. It was at this time, in the late stages of the last glaciation, that humans first moved through Beringia and populated the continents now known as North and South America, a biogeographical movement that was to have vast and unpredictable effects on all of the biomes of those lands.

■ TROPICAL FORESTS IN THE LAST TWO MILLION YEARS

The massive and rapid climatic changes experienced by the Earth in the last two million years have affected the distributions of all of the world's biomes, including the tropical forests. Pollen analysis, or *palynology,* (see sidebar "Pollen Analysis," page 179) of lake and peat sediments in various parts of the world has revealed that the zones occupied by the different biomes were pushed toward the equator as the ice sheets and glaciers expanded. In North America, for example, the Great Lakes region was covered by the ice of the Laurentide ice sheet, and was bordered to the south by a band of tundra. Boreal coniferous forest occupied the bulk of what is now the United States, giving way to temperate forest only in the southern states around the Gulf of Mexico. Thus, all the zone boundaries had moved south.

Pollen evidence is abundant in North America because of the frequency of lakes and peat bogs suitable for analysis, so it has been possible to build up a very detailed picture of vegetation history over the past 20,000 years. Things have not proved to be nearly so easy in the Tropics, however, for a number of reasons: the lack of suitable lake sites, especially in the lowlands; the greater technical difficulties in identifying pollen grains; the lack of wind-pollinated tress that produce large quantities of pollen; and the scarcity of research workers in the region. As a result of all these factors, the history of vegetation in the Tropics is still hazy.

In the Amazon region of South America, most of the available lake sites suitable for pollen analysis have been in the foothills of the Andes Mountains. These have shown that forests, including the rain forests, on the eastern side of the Andes chain retreated down to lower elevations during the height of the last glaciation and were confined to levels below 6,000 feet (2,000 m). As the climate warmed, starting 15,000 years ago, the forests expanded upward until they reached their current level of 10,400 feet (3,400 m). The limited evidence of pollen from lowland Amazonia suggests that during the period of cold the forest became dominated by trees currently associated with higher altitudes. Thus the nature of the Amazon forest was likely closer to the current montane forests of the Andes.

There has been much speculation among biogeographers about whether the Amazon forest was fragmented into smaller units during the times of cold. Some have suggested that the forest failed to remain intact during the glacial episodes of higher latitudes, and that much of the lowland region became occupied by savanna woodlands and grasslands, effectively splitting the forest into a series of smaller refuges where the tropical forest plants and animals survived the climatically unfavorable times. The evidence for such fragmentation is indirect and widely disputed. Zoologists have drawn attention to the fact that certain parts of the South American tropical forests are richer in certain organisms than others. There are locations where endemic species are concentrated, making them particularly biodiverse, and this has been observed among birds, butterflies, lizards, and certain families of trees. The critical question, of course, is whether these different groups of organisms all show the highest levels of endemism and diversity in the same locations, and the answer seems to be that they do. When the patterns are mapped for all the groups studied, there are about a dozen localities in which all overlap in their diversity. This overlap suggests that these areas are the refuges in which forest communities survived the times of cold. Direct evidence from pollen analysis for the invasion of lowland Amazonia by savanna vegetation is still lacking, but it is possible that the fragmentation was less dramatic than that suggested by this model. Perhaps the forest changed its composition to a more montane or to a drier character and the lowland rain forest communities became fragmented within this simpler type of forest ecosystem.

In Africa, biogeographers have revealed a pattern of patches of endemism and diversity similar to found in the Amazon. In West Africa such a region is in the extreme west in Liberia, and a few locations exist in the western part of the Congo River basin. There is also a rich area in the east of Africa in the Great Rift Valley and the upper reaches of the Congo River. Many rain forest species of plants and animals are confined to these general areas within central Africa, forming separate, *disjunct* populations. The gorilla is one example of such a disjunct distribution, having lowland populations in the west and an upland population in the mountains of Uganda in the east (see illustration). Biogeographers suggest that this and many other species became fragmented during the cold period 20,000 years ago and have never managed to recolonize the intervening forest that has developed in the current warm episode of the last 15,000 years. As in the case of South America, the argument

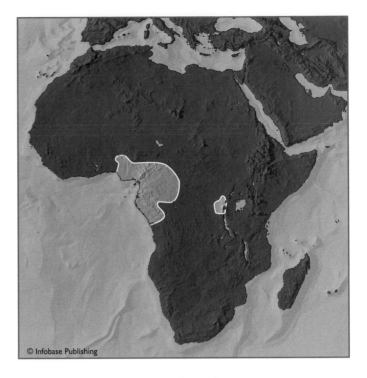

© Infobase Publishing

The distribution of gorillas in Africa. There are two separate populations of this great ape, forming a disjunct distribution. It is possible that the two populations were separated as a result of the climatic changes of the last ice age, when forests became fragmented. Although forest has become reestablished over the last 10,000 years since the Ice Age, the gorilla populations have failed to link up again.

is based largely on indirect evidence in the form of present distribution patterns of organisms.

Paleoecologists have carried out pollen analyses on various lakes and swamps in East Africa and have come to the conclusion that the development of glaciers on the East African mountains was accompanied by a time of cold, relatively arid climatic conditions. The altitudinal limits of montane forest was lowered by about 1,500 feet (500 m), which means that this vegetation type was much more extensive in East Africa in the past. They speculate that dry vegetation extended through central Africa at that time, probably linking the northern and southern savanna vegetation belts of the present day and pinching out the regions of rain forest in the main part of the Congo River basin. As in the Amazonian situation, this vegetation may have been dry forest rather than open grassland. Geologists, however, have found deposits of dune sands in Central Africa that suggest considerable aridity during the glacial.

In West Africa there are several sets of pollen analytical data that help clarify the environmental history of the past 20,000 years. These indicate that savanna vegetation dominated much of the region throughout glacial times, when the fossil vegetation indicates that the average temperature was probably 9–14°F (5–8°C) lower than that at present. Montane forest trees, such as the montane olive (*Olea capensis*), were present at much lower altitudes than where they are found now. Biogeographers have found the fossil remains of termite mounds built by species typical of savanna environments within the rain forests of southern Nigeria in West Africa, a finding suggesting that savanna prevailed in the area in the past. Overall, there is much stronger evidence for the total disruption and fragmentation of the African tropical forests than for such conditions in South America. It is possible that this history of climatically induced disturbance accounts for the much lower biodiversity of the African forests compared to that of South and Central America and Southeast Asia.

Tree lines began to move up the mountains of East Africa by 15,000 years ago with the arrival of the present interglacial, and in West Africa the rain forest began to expand northward. Between 10,000 and 5,000 years ago, rain forest extended to about 16°N, deep into what are now the dry lands of the Sahara. But after 5,000 years ago, aridity set in once more and the forests have subsequently retreated southward to about 8°N. The early part of the current interglacial in Africa, therefore, was evidently a time of warmth and wetness, allowing rapid expansion of the tropical forests.

Evidence for increased rainfall in Africa from 10,000 years ago has been found in an unexpected source, the sediments of the eastern Mediterranean Sea. Sediments that accumulated at this time have been recovered by taking deep cores from the bottom of the sea, and they show thick layers of black ooze, which was clearly deposited under anaerobic conditions. The lack of oxygen in the waters of the eastern Mediterranean at that time was due to the massive inflow of freshwater from the Nile River, draining the eastern regions of Africa. The freshwater was less dense than the salty waters of the Mediterranean and formed a layer that floated on top of the saline waters. In this stable and stagnant condition, there was little mixing between the two water bodies and consequently little diffusion of oxygen from the atmosphere to the deeper water. Bacterial decomposition was curtailed, and black anoxic mud was formed as a result. The black mud band of the eastern Mediterranean, formed 10,000 years ago, thus testifies to the increasing wetness of the African climate at that time.

In Southeast Asia, sea level changes were particularly important because this part of the world saw the level of the oceans relative to land fall by as much as 370 feet (120 m), leaving many of the islands and landmasses joined in one large area, which has been called Sundaland. Pollen analyses from this region have shown that some areas, such as the Malaysian Peninsula, certainly became dry and cool during the ice age and supported savanna vegetation, but other regions, such as northern Borneo, remained covered by tropical forest throughout the cold episodes. Fragmentation may not have been quite as widespread as that found in Africa, which could account for the higher biodiversity of this region, but savanna is likely to have occupied much of the inland regions of Sundaland. Another approach to the study of past vegetation patterns in Southeast Asia, adopted by some biogeographers, is the study of termite distribution patterns. Some termite species are poor dispersers, so they tend to remain in one region. This means that if their populations are split by habitat fragmentation, they remain in the sites of refuge long after the habitat has recovered and lines of communication for most organisms have been reestablished. By studying modern termite distribution patterns, biogeographers have deduced that northern and eastern Borneo and northern and western Sumatra were the likely locations of rain forest refugia through the height of the last glacial event.

In New Guinea some researchers estimate that the rain forest was reduced to just 75 percent of its current extent, mainly because of the loss of mountain forests, where lower temperatures resulted in a descent of the tree lines. What is not known is how much of the land now lost below the oceans was occupied by rain forest at this time. It seems entirely likely that the forest expanded out into the newly exposed lands as the sea levels fell during the high-latitude glaciations.

In Australia, which was linked to New Guinea by the exposed land during the glaciations, the rain forest is now mainly found in the northern regions, especially Queensland. Pollen analyses of lakes from this region have produced some very detailed information about the recent history of vegetation. They show that the rain forests vanished completely during the time when the glaciation of the high latitudes reached its peak. At that time, dry woodland and forests of conifers, such as *Araucaria*, occupied the region. Some of the lake sediments date from earlier times and record the vegetation of the previous interglacial of 120,000 years ago, when, once again, rain forest occupied the region.

The general conclusion from the studies of rain forest history around the world is that they were greatly reduced in extent during the glaciations of the Pleistocene, being replaced over much of their range by dry woodland or savanna grassland. But in all three of the major rain forest regions of the world, Central and South America, Africa, and Southeast Asia, patches of rain forest persisted and provided habitats in which the rich forest flora and fauna could survive.

■ CONSEQUENCES OF FOREST FRAGMENTATION

One of the general principles of biogeography states that large areas of a particular habitat tend to contain more species of animals and plants than do small areas of the same habitat. A large area of forest, for example, will contain more species of trees, herbs, birds, and mammals than will a small area. One of the main reasons for this difference is that a small patch of forest will have a longer perimeter per unit area than a larger patch, and in the marginal areas the conditions are least favorable for survival and the dangers of predation are greatest. Conservationists regard large areas of an ecosystem to be more valuable than small areas (see "The Problem of Evaluation," pages 199–200), and the *fragmentation* of the rain forests that took place during the last and earlier glacial maxima could be viewed as a very unfortunate event from the point of view of biodiversity. Yet the number of species within tropical forests has remained high, so is it possible that fragmentation was harmless, or even beneficial?

One theory is that the separation of populations of organisms into isolated units may have encouraged the process of evolution and speciation. When Charles Darwin set out his ideas about evolution by natural selection, he suggested that any kind of isolation led to a breakdown in interbreeding so that the isolated populations might develop along independent lines that related specifically to the pressures of selection within their particular areas. The fact that there are so many spatially separated endemic species in the Amazon Basin, for example, could mean that each of these species has arisen from a common ancestor during a period of breeding isolation. In the case of the toucanets (*Selenidera* species), a group of birds closely related to the toucans, for example, there are five species occupying the Amazon Basin, all of which have non-overlapping distributions. It is likely that these species had a common ancestor, and it is possible that they split off from one another genetically as a result of fragmentation of the original population and development of each subpopulation in isolation. Is it possible that this took place during the height of the last ice age, when the rain forests were fragmented?

This sounds like a very logical proposal, and it has received a great deal of support from biogeographers in the past, but there are now many doubts about whether it would work. Considering the final cold phase of the Pleistocene, this lasted for only about 80,000 years, and forest fragments would not necessarily have been completely isolated for the whole of that time. The climate fluctuated considerably during each glacial episode, leading to an alternation of very cold times with relatively mild spells. So, full fragmentation may have occurred for only a few thousand years. Also, the supposed fragmentation of the tropical forests may not have been as great as has been supposed. The Amazon rain forest, for example, may have become interspersed with dry forest rather than grassland, so many forest species may have become less isolated than has been believed.

Molecular biologists have also come up with a new approach to the question of evolutionary relationships and the speed at which speciation can occur. By comparing the chemistry of DNA between different organisms, it is possible to determine how closely they are related and, in some cases, how long ago they are likely to have drifted apart in their genetic constitution. Evolution causes the molecular clock to tick, gradually leading to increasing differences between the genetic constitutions of species. Studies of many organisms have shown that evolution is relatively slow and is unlikely to have taken place in the time framework of the last glaciation. A time span of 100,000 years is very short for significant differentiation of populations in any organism that lacks a very short generation time. Bacteria can manage such rates of evolution without any difficulty, but birds would find it difficult.

Fragmentation during the last glaciation as a mechanism for enhancing the biodiversity of the rain forest is thus a theory that has lost support over recent years. It is nevertheless possible that earlier or repeated fragmentation during former glacial events during the Pleistocene may have contributed to the genetic aspects of forest biodiversity.

Fragmentation can also have detrimental effects on biodiversity, especially if the fragments are very small and isolated. A small fragment of a habitat may contain a limited number of individuals of any population, left behind as *relicts,* and this means that they will be forced to breed among themselves. It is well known that inbreeding can lead to genetic weakness because unhealthy, recessive genes build up in the population and can express themselves in weak and unfit individuals. All populations benefit from an input of new genes from outside populations, bringing new strength and vigor to the community, and if genetic exchange between isolated populations is limited, the ability of organisms to survive and adapt is impaired. Species that exist in a series of isolated groups are called *metapopulations,* and studies by ecologists have shown that small groups within the metapopulation are very liable to local extinction. It can be argued, therefore, that any fragmentation of the rain forest that occurred during the glacial episodes could have had harmful genetic effects rather than stimulating evolutionary divergence and speciation. The crucial question here is the size of the isolated populations. If they were large, then genetic impoverishment is unlikely, but if they were small, then inbreeding and loss of fitness are more probable.

Whether it has been a bad or a good thing for biodiversity, fragmentation is a problem that has been faced by the forests in the past and is a stress that they have survived. With the arrival of human cultures on the planet, however, fragmentation is being renewed as forests are felled. It remains to be seen whether the forest and its biodiversity can survive the next and even more severe onslaught.

■ CONCLUSIONS

During the 4.6 million years of Earth history there have been great changes in global climate. Times of warmth have occasionally, though somewhat rarely, been interrupted by times of cold, when parts of the Earth, usually in high latitudes or at high altitudes, have been covered by ice. There have been at least eight of these ice ages, and possibly several more, particularly in the earlier stages of Earth history. Ice ages seem to occur regularly at intervals of about 150 million years, but no one has yet been able to explain the causes that underlie this cycle. The Earth has been experiencing one of these ice ages for the last two million years.

Tropical forests, particularly the rain forests, require heat and moisture, so it is reasonable to suppose that they have been favored by the warm times of Earth history, as long as the warmth was not associated with drought. But tropical forests can survive in low latitudes even when the world is in an ice age, as the current ice age has demonstrated.

Forests are composed of trees, and the tree life form did not evolve until late Devonian times, about 400 million years ago, so the earliest forests date from that time. The development of rain forests is well recorded in the geological sediments of the succeeding Carboniferous period, when the great coal-forming swamps were formed. Flowering plants and mammals did not exist at that time, so the vegetation consisted of trees related to ferns, horsetails, and club mosses, and the fauna consisted of insects and amphibians. By the end of the Carboniferous, the Earth had cooled and entered an ice age.

Throughout the Cretaceous period, vegetation was dominated by conifers and cycads and the giant reptiles roamed the Earth. This period came to an abrupt end about 65 million years ago when a bolide struck the planet. The subsequent Tertiary subera saw the development of "tropical" rain forests in such unlikely locations as Alaska and England, indicating that global temperature was high once again. The trees of these forests were closely related to modern tropical groups, as were the animals that inhabited them, so the modern tropical forests can be regarded as the descendents of these Tertiary forests.

The temperature of the Earth fell in a series of steps through the Tertiary, and by two million years ago ice sheets were once again developing. The geological records are much more complete for this recent ice age, so geologists have been able to work out the sequence of events more fully. It is evident that there have been a series of glacial advances and retreats in these recent times as cold episodes have alternated with warm, interglacial times. Overall, the cold has dominated and warm intervals have been short. The most probable cause of the alternation in climate is the pattern of movement of the Earth around the Sun and the variations in the tilt of the Earth on its axis. When these cycles are superimposed, they produce a model in which cold episodes should occur roughly every 100,000 years, which agrees closely with geological observations. Glacial cycles are thus controlled by astronomical factors.

Changing climatic conditions are evident in the geological record of rocks, and one of the most useful techniques for recording these changes has been the ratio of oxygen isotopes in the ice sheets and in ocean sediments. From these, it has been possible to piece together a very detailed picture of changing global temperatures over the past two million years. One important fact that has emerged from these studies is that the climate can change very rapidly, sometimes by several degrees over a matter of decades.

The effects that these climatic changes have on vegetation and animal life are evident in the fossil record. Vegetation leaves a particularly clear record in lake sediments and peat deposits in the form of pollen grains, which are produced in very large quantities, preserve well in wet conditions, and are recognizable on microscopic examination. Pollen analysis is a process in which the pollen content of stratified sediments is examined and recorded, resulting in a reconstruction of past changes in the vegetation of a region. Using this technique, ecologists have been able to follow changes in the boundaries of the Earth's great biomes during times of climate change.

The history of tropical forests during the last two million years of the Pleistocene has been less easy to reconstruct than that of biomes from higher latitudes. In the high latitudes, there is a greater abundance of suitable lake and bog sites and a higher density of research workers. The nature of the tropical vegetation, by contrast, with an abundant variety of insect-pollinated trees that produce little pollen, has made the task of vegetation reconstruction very difficult.

The general picture that has emerged from tropical forest studies, however, is that the glacials of the high latitudes was accompanied by a considerable drop in temperature even in equatorial regions. The result has been a change in forest composition and often a fragmentation of the forests, particularly the rain forests, into isolated patches, called refugia. In the Amazon Basin, these refugial patches may have been separated from one another by areas of dry montane types of forest. In Africa, dry savanna vegetation probably dominated much of the Congo Basin, leaving patches of rain forest in the west and in the east of Africa. In Southeast Asia there was also a contraction of the rain forest, and large areas became dominated by savanna. Northern Queensland in Australia saw the development of dry *Araucaria* forests during the times of cold.

The warming of the Earth's climate over the past 15,000 years has resulted in the spread of the tropical forests once more, but the effects of fragmentation are still apparent among some species, which have remained isolated in patches. The gorilla in Africa is a good example, with populations present in the lowlands of the west and the highlands of the east. Rain forest has spread from its refugia into more northerly and southerly latitudes, especially in the early part of the current interglacial, 10,000 to 5,000 years ago.

A history of successive fragmentation for the rain forest could have affected its composition in a number of different ways. If a species is split into isolated populations for a long period of time, it may develop along different lines in the various locations, and these difference may persist even if the populations are eventually brought together once again. Animals, for example, may develop different patterns of courtship and breeding behavior, so may remain separated even though they occupy the same area. Fragmentation could thus result in speciation and the enhancement of biodiversity. The idea is quite convincing, but evidence is not consistent, and many biologists argue that the time needed for such developments has not been available in the Pleistocene glacial cycle.

Fragmentation could also have harmful effects on biodiversity because small, isolated populations are in danger of inbreeding and consequent genetic uniformity, weakness, and lack of adaptability. This possibility depends on the size of the refugial fragments of forest and the density of populations of species within them.

The tropical forests have thus had a very checkered history. There have been times of great success and widespread distribution, as well as occasions when they have been reduced to scattered fragments. They have proved remarkably resilient in their survival through times of great climatic stress, but they now face a new set of problems imposed on them not by climate but by a newly arrived species on the face of the Earth—humankind.

9

People in the Tropical Forests

Human beings are part of tropical forest biodiversity. *Homo sapiens* is a species that occupies most of the forested regions of the planet and plays a part in the forest ecosystem. People living in small groups and at low densities in the tropical forest form part of the forest food web. They prey upon certain animals and collect fruit, fiber, and timber from some of the plants. They thus modify the structure of the ecosystem and alter the pattern of energy flow, but are integrated within the system and may even contribute to habitat diversity and ecosystem stability. High human population densities, however, place unsustainable demands on any ecosystem, and the tropical forests are no exception. Temporary, small-scale clearance for settlement and agriculture, followed by abandonment and relocation, is easily accommodated by the dynamic forest. But large-scale deforestation and the conversion of areas of forest to grazing land create conditions that may place the ecosystem beyond recovery.

The role of humans in the tropical forest, therefore, is complex. The ecosystem is a relatively robust one and can cope with considerable pressure and disturbance, but there comes a point when the demands of people on the forest become excessive and bring the entire system to a point of collapse. People have recently become a problem for the forest, expanding their populations until they have reached a level that is difficult to sustain.

■ THE EMERGENCE OF MODERN HUMANS

Primates, the animal group to which biologists assign the human species, are largely tropical forest animals, although some members have proved very adaptable to other types of habitats. The primates are divided into Old World and New World groups; these are thought to have separated some 40 million years ago as the Atlantic Ocean widened, separating Africa from South America. By seven million years ago, the Old World group had split into different lines of development, including one that led to the great apes and another that led to humans. These organisms were still essentially forest creatures, and it was not until about four million years ago that the ancestors of humanity showed an undoubted bipedal stance and adaptations to the hand that indicate life in a wooded grassland, or savanna habitat.

The genus *Homo*, of which we are a member, is first found in the fossil record from deposits of about 2.5 million years ago, and our own species, *Homo sapiens*, first occurs in the geological record at 260,000 years ago. Paleoanthropologists, scientists who study the early history of humans, still argue about the place of origin and the pattern of spread of the human species. One member of the *Homo* genus, *H. erectus*, had spread out of Africa and into southern Asia by one million years ago; this species is regarded as the direct ancestor of *H. sapiens*. Neanderthal man, *H. neanderthalensis*, was a species that diverged from the main line of ancestry about 130,000 years ago, but did not survive through the final cold episode of the last ice age. It was during that final ice age, around 50,000 years ago, that our species began to develop new techniques and habits, such as the production of more refined stone tools, the creation of artistic images depicting possible religious rites and deities, and the wearing of decorative jewelry. Modern human cultures began to emerge.

The new surge in human cultural development was accompanied by geographic spread through Europe, Asia, and by 40,000 years ago into Australia. The New World remained unoccupied throughout this time. The invasion of Australia is particularly remarkable because it must have involved sea travel, and there is no direct archaeological evidence for boats until relatively recent times—13,000 years ago in the Mediterranean region. The invasion of Australia and the numerous islands of the Southeast Asian region may well have taken place by chance dispersal on rafts. From a biological and ecological point of view, this dispersing species was particularly remarkable for its adaptability. In Africa it met with semi-arid scrub and desert; in Europe it found temperate forests of deciduous and evergreen character; in

India it found savanna; and in the southeastern regions of Asia it discovered both dry deciduous and tropical rain forests. No other species displayed such a capacity to cope with the challenges of so many different habitats and was able to prove so invasive.

The tundra of the far north proved less attractive, but the prospect of hunting large mammals, such as mammoth, giant elk, and caribou, drew the expanding populations of humans even into these testing regions. By 20,000 years ago, people had spread into northern Europe and Asia, including the far eastern corners of Siberia. Global sea levels at that time were low because so much of the Earth's water was locked up in the ice sheets and glaciers, so there was a land bridge between Siberia and Alaska. The environment of this bridge was harsh, unforested tundra, but it was not covered by ice, so mammals, including humans, were able to cross into the New World on foot. Exactly when this took place is still uncertain, but Alaska was certainly occupied by humans by around 12,000 years ago. The subsequent spread of people through the Americas is remarkably rapid. By 11,000 years ago they had reached what is now Mexico and had spread right through North America, and within a few hundred years were present in Amazonia and all the way to Patagonia.

Some anthropologists feel that this view of the invasion of the Americas is too simple to explain the facts. They claim that the people arriving in the New World were not all from eastern Siberia, but that some came from southern Asia and the southern Pacific Rim. Archaeologists have examined some of the most ancient American skeletons and feel that they more closely resemble the southern Asian or even early Australian people than the folk of Siberia. Perhaps there were different waves of invasion involving different groups of people, and perhaps the Native American populations of the Baja and Mexican regions had a different origin from those of the north. There are also arguments about dates, because some South American settlements have been dated to 12,500 years ago, which does not fit in with the pattern of spread previously accepted. The arguments will undoubtedly continue until additional evidence is available, but one thing is certain—by soon after 11,000 years ago the peopling of the Americas was complete.

The spread of human cultures through North and South America again illustrates the cultural adaptability of the species to very different climates and environments. The people who had entered Alaska were of Siberian descent and had lived in a tundra environment. Some of these people remained within the Arctic tundra of northern Canada, but others traveled south and encountered coniferous and temperate deciduous forest, prairie grasslands, dry scrub, and deserts. In Central America they first met up with the tropical forests and from there spread rapidly into the tropical rain forests of Amazonia. Within relatively few generations, they had made the change from living in tundra to living in the rain forest. It is possible, if the South Asian theory of origin is correct, that some of the people involved in the invasion of the Americas originated from tropical regions, in which case some of the skills needed for a forest existence were already present in their culture.

People were thus present throughout the rain forest regions of the world as the climate became warmer at the close of the last ice age, from 13,000 years ago onward. They were present in the Amazon Basin, the lowlands of the Congo, and the island of New Guinea during the post-glacial times, when hot and wet conditions developed, leading to the gradual replacement of the drier forests of the ice age with the rain forests of today. Humans have thus been a member of the tropical forest ecosystem ever since the post-glacial recovery of this biome, and they have played an important part in its development and biodiversity.

■ EARLY FOREST PEOPLE

The people inhabiting North America 11,000 years ago have been named "Clovis people," after a town in New Mexico where their archaeological remains have been described in detail. They are characterized by their finely chipped stone spear points and other tools, and are often associated with the bones of large mammals, such as mammoths, on which they preyed. Such a lifestyle would have been appropriate in the alpine tundra habitats of tropical mountains, and one can imagine that these people were able to advance southward down the mountain chain of the Andes, hunting game. Open savanna and prairie habitat would also have been suitable for this type of hunting, but the tropical forests must have presented a considerable challenge. Until relatively recently, anthropologists believed that the early colonists of South America found the Amazonian forests a barrier that was impassable, so they concentrated their populations and movements on the surrounding mountain masses.

It is very difficult to conduct archaeological surveys seeking early settlement in a habitat as complex as a tropical rain forest, and the lack of evidence for early settlement could be a consequence of weak survey data. Researchers from the University of Illinois have recently opened up a new chapter in the history of the Amazon rain forests by discovering and excavating a cave settlement in the Lower Amazon region of Brazil about 400 miles downstream from the city of Manaus. In the sandstone hills of Monte Alegre, close to the course of the main Amazon River, the research team discovered a cave containing stone artifacts, together with paintings on the rock walls. The floor of the cave was littered with thousands of seeds, together with the skeletal remains of fish and small mammals, as well as charcoal from a range of rain forest trees.

It is clear that these people, who settled in the cave over 11,000 years ago, lived by hunting small forest creatures using spears and arrows, and gathered the fruits of the forest to supplement their diet. Many of the forests fruits they consumed are still valued as food today by the Amazonian peoples, including the Brazil nut (*Bertholetia excelsa*), which is a large seed borne by a tall rain forest tree native to the Amazon and is highly regarded as a nutritious source of food. Some of the seeds found, such as those of *Vitex cymosa*, are currently used for fish bait, so it is possible that the original inhabitants of the Amazon also used them for this purpose. Palms seeds were also frequent in the cave deposits, probably reflecting the use of these plants as food. Many of the plant species represented in the remains belong to species favored by forest disturbance. It is possible that the early settlers concentrated their gathering activities in natural clearings and riverside habitats, but the observation also raises the intriguing question of whether these people had begun to manage their forest habitat, possibly using fire or wood cutting to open up the habitat and encourage the growth of their preferred food suppliers.

The animal remains found in the Monte Alegre caves included rodents, bats, turtles, snakes, tortoises, turtles, amphibians, fish, mollusks, and birds. The fish varied from small ones of less than two inches (5 cm) to large ones over five feet (1.5 m) in length. The cave was continuously occupied for a period of about 1,200 years, after which it was abandoned and the ancient deposits became buried beneath layers of sand. In later years, others came to use the cave, first fishermen and later farmers who had begun to clear the forest. The results of this excavation illustrate that a tropical forest culture existed at the same time as when the Clovis big-game hunters dominated North America. They lived 5,000 miles (8,000 km) south of the Clovis people, and although their stone tools resembled those of the Clovis in their skillful construction, they were quite different in form and clearly adapted for a life of small-mammal hunting and fruit gathering in the rain forest.

Archaeological evidence from other tropical forest regions of the world indicates that people were present 11,000 years ago, when the final retreat of glaciers was taking place in the higher latitudes and altitudes. In New Guinea, for example, there was a small population of hunters and gatherers foraging for the fruits of the *Pandanus* tree and collecting mollusks, especially around the coasts. In northern Australia, the early inhabitants encountered dry *Araucaria* forests. Their hunting activities, particularly the use of fire in driving game, may have held back the development of wetter rain forest in the post-glacial period, encouraging the growth of *Eucalyptus* forest instead. In East Africa, the change to more humid conditions at the end of the ice age is associated with alterations in the diet and behavior of the human occupants of this region. The people turned from big-game hunting in savanna environments to smaller animals associated with the forest.

It is well to remember that the rain forests of the Amazon, Southeast Asia, Australia, and Africa were all occupied by human cultures throughout their post-glacial development. It is entirely likely that people have had a hand in shaping the structure and ecology of the tropical forests throughout that time, and some of the current biodiversity of the area may well be a consequence of human settlement and activity.

■ DOMESTICATION AND AGRICULTURE IN THE FOREST

The cultivation of plants and the domestication of animals was an enormously important development in human cultural evolution. In addition to gathering wild fruits and hunting jungle animals, humans devised a means of controlling food production and alleviating some of the problems associated with fluctuations in food supply within the forest. This practice was not entirely new; termites already cultivated fungus gardens, and some ants used a system of slavery whereby they exploited other species in their service. But agriculture, in its widest sense, provided humans with new opportunities to expand their populations and develop more sophisticated social systems.

Agriculture did not arise only once in the course of history, but at many times and in many different places. Some centers of agricultural innovation were in the grasslands and open woodlands of the Middle East in Southwest Asia, and others in the Far East of continental Asia. But agricultural ideas also entered the minds of the early settlers of the New World Tropics. Archaeologists excavating a series of caves in Mexico in the 1950s discovered deposits evidently associated with early human settlement, and among the detritus they found seeds, fruits, and stalks of an edible plant, the squash (*Cucurbita pepo*). In later human history, this plant was to give rise to various important domesticated crop plants, such as pumpkins and summer squash. Two important questions faced the archaeologists who discovered them, however: Were these seeds simply wild squash that the inhabitants of the cave had gathered, or were they evidence of early domestication? If they were evidence of the latter, then how early did this occur? The answer to the first question was revealed by careful measurements; the size of the seeds was greater than that of wild squash, so these plants had been selectively bred for agricultural purposes. The second question has been disputed for many years because the radiocarbon technique of dating has certain errors associated with it; nonetheless, the most recent dates for these Mexican squash seeds lie between 8,000 and 10,000 years ago. These deposits are thus the oldest record of agricultural activity in the New World and prove that domestication was taking place in Central America at roughly the same time as in the Middle East.

The tools used by early farmers can also provide information about the systems of agriculture that were used and the types of crops that were grown in the Tropics. Milling stones, for example, can be recognized because they are generally large cobbles that have one of their faces scarred by the activity of constant grinding. Their presence suggests that seeds or other plant tissues are being cultivated and then crushed to extract the valuable food content, often starch. Archaeologists usually rely on finding some intact plant material to identify the nature of the crop being milled. Macerated seeds or grains of a crop like corn are often easy to identify, but some tropical food plants retain starch in their roots or tubers and thus do not survive well as fossils. Botanists have studied the structure of the starch grains stored in a wide range of root and tuber crops and have found that the different species can be identified, and that these starch grains may survive for thousands of years on the surface of milling stones. In one study of stones from a settlement in Panama dating from about 7,000 years ago, archaeologists discovered starch grains of manioc (*Manihot esculenta*), yam (*Dioscorea* species), and arrowroot (*Maranta arundinacea*), revealing evidence for very early cultivation of root crops in the lowland forests of Central America. Manioc, or cassava, has become an extremely important crop in tropical countries throughout the world, and is likely to be of South American origin, as is arrowroot. It is possible that even as long ago as 7,000 years domesticated plants were being collected, transported, introduced, and cultivated in different parts of the tropical forest biome.

In the Southeast Asian forests there is some evidence for the cultivation of sugar cane (*Saccharum officinarum*) in the highlands of New Guinea about 5,000 years ago. The pig (*Sus scrofa*) was also domesticated at this time. Fossil remains of pigs have been found associated with human settlements and were almost certainly introduced to the island by humans. By 3,000 years ago, agriculture was widespread, as people of the forest had a range of stone tools and digging sticks to assist them in forest clearance and cultivating the cleared areas. People have thus been creating openings in the forests of Southeast Asia for at least 5,000 years. The effects on the forest composition can be detected in the pollen sedimenting in the lakes of the area (see sidebar "Pollen Analysis," page 179), the pollen record being sensitive enough to provide records of forest disturbance in the form of pollen from trees usually associated with secondary successions (see "Secondary Forests," pages 62–65). It is often difficult, however, to be sure that the disturbances were caused by human activity rather than by natural catastrophes. Some types of pollen seem to be more closely associated with human activity and have been termed *anthropogenic* indicators. In the

Southeast Asian region, *Pandanus* is a good example of an anthropogenic indicator. It is a genus of trees that provides nuts popular for human consumption, and it is likely that early human settlers cleared areas of dense forest to encourage the growth of this plant, which prefers more open conditions. In time, people domesticated *Pandanus* and selectively bred strains that produced higher yields of nuts. Its pollen grains are distinctive, and when they are present in greater abundance at certain depths in the sediments of lakes, it is a reasonable indication that human agricultural activity took place at that time.

In tropical Africa, plant domestication took place about 4,000 years ago with the cultivation of plants such as sorghum (*Sorghum vulgare*) and millet (*Pennisetum typhoideum*). Pastoralism had been practiced in the drier regions bordering the Sahara, but cattle, sheep, and goats were not well suited to the forest environment. The introduction of domesticated grazing animals to the tropical forest regions of Africa inevitably led to forest clearance and regular burning to maintain savanna woodlands and grasslands.

The early settlers of the tropical forests were innovative and experimental in their exploitation of the local plant and animal resources. By 3,600 years ago, the people of Central America had discovered that the sticky latex, exuded by some trees when their bark is damaged, hardens on contact with air, and when mixed with other materials can produce an elastic, rubbery material. The tree *Castilla elastica*, for example, which is native to Central America, produces a sticky white liquid when injured, which dries to a brittle material. But if the latex is mixed with juice from the morning glory vine (*Ipomoea alba*), it remains more malleable and can be moulded into rubber balls, bands, and even figurines. Balls were used by these people for recreation, thus the people of the Central American forests were probably the first inventors of ball games played in designated courts.

By 2,600 years ago, members of the early Mayan civilization of Central America had learned how to manufacture a chocolate drink using the beans of the cacao plant (*Theobroma cacao*). Archaeologists have discovered vessels from that period that have the appearance of teapots with long spouts and which contained residues of chocolate. The chocolate drink was boiled in the vessel and the foam was poured from the spout. This practice was still found in Central American society right up to the Spanish conquest in the sixteenth century.

The people who settled in the tropical forests were essentially "stone age" in their culture, depending entirely upon stone, wood, and bone tools to assist their survival in the forest. Some tropical forest societies have remained in this condition right up to the present day, whereas others learned the art of iron forging, providing them with a more sophisticated system of tools for felling trees and managing the forest. In the south of India and Sri Lanka, for example, the local rocks are

Easter Island

Easter Island lies in the South Pacific in a subtropical area of Polynesia. It is a small island, only 66 square miles (170 square km) in size, and is world famous because of its massive statues of elongated human heads, planted in the sides of its hills. The landscape is bare of trees, but this was not always the case, for pollen analysis of lake deposits have shown that it once bore dense forests of giant palm trees of a species now extinct. Polynesian people, migrating over the South Pacific in small boats, did not arrive in this tropical paradise until about 900 C.E., which was rather late in the Polynesian expansion. The people were farmers and brought with them cultivated crop plants, such as sweet potatoes, yams, bananas, and sugar cane, together with chickens. They cultivated the land and their population expanded, possibly reaching 30,000, or 450 people per square mile (170 km^{-2}). This expansion placed great stress on the environmental resources, especially on the forests of palm trees, which people used for construction, fuel, and food. They quarried the massive statues from local stone and somehow managed to transport them from the quarry to the site of erection, despite their size and weight. Some were 70 feet (21 m) long and weighed 270 tons (273 tonnes). The most likely means of transport was to use the trunks of the palm trees as rollers, which must have involved felling yet more of the forest. Most of the statues were erected between 1100 and 1600 C.E. Statue building then abruptly ceased and the entire society went into a disastrous decline, which has posed a great enigma to archaeologists. The explanation for this decline came with the use of pollen analysis from ancient lake deposits on the island, which revealed that the forests of Easter Island finally vanished at the time of the collapse of the civilization. The giant palm tree became extinct along with 21 other species of tree at about 1600 C.E., so the society had evidently outgrown its resources and failed to conserve the forest so that it could be exploited at a sustainable level. The case of Easter Island demonstrates that primitive societies were just as capable of destroying their environments as modern societies, given the technical means and the appropriate population density.

---

often rich in oxides of iron, these resources having been put to use for at least 2,000 years. One of the problems in extracting iron from its ore is maintaining a high enough temperature to melt the metal. Archaeologists working in Sri Lanka have discovered ancient furnaces, built on the west-facing sides of hills, that used the natural wind flow as a kind of bellows, forcing air over burning *charcoal* to raise the temperature of the furnace to the required level. Similar furnaces have been found in Burma; iron smelting also took place in East Africa and West Kalimantan, Indonesia. So the forest people developed new materials that gave them greater control over the management of their habitat.

The great danger with advances in culture is that they bring the possibility of environmental problems, and this has certainly occurred in some parts of the tropical forest habitat. Perhaps the most disastrous account of environmental destruction resulting from prehistoric mismanagement is that of Easter Island (see sidebar above). It is easy to assume that primitive cultures in a productive tropical habitat lived in some kind of idyllic harmony with nature. But their population growth and their life spans were undoubtedly limited by their cultural restrictions. When these restrictions were relaxed, for example, by the development of more advanced technology or social organization, populations expanded and the environment suffered.

In the case of Easter Island, the outcome was the collapse of the entire society.

■ HUMAN POPULATIONS AND DISEASE IN THE TROPICAL FORESTS

The people of the tropical forests, as the example of Easter Island demonstrates, were quite capable of environmental misuse and overexploitation when the conditions allowed such developments. On the whole, the populations of jungle-dwelling people were controlled by their environment. Among many forest cultures, infant mortality remains high, sometimes resulting in only 50 percent of babies surviving for two years. Disease, parasitism, and starvation have long been the controlling factors on population expansion among humans. Intertribal strife and warfare have also placed constraints on the growth of populations.

Malaria (see sidebar on page 191) is one of the most widespread and dangerous tropical diseases that has long proved a limitation to population expansion. Written records

Mosquitoes and Malaria

Mosquitoes are found in many of the world's wetlands, including those of the Tropics. Not all are involved in the life cycle of the malarial parasite, however. Scientists have recognized about 3,300 species of mosquito, of which 410 belong to the genus *Anopheles*. Within this genus, 70 species of *Anopheles* are known to transmit the malarial parasite, the most dangerous of which is *Plasmodium falciparum*. The larva of the mosquito is aquatic, hatching, spending its early life, and pupating in wetland pools or even the hollow tree stumps of the forest (see illustration). The adult insect, however, can disperse widely through the air. The males and females form courtship swarms at sunset, the female not able to lay eggs until she has consumed protein. She obtains this from the blood of mammals (although some species parasitize birds or reptiles) through use of mouthparts including a long, piercing tube with which she can cut into skin. She then forces the tube into a blood-carrying capillary beneath the skin and injects saliva containing anticlotting agents to ensure that the blood maintains its fluidity while being withdrawn. The mosquito takes only a small amount of blood, usually about 0.0002 ounces (0.001 g), but this provides enough protein to manufacture up to 500 eggs.

If the mosquito has not previously taken a meal from a host contaminated with the malarial parasite, the effect on its victim is more of an irritation than a problem. But if the mosquito has fed on a person with malaria parasites present in the blood, it will have taken malarial parasites into its stomach, where the parasite completes its sexual reproduction and the zygotes take up residence in the salivary glands of the mosquito. When the female takes its next blood meal, these will be injected into the host's bloodstream with the saliva and the malarial parasite has successfully moved to a new host.

Once in the host's bloodstream, the parasite is carried to the liver, where it infects the liver cells. From here it enters red blood cells and replicates, eventually causing rupture of the blood cells and the release of parasites into the blood stream. This release takes place every two to three days and causes the feverish

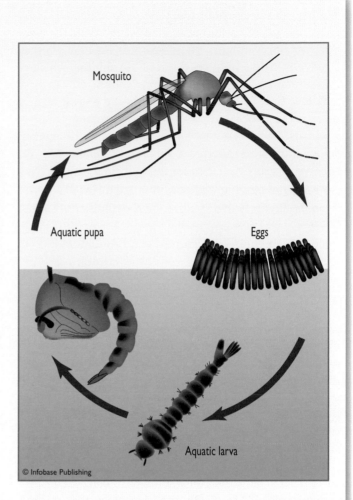

The life cycle of the mosquito. Female mosquitoes take a blood meal before reproduction. They lay eggs in aquatic situations, often in small temporary water bodies, and the larva and pupa spend their lives below water.

symptoms of malaria to become apparent in the host. The disease cannot be passed from human to human except by mosquito transfer, so control of the disease has often focused on control of the vector, often involving the destruction of its wetland habitat. The human immunovirus (HIV) cannot be transferred by mosquito because the disease agent is not able to survive in the insect, but mosquitoes can act as vectors for some other diseases, such as yellow fever, filariasis, and dengue fever.

of the disease go back almost 5,000 years, from descriptions found in early Chinese documents. It is a widespread disease, found in the Tropics and Subtropics throughout the world, and affects about 500 million people.

The people of the forest have thus lived with malaria for very many generations, and, as in the case of most diseases, they have sought treatments for the condition in the plants found growing there. Some of the inhabitants of the

Amazon forests developed a treatment based on the extract of the bark of the cinchona tree (*Cinchona officinalis*). This remedy was taken to Europe by Jesuit priests in the 17th century and proved most efficacious. It is now known that the active principle of the bark extract is quinine, which is still used in malaria treatment and prevention.

Yellow fever is another mosquito-borne disease that is widespread in Africa and the New World Tropics. This disease is caused by a virus that infects a wide range of monkey species but can be transferred to humans via mosquitoes. As in the case of many viral diseases, an individual who survives an attack can develop a degree of immunity. Native populations of the forest, therefore, often carry such immunity, and the disease has proved particularly lethal to people from other parts of the world entering the forest biome. Dengue fever is caused by another virus that resides in the monkey population of the forest and can then be transferred to humans by means of mosquitoes. Like yellow fever, it can also be transferred between humans as a result of successive mosquito bites. It is also found throughout the tropical areas of the world, but is particularly prevalent in Southeast Asia. Medical scientists working in the tropical forests of Malaysia found that the dengue fever virus was present in approximately one quarter of the monkey populations of mangrove swamps, bog forests, and primary rain forests of the region. This is a natural reservoir of the disease that can never be eliminated, so there is little hope of ever totally eradicating the virus.

Lymphatic filariasis, sometimes called elephantiasis, is caused by a parasitic nematode worm that is also carried by mosquitoes. It affects more than 90 million people, mainly in Africa and Asia, but also to some extent in South America. Infection by the worm causes gross enlargement of the affected organs, especially the legs or the scrotum, and sculptures show that the disease occurred 5,000 years ago in Ancient Egypt. It is not normally a fatal disease but is extremely disabling and disfiguring. Parasites are generally more common in the Tropics than in temperate regions, and many are found within the people of the forests. Hookworm disease, like filariasis, is caused by nematodes that live in the small intestine of more than 900 million people in the Tropics. The eggs are shed in feces, hatch in the soil, and burrow through the skin of the feet of passing people. They then migrate through the bloodstream until they reach the lungs of their host, and from there they are coughed up into the mouth, which enables them to reach the gut, where they take up residence in the intestine wall. They feed on the blood of their hosts, once they are established in the gut, causing considerable blood loss and can result in anemia. Unsanitary, cramped living conditions, poor hygiene, and barefoot visits to latrine areas all contribute to the spread of this debilitating parasite.

Fungi are extremely successful as parasitic organisms of humans in the Tropics. Most fungi are favored by warm, wet conditions, so the rain forests provide an ideal environment for their spread. Many of the parasitic fungi cause superficial infections of the skin, such as ringworm. This fungus is extremely prevalent, affecting about 35 percent of the people living in the rain forests of the world. Other fungi live just below the surface of the skin, the subcutaneous mycoses, causing inflamed abscesses. They often gain entry to the body as a result of small injuries, such as the penetration of a thorn into a limb. Superficial and subcutaneous fungal infections are rarely life threatening, but systemic infections of fungi can be extremely serious. These usually infect their host through the lungs and respiratory tracts and are particularly common and dangerous among those with low immunological response, such as people suffering from AIDS.

The human immunovirus (HIV), the causal agent of AIDS, had a tropical origin. The first human death from AIDS was that of a man from the Congo Basin of West Africa in 1959. Medical research then demonstrated that the virus causing this disease was a strain of the simian immunovirus (SIV), found in other primates, such as the chimpanzee and the gorilla. Gorillas are particularly prone to SIV and zoologists have surveyed the prevalence of this virus in gorilla populations by analyzing fecal material. The use of feces in this way provides a technique that avoids disturbance and distress to the wild populations because there is no need to locate, sedate, and take blood samples from the animals. Some researchers have speculated that the mutation and transmission of this virus to humans was a direct result of increased forest clearance in West Africa, leading to increased hunting and contact between humans and chimpanzees. Butchering the chimps and gorillas would provide frequent opportunity for their blood to come into contact with minor wounds on the hands of the hunters, and this mixing would give the virus an opportunity to adapt to a new host. The cost has been enormous, with more that 50 million people now infected with the HIV virus. Some tropical countries, such as Botswana and Zimbabwe, have consequently seen life expectancy fall from around 70 years to less than 30. Gorillas also carry the Ebola virus, which in some areas has caused gorilla populations to fall by as much as 95 percent. This virus is deadly to gorillas and also to people if they come into close contact with the animals.

AIDS is a very recent scourge of the tropical peoples, but many other diseases have preceded it, often brought into the region by colonists from other parts of the globe. When Europeans first sought out and slaughtered the dwellers of the South American jungles, they brought with them far more insidious sources of death than the gun and the sword—disease. Many of the diseases common in Europe, such as measles and influenza, proved fatal to the South American populations, who had no natural immunity to these viruses. Other diseases, such as tuberculosis and smallpox, were even more devastating, and the native people of the New World suffered population crashes from which they never recovered.

There are many reasons, therefore, why the natural population levels of humans in the tropical forests are low, even before the advent of Europeans. Food supply, especially animal protein, is limited and not easy to harvest. The opportunities for agriculture are fewer than in the temperate zone and the culture of domesticated plants and animals is highly energy intensive, involving the felling of forests and the maintenance of open clearings. Parasitism and disease are rife, lowering the life expectancy of forest dwellers and putting a brake on population growth. As a result, the population density of the deep tropical forest regions of the world is extremely low. The Amazon forests currently sustain only about five people per square mile (2 km^{-2}), while the forests of Indonesia have densities of about 13 people per square mile (5 km^{-2}). The forests of equatorial Africa are slightly more densely populated, with about 50 people per square mile (20 km^{-2}). Even this, however, is sparse compared with more than 780 people per square mile (300 km^{-2}) found in western Europe, India, and China. Things are changing, however, and many of the tropical forest regions are now seeing population growth rates of up to 4 percent, leading to regions such as West Africa, Sri Lanka, Thailand, and Malaysia rapidly approaching the population densities of the more crowded parts of the world.

The forest dwellers of the world are generally small in stature, which is certainly in part a result of limited protein intake and generally a low-calorie diet. The effects of diet are most apparent in pregnant females, who do not gain weight in the same way as those of developed Western societies. A woman from the United States will normally increase in weight by about 28 pounds (13 kg) during pregnancy, but a woman from the forests of the Congo will on average gain only 11 pounds (5 kg). Consequently, the birth weight of infants is low. While being fed on breast milk, the baby receives a nutritionally balanced diet, but this is dependent on the health and welfare of the mother, who is placed under nutritional strain by the demands of the baby. Once the baby is weaned, poor diet often ensues and infant mortality is consequently high.

One positive aspect of forest dwelling is that blood pressure problems are relatively rare among forest people. Among the people of the developed world, a blood pressure of 120/75 would be regarded as healthy and normal, but many people live with blood pressure elevated above 150/90. To assess blood pressure, measured in millimeters of mercury, two readings are taken, the first (systolic) is the pressure when the heart contracts, and the second (diastolic) is the pressure when the heart relaxes. An inhabitant of the Amazon forest would generally have a blood pressure closer to 90/60. This lower blood pressure may in part be due to the scarcity of salt, sodium chloride, in the heart of the forest. The people of the Amazon often use potassium salts rather than sodium salts for flavoring, and such use may well have beneficial effects. Despite these advantages, however, the life expectancy of a forest dweller is only about 40 years.

Many studies have been carried out to try to determine the carrying capacity of tropical forest habitats for human beings. Just how many people could a tropical forest support if it were managed properly? The outcome of such calculations suggests that between 65 and 260 people could be accommodated per square mile (25 to 100 km^{-2}), but such densities would involve great changes in forest structure and biodiversity.

■ MODERN FOREST PEOPLE

The people of the forest once existed entirely by hunting animals and by collecting plant products, such as fruits, roots, and tubers. This way of life is called hunter-gatherer subsistence, and it is still practiced in some forest societies. On the island of Mindanao in the Philippines, for instance, there are still members the Tasaday tribe, which occupies a remote rocky outcrop on the south of the island. They remained unknown to the Western world until 1966, when they were first "discovered." Very few members of this tribe remain; they survive in a stone-age type of society, living in caves and having only stone tools. Their diet is mainly vegetarian, and they are reputed to know the names and properties of 1,600 different types of plant that they find useful. They eat nuts and fruit, yams, and palm pith, but also catch small fish, frogs, and crabs that provide them with additional protein. They are an extremely peaceful society, possessing no weapons and having no word for war in their language. One might expect life to be hard when depending on collecting forest produce for food, but in fact each member of the tribe spends only a few hours each day collecting food, so the tribe is not short of leisure time. A hunter-gatherer society, therefore, requires low levels of energy expenditure to maintain its food supply, but is only able to sustain very low densities of population.

Most modern forest societies practice some form of agriculture. Domesticated plants and animals have a long history and provide a reliable source of food that is less likely to fail when conditions in the forest become testing, as in the case of drought or extensive storm damage. But an agricultural society needs to modify the environment to make it more suitable for the survival and productivity of domesticated organisms. Plants need the appropriate soil and light conditions, and animals need to gain access to suitable food supplies. On the whole, forest habitats, with their complex structures and high canopies, are not optimal for the growth

Shifting Cultivation

The agricultural practice known as shifting cultivation has also been called *slash-and-burn*. The nomadic people of the tropical forests select an area they consider suitable for cultivation, which usually means sites with relatively dry soils rather than the low-lying alluvial and wetland sites; selection of the area depends partly on the crop that is to be grown and partly on the relative difficulty in clearing forest. Dry sites often carry less biomass and are more fire-prone and are thus more easily cleared, but some crops, such as rice, need wet soils, so the drier sites are unsuitable. Clearance of forest is difficult and involves intensive labor with primitive tools. Larger trees may prove impossible to fell, but these can be killed by cutting a ring around their base, stripping off the bark and the living cells beneath and thus preventing the movement of food materials from the canopy to the roots. As the tree dies, it becomes more susceptible to fire and can be burned. Once the canopy is opened and the light supply to the ground is improved, the soil must be tilled by means of digging sticks and other tools. It is then possible to plant crops as well as forest plants used for medical purposes or food production. The edges of the clearing are often particularly suitable for encouraging the growth of useful native plants, and the dense growth of vegetation that takes advantage of the increased light levels also provides a source of food for foraging animals, such as pigs.

Forest soils are poor in the nutrient elements required for plant growth, so the quantity and quality of crops diminishes year by year as the soils are exhausted. After about five years the site is abandoned and the surrounding jungle is invaded once more as the people move on to a new site. The cultivators use no fertilizers, except perhaps some animal manure, and apply no pesticides. As a result, the clearings are often densely covered with weeds, and crops are often ravaged by local herbivores, but the high rates of plant growth and productivity make the cultivation exercise worthwhile. By growing several different plants together, the forest agriculturalist can also avoid some of the problems of pest abundance that occur in the extensive monoculture farming practiced in the developed world. Farmers often use fire to assist in clearing forest, especially in the drier sites, liberating some of the elements contained within the trees that have been felled and causing a surge of nutrient supply to the soil. Heavy and persistent rain, however, soon takes most of this input of nutrients through the soil and on into the rivers. With the abandonment of the clearing, secondary succession takes place (see "Secondary Forests," pages 62–65) and the forest gradually regains its complex structure.

---

of crops or the grazing of domestic stock, so trees need to be removed and the soil must be tilled. This requires a great deal of effort, and even when an area is cleared, the soil often proves very limited in its nutrient content and cannot support crops for very long (see "Soil Formation and Maturation," pages 46–50). As a result, a distinctive type of agriculture has evolved in the forest regions called *shifting cultivation* (see sidebar).

The forest inhabitants who practice shifting cultivation usually continue to hunt and gather from the unmanaged forest, harvesting fruit and "bush meat" to supplement their diets. In the clearings they have constructed, they can cultivate a range of different crop plants. In the forests of Columbia, for example, banana, yam, and manioc are often planted beside one another, in among the dead remains of roots and stumps from the felled trees. Each clearing provides sustenance for a small village, usually ranging from 25 to 110 inhabitants, but the productivity of the poor soils of the Amazon Basin is very low. In the first year, a manioc crop may reach as much as seven tons per acre, but this is

likely to fall to five tons per acre in the second year, and only four tons per acre in the third year. At this low level of productivity, the crop is hardly worth the effort expended, so the site is usually abandoned after three years and the people clear a new area. After about five or ten years, all of the land within easy reach of the village has been cleared, and the entire group moves on to a new location.

The soils of the drier regions of the Amazon are particularly poor in nutrients, but the wet regions are supplied with a new input of silt as the rivers regularly overflow their banks. The people of these wet regions benefit from the higher productivity of crops when the waters recede, but they undergo periods of famine when they have to move to higher ground during the floods and wait until they can begin cultivation again. The people of the wet forests, however, do not need to move into new regions because their soils are never exhausted, and the size of the village groups is often larger than that of the dry-land people, sometimes becoming as large as 2,500. Like the dry-land tribes, they cultivate manioc but also use corn and wild rice in their diet.

The timing of planting and harvesting, of course, is critical. They need to wait until the soil has dried after the retreat of the floods, otherwise the newly planted crops will rot, but if they leave it too late the waters will return before they have completed their harvest. The wetland people have one other advantage, namely the abundance of fish and other wildlife, such as tapirs, peccaries, and capybara, with which to supplement their diet. They shoot larger animals, including large catfish, using a bow and arrow, and they catch fish by crushing the leaves of some of the local plants rich in narcotic drugs and spreading the material in the water. Fish are stunned, float to the surface, and can be collected by hand from canoes.

One of the most distinctive groups of people from the West African forests are the pygmies, who occupy the rain forests from the west coast through to Uganda and Rwanda. The Mbuti people of the eastern forest are among the smallest, with an average male standing at about four feet six inches (1.37 m); females are even smaller. They are skilled hunters, using bows and arrows, spears, and nets with which to entrap animals. They are even able to kill elephants using these primitive weapons. They live in small bands, up to 100 strong, and create camps with shelters made from branches and leaves. The pygmy people trade with neighboring societies, especially for manioc, bananas, beans, and salt.

The Dyak tribe of Borneo relies mainly on rice cultivation for subsistence. They use complex astronomical calculations to determine the best time for planting and harvesting. They practice shifting cultivation, and normally use a cleared area for only three years before moving on to the next site. It takes the men about three months to clear an area of four or five acres using small iron hand axes. Another people of this region of Southeast Asia, the Punans, live in the deepest forest and are extremely adept hunters, using blow pipes as well as guns. Some clear areas for cultivation, but others hold religious beliefs that prevent them from felling trees, so they exist entirely by hunting. The tribal groups are extremely warlike and have long practiced headhunting, hanging their prizes in the roofs of elaborately constructed longhouses in their villages.

The people of the forest have thus maintained a hunter-gatherer lifestyle, with many adopting simple agricultural systems that involve the cultivation of crops, usually in mixed assemblages, and the domestication of some animals, particularly pigs. The impact of this way of life on the forest varies with the density of humans. In areas of low density, the abandoned clearings form part of the general mosaic of the forest, creating a patchiness that is itself a contribution to biodiversity. In a study of the West African rain forest, ecologists found that the effects of shifting cultivation were still apparent in the composition and structure of the forest an estimated 400 years after human clearance, so the forests that appear to be pristine and untouched by human hands are all likely to bear some subtle human imprint in their composition. People are thus a component of the tropical forest ecosystem and have been for many thousands of years. They may be regarded as an essential part of its biodiversity.

■ FOREST RELIGIONS, MYTHS, AND RITUALS

Living as component parts of the forest ecosystem, as well as partaking in the day-to-day function of the forest, brings the forest people very close to nature. Thus their philosophy and religion relate closely to the natural world. It is perhaps difficult for a society that collects its daily needs from a store to appreciate the dependence on the natural world that is only too obvious to the inhabitant of a forest. Children are born on the mud floor beneath a leaf-covered shelter; water is obtained from a murky river; food is obtained by harvesting insect-infested crops, gathering fallen fruit from trees, or stalking and killing a forest animal that fails to escape capture. Disease is treated with highly toxic herbs in quantities determined by a long experience of trial and error, and death is accepted as the inevitable conclusion of a short and hard life.

Humans are capable of abstract thought, and are spiritual creatures often conscious of forces beyond rational analysis and explanation. To the forest dwellers, the natural world is alive with spirits, both good and evil, that must be encouraged, placated, or warded off, depending on their benign or malign natures. To these ends, many rituals are found among the people who cling to these animistic beliefs. Totemic, decorative items of apparel are very common among forest people, sometimes taking the form of headdresses or body painting, and sometimes involving scarification of the skin, where patterns or totemic devices are carved using a sharp implement. Conducted under unhygienic conditions, such skin carvings often become infected and can lead to septicemia.

The spirits of the dead are particularly to be feared, especially by those who may be responsible for the death through acts of war. The Dyak people of Borneo protect themselves from the spirits of warriors they have slain in battle by removing the heads of their victims and hanging them on their houses or even on belts around their waists. Headhunters would sometimes stitch together the lips of the trophy heads, thus ensuring that their spirits are imprisoned. By abusing the mortal remains of their victims, the victors proclaimed their mastery over both their bodies and their souls, thus protecting themselves from revengeful spirits that wander the world. Consuming the flesh of victims, or cannibalism, is an additional means of establishing physical and spiritual superiority and is regarded as a way to assimilate the courage and the strength of the fallen warrior. In some South American tribes, cannibalism was extended to the immediate family. By eating part of a dead relative, one displayed affection and admiration for their attributes and acquired some of these in the process.

Tokens of good fortune are also found among the Dyak people, as in the case of a crocodile tooth, which is regarded as a protection for a woman during her pregnancy. The newborn baby is also given good luck tokens, such as a piece of thorny bush that is placed above the baby's head to protect it from the evil spirits as they try to descend into the new soul. The Dyaks also worship a number of gods, including Gana, who protects the hunter and is greatly impressed by elaborate headdresses that they wear. The supreme god is Sang Huang, the ruler of the seven heavens, creator of the universe, who rules over the lesser gods, each of which has control over the various levels of heaven. The supreme god of the Semang people of Malaysia is Ta Pedn, who controls a heaven in which fruit trees grow and where people reside after death. The ministering spirits of this supreme god inhabit trees and flowers, and they are able to commune with human witch doctors through these earthly symbols.

Creation myths abound among the forest people, such as the inhabitants of the Andaman Islands in the Indian Ocean, who believe that a storm goddess, Paluga, created the Earth, which now rests upon a massive palm tree. They fear that if Paluga becomes angry she may shake the tree and cause the Earth to fall into chaos, a very likely event in the case of tsunami, which is an ever-present fear among the people of this region.

The Mbuti pygmies of Africa regard the forest itself as the great source of all life, which determines all that happens from day to day and protects them from evil. Sometimes the forest seems to lose concentration and forget them, as when disaster falls in the form of drought, famine, disease, and death. On such occasions the pygmies believe that a noisy festivity is required to wake up the slumbering forest. In Brazil, the Xingu Indians have a vast repertoire of just-so stories explaining all of the minor features of the forest, such as why birds sing and who first discovered fire. This latter story involves the king vulture, who alone possessed the secret of fire. A cunning warrior drew a picture of a dead deer in the soil, and when the vulture descended to feed upon it he caught it and would release it again only in exchange for the gift of fire.

Some of the more complex societies that arose from the people of the forest, such as the Mayan civilization of Central America, developed more complex religions and rituals. The Mayans were strongly dependent on the arrival of the spring rains, and they went to great lengths to placate and stimulate the rain god Chac. If the rains were delayed, they erected a great pyre of rubber trees and burned it to create large clouds of black smoke. This evidently got up Chac's nose, and he would bring rain clouds to quench the fire. Some scientists have speculated that the particulate matter in the smoke formed an *aerosol* and may have acted as nuclei for the formation of water droplets and created rain clouds, in which case the Mayan practice had a very direct impact on the weather.

The rituals and myths of the forest people thus developed along lines that were very close to nature and seemed very strange to the people of European civilization who first encountered them.

■ FOREST EXPLORATION

Five hundred years ago the tropical forests were practically unknown to Western civilization. There were rumors and exaggerated tales of mountains and lakes deep in the heart of Africa, based on ancient accounts from Greek explorers in the fifth century B.C.E. and derived from the stories brought back by Arab traders and slave traders on the east coast of the continent. In 1493, Christopher Columbus became the first European to describe a tropical rain forest on the island of Antigua in the Caribbean (see "Introduction," page xvii). A Spanish navigator, Vicente Yáñez Pinzon (1460–1542), who accompanied Columbus on his 1493 voyage, sailed back to the New World in the year 1500, and passed south along the coast of South America. On reaching the mouth of the Amazon River, he was greatly impressed by its size, regarding it as the entrance to a vast inland lake. The impenetrable jungles of the Amazon Delta, however, defied exploration and the largest rain forest on Earth remained a mystery.

During the course of their conquest of South America, the Spanish found the Andes Mountains to be the most promising area, in part because of the great wealth in gold to be plundered from the Incan people, but also because the mountains proved more suitable for horse transport and access. In 1541, Gonzalo Pizarro (1506–48) led an expedition along the Andes chain, setting out from Ecuador and traveling south through what is now Peru. There the expedition became entangled and lost in endless tracts of montane forest and they began to run short of food and supplies. At this point the group split up, one part of the group heading back over land, while the other, under the leadership of Franciso de Orellana (1511–46), built a boat and sailed down a river that they hoped would lead them to a village that might replenish their food supplies.

Orellana's voyage downriver has become one of the most remarkable of all tales of exploration. He soon discovered a village with abundant food, but decided not to return to Pizarro but to continue on down the river, eventually covering over 2,000 miles (3,200 km). He encountered many different tribes, some of whom were far from friendly; one particularly warlike group seemed to consist entirely of pale-skinned women. This encounter brought to mind the ancient Greek legend of a tribe of warrior women in Asia who bore the name Amazons. When the tale was repeated on returning to civilization, the name "Amazon" became associated with this great South American river. It is still not known whether the tale of the women warriors was correct, or if the sailors were confused by long-haired males, but the story added glamour and mystery that was greatly prized by early explorers. Orellana reached the mouth of the Amazon and returned directly to Spain, avoiding the displeasure he would undoubtedly meet from the disgruntled Pizarro, who accused him of desertion. Meanwhile, Pizarro had given up waiting for the return of Orellana and succeeded in his overland return. But only 80 of his 220 Spanish soldiers survived the expedition, and none of the 4,000 Indians. Orellana had sailed the entire length of the Amazon River and was thus the first European to travel deep within the tropical rain forest.

The African forests were first approached from the east, as Christian missionaries and other explorers penetrated the Dark Continent from the coast of the Indian Ocean, following in the steps of Arab slave traders. In this way, these pioneers crossed the savannas of the east and encountered the forests and the swamps of what is now Uganda on the west side of Lake Victoria. But the main body of the African rain forest lies father to the west, in the basin of the great Congo River. This region remained uncharted until an American explorer, Henry Morton Stanley (1841–1904), penetrated the region. Stanley was Welsh by birth, making him eligible for British knighthood that he eventually received in 1899 from Queen Victoria. But he had become an American citizen and made his living by exploring and reporting his travels. His most famous discovery was David Livingstone (1813–73), who had disappeared into the heart of Africa and whom Stanley eventually found in 1871. In the course of his wanderings, Livingstone made an important discovery, the Lualaba River, which ran northward from the country to the west of Lake Tanganyika. In Livingstone's view, this could be the source of the Nile River, which was the center of great public interest at the time.

In 1874 Stanley obtained financial backing from newspapers in New York and London, enabling him to launch an expedition to explore further the central regions of Africa. Stanley located Livingstone's Lualaba River and set off by boat, firmly expecting to end up on the Nile in Egypt. Instead, he encountered many trials, including battles with local people, near starvation, disease, and all the stresses associated with a permanently wet tropical environment. But eventually he found himself in the mangrove swamps at the mouth of the Congo on the Atlantic coast of West Africa. He had set off with 356 companions, but only 114 survived the journey, which had taken a total of three years to com-

plete. Stanley thus became the first man of European origin to traverse the rain forests of Africa.

The forests of Southeast Asia are largely scattered on different islands, so their exploration has been more piecemeal than that of the Amazon and Congo basins. They have, however, proved just as impenetrable to explorers, and new tribes and peoples were still being discovered in the 1960s. The extent and distribution of tropical forests are now well known and accurately mapped using satellite observation, but this does not mean that all of the forests' secrets have been revealed. Close scrutiny of the tropical forest is still urgently required, and scientists have become the new generation of explorers as they document the myriad inhabitants of the rapidly disappearing forest.

◼ CONCLUSIONS

Human beings arose as part of natural ecosystems, including the tropical forest. The species *Homo sapiens* first appears in the fossil record about 260,000 years ago and by 50,000 years ago had developed a complex culture involving the production of stone tools and decorative works of art and jewelry. The species spread out of Africa and expanded through southern Asia and reached Australia by 40,000 years ago. The final ice age limited the spread to the north, but as the ice retreated after 20,000 years ago, humans moved into northern Asia and over the land bridge into North America. By 12,000 years ago, people were spreading south through the New World into Central and South America, and all of the major continents, with the exception of Antarctica were occupied.

Cave-dwelling people occupied parts of the Amazon Basin 11,000 years ago and lived by hunting small animals and gathering fruits from the forest. Rain forest was spreading at this time as the climate in equatorial regions became warmer and wetter, and people occupied the forests of Southeast Asia and Africa. The idea of plant domestication arose independently in several different parts of the world. The first evidence of squash domestication and cultivation in Central America may date back as far as 10,000 years, at the same time as domestication arose in the Middle East. Certainly, by 7,000 years ago manioc and yam were being cultivated and harvested in the New World Tropics. Sugar cane was domesticated in Southeast Asia by 5,000 years ago, and sorghum and millet were cultivated in Africa by 4,000 years ago. By 3,600 years ago, cacao was being used in Central America. So the history of cultivation within the tropical forest regions of the world is very ancient.

Early techniques of farming were primitive, based on stone tools, but the discovery of iron brought more sophisticated tools, and these have been present in the forests of India for at least 2,000 years. Extensive forest destruction then became possible, but catastrophe was rare and restricted to island situations where resources were limited and populations expanded too fast. Easter Island in the South Pacific is an example of how overexploitation of limited resources can lead to the collapse of a culture in the Tropics.

Human population levels in the forest were held in check by several factors, one of the most important being disease. Malaria is very widespread through the world's tropical areas and is likely to have limited human population growth throughout history. It currently affects over 500 million people in the world and is carried by mosquitoes, as are yellow fever and lymphatic filiariasis, a parasitic disease that causes massive swellings of limbs. Dengue fever is a mosquito-carried virus disease, which humans share with monkeys, and HIV is another virus affecting primates that probably arose initially in wild populations. These diseases have a heavy impact on human populations through the reduction of life expectancy. The spread of AIDS in Botswana, for example, has reduced life expectancy to less than 30 years.

Population densities in the tropical forests varies, being least in the Amazon jungle, with fewer than five people per square mile (2 km⁻²), and highest in West Africa with about 50 people per square mile (20 km⁻²). Some scientists believe that populations up to five times as great could be supported in these areas, but it would undoubtedly have a severe impact on the forest structure and biodiversity.

Some of the people of the forest still live in a stone age culture and exist by hunting and gathering. Many tropical forest regions, however, are occupied by people who use shifting cultivation—clearing an area, cultivating it for a few years, then abandoning it and moving on. All forest people, however, supplement their agriculture with hunting and gathering, harvesting bush meat from their local forest surroundings.

The people of the forest are strongly dependent on their local habitat for their food production, and most cultures have developed rituals and religious systems that involve the animals and plants that surround them. Most believe that the forest is occupied by spirits, some good and some evil, which must be placated or warded off to avoid disaster. Personal decoration and various rituals have evolved as a means of pleasing the gods. Some of the religions have a belief in a supreme creative being, but often as a commander of lesser gods.

The tropical forests were unknown to Western civilization until the period of global exploration in the late 14th and 15th centuries. The Spanish conquests of South and Central America revealed the vastness of the forests of the New World, particularly those of the Amazon Basin. In Africa, the deep forests of the Congo Basin remained a mystery to European and North American cultures right up until the late 19th century, when geographical exploration once again captured the public imagination. With the arrival of new peoples, cultures, technologies, and ideas, the original residents of the tropical forest have begun to face large and inevitable changes in their way of life.

10

The Value of Tropical Forests

When people hear the word value they immediately think in financial terms. How much money would this particular object or commodity fetch in the market place? In many areas of life this approach is perfectly appropriate, but there are other areas where it is not; there are some things that money simply cannot buy. The problem faced by conservationists and ecologists is that they are often arguing the case for the protection of parts of the environment that could be exploited in alternative ways, and the economic arguments for exploitation are often very convincing. The question of evaluation in environmental matters is therefore one that requires careful thought and consideration.

■ THE PROBLEM OF EVALUATION

Conservation is more than simply preservation. It involves the active management of an ecosystem to ensure that it continues to function, and sometimes this means that human activities need to be sustained. Landscapes have often been modified by a long history of human intervention, as in the use of fire, or farming, or the harvesting of certain products, including hunting. The conservationist is not always averse to the sustainable use of an ecosystem for human support or recreation, but is opposed to activities that will damage the ecosystem and reduce its biodiversity. It is thus possible that conservation can accommodate economic requirements in the form of a sustainable harvest from a managed ecosystem. In the case of forests, for example, the extraction of timber is the most obvious means of obtaining economic gain from the ecosystem. Clear felling is the simplest approach, which is likely to produce the maximum immediate economic advantage, but clearly results in the destruction of the structure and a loss of biodiversity. There are alternative ways of harvesting the products of the forest that are not quite as attractive as immediate sources of revenue but could result in long-term, sustainable income from the ecosystem.

In a study of an area of tropical rain forest in the Amazon Basin, economists calculated that the clear felling and extraction of timber would yield approximately $400 per acre (almost $1,000 per ha). They also calculated that the removal of other forest products, such as fruits, herbs, and latex (see sidebar "Rubber," page 205), could generate $280 per acre per year ($700 per ha). If these figures are correct, then clearly the regular extraction of renewable materials would very soon exceed the immediate financial gain of timber extraction. Other economists, however, have questioned these figures on the basis that the values depend on whether there is a local market for the products, or whether the cost of transport to the marketplace would make the whole exercise nonviable. Many conservationists and ecologists are very cautious about resorting to economic justification for their recommendations because they recognize that the real value of conservation is very difficult to evaluate in this way. They look instead for other criteria on the basis of which they can present a conservation argument.

Biodiversity is itself extremely valuable. Every species of organism on the planet has its own unique genetic makeup that has been honed over millions of years to fit the environment in which it lives. Even within a species there are variations in genetic constitution that confer adaptability to the organism, and all of these variants are of value because they could not be reconstructed if lost. Humans place great economic value on certain works of art because they are unique and could never be replaced if lost, and species of plants, animals, and even microbes, are similarly irreplaceable. Extinction is indeed forever.

But are all species of equal value? Is the giant panda worth the same as the smallpox virus? This is a difficult question to answer, but until the possible value of each species is known, it is wise to avoid steps that cannot be retraced. Many obscure and unlikely species have proved unexpectedly valuable as the sources of drugs and chemicals or as laboratory research tools, so discarding species before they are fully appreciated is extremely unwise. There is also an ethical argument. Some believe that species apart from humans have certain rights, the right to survival being basic among these. Even if the dodo or the passenger pigeon had little economic value, their loss at the hands of human mismanagement is something that most people would regret. The world is a poorer place without them.

The same argument can be applied to whole ecosystems and landscapes. Many people feel uplifted by natural environments and wilderness. For those who may never have the opportunity to visit truly wild places, it is still reassuring to know that they exist. Tourism to the world's wildernesses (see "Ecotourism," pages 219–220) is greatly on the increase, and television wildlife documentaries are a source of pleasure to many. So the existence of global biodiversity can have a positive impact on human health and well-being, even though this is difficult to evaluate in economic terms.

Many of the Earth's great ecosystems, the tropical forests among them, have significant effects on the global environment, including climate (see "The Forest as a Carbon Sink," pages 209–211). Scientists have trouble quantifying such effects because of the scale of the exercise and the difficulty in obtaining precise measurements. Most agree, however, that conservation is necessary to ensure the continued function of global processes, including the interaction between the atmosphere and the biosphere. The economic implications of this argument are so vast as to be incalculable.

The global aspects of conservation are vast and may be difficult to conceive, but at a local level it is often possible to appreciate more easily the importance and value of natural ecosystems to the people who are a part of them. In the case of the tropical forests, human populations are often quite low, but these people are nonetheless dependent on the continued productivity of the forest. Any assessment of the value of tropical forests must begin with the needs of the local people. People are part of the forest and sustaining the forest in the future will require carefully planned *management*.

■ TIMBER PRODUCTION AND FORESTRY

Timber is the most evident of the forest's resources. The enormous bulk of wood present in the trunks of the forest giants holds great economic attraction to those whose sole interest is the exploitation of the environment. For the forest people, timber has been a source of building material for their homes and shelters and has been an obstruction to the development of agriculture. The opening of forest clearings by felling smaller trees and ring-barking larger ones (see "Domestication and Agriculture in the Forest" pages 188–190) was an immediate requirement for those forest people who adopted an agricultural way of life.

In addition to timber providing materials for building, wood is also useful as a fuel, and many people who live in the tropical forests of the world still use wood as a major energy resource. Wood, especially fresh green wood, is not a very efficient fuel as it is difficult to generate high temperatures with a wood fire. But charcoal, produced by the partial combustion of wood under conditions of limited oxygen supply, is much more efficient. Charcoal burning (see illustration on page 201) is still a widespread industry among tropical forest peoples.

As commercial interests brought timber merchants from developed countries, selective logging began to take hold. Not all forest timbers are of equal value because their properties, such as hardness, durability, and color, vary with the species. Early commercial exploitation concentrated on attractive woods for cabinet making, including ebony (*Diospyros ebenum*), satinwood (*Chloroxylon swietenia*), mahogany (*Swietenia* species, see sidebar "Mahogany," page 115), and teak (*Tectona grandis*) (see sidebar). The timber seekers would locate these trees within the forest and fell the individuals that interested them, extracting them by buffalo or mule, dragging them through the forest. This created small openings in the forest and was less damaging than clear felling, but selective logging and extraction of this type does result in some damage to surrounding trees. Ecologists have recently studied the effects of selective logging in Guiana and found that the number of trees removed was just 1.2 per acre (3 per ha), but that in the process of felling and removal the loggers damaged 38 percent of the remaining trees. They measured the impact of the logging on biodiversity by checking bird numbers and found that these had decreased by about 30 percent, which they attributed to the loss of deep shade habitats. When they revisited the sites 10 years later they found that the forest had still not fully recovered its structure or diversity. Although selective logging may seem like an attractive option, it does have lasting negative effects on the forest.

During the 1950s two great changes took place in the exploitation of tropical forests for timber. First, there was the invention of the one-man-operated chain saw, which allowed a single forester to fell even a massive tree in just half an hour. Then came an increase in world demand for timber, including the less durable and lighter timbers that had previously been neglected. Selection was no longer needed or desirable, and extraction of timber was simplified by clear felling large areas of forest and building roads to allow motor vehicles to drag out the felled trees. This has allowed timber harvesting to proceed at much faster rates and with far greater destructive impact than was previously the case. Clear felling, especially if followed by fire, leaves soils exposed to the heavy rainfall, so that leaching and erosion take place and the regeneration of forest is slower. Land managers, especially in South America, often use the cleared

Wood from an area of cleared tropical forest is being burned under conditions of low oxygen supply, and this results in the formation of charcoal, which is a fuel that produces a high temperature when combusted in air. *(Peter D. Moore)*

land for grazing cattle, a practice ensuring that the forest does not return.

Sustained forestry in the rain forest is possible, however, and biodiversity loss can be minimized if the harvesting is properly managed. Instead of clear felling large blocks of forest, the harvest should be confined to strips along the contours at the base of hills (see diagram on page 202). Any nutrients leached or soil eroded from this strip is carried downhill into the valley forest, which is left untouched. Timber can be extracted along the line of the contour to avoid damage to the forest above and below the harvested strip. When this line of forest has begun to regenerate, or when new trees have been planted, the next strip to be harvested is the contour line immediately above. Nutrients are leached from the newly harvested area and are carried down slope causing soil *entrophication* or fertilization. This enrich-

Teak

Teak (*Tectona grandis*) is among the most prized of the tropical hardwoods. It is native to India, Burma, Thailand, and Vietnam, and grows in the forests neighboring the Himalaya mountains in the monsoon forests. It is a deciduous species that requires an annual dry season so is not found in the true rain forests. When the monsoon arrives in India in July, the teak tree puts out its new set of leaves and comes into flower. It is a tall tree, growing to a height of 150 feet (45 m), and old specimens may have a basal diameter of eight feet (2.5 m). As in many of the tall forest trees, the base of the trunk is supported by a series of buttresses.

The wood of teak is particularly attractive, having a golden or reddish brown color, often streaked with darker colors. When harvested and cut into planks, it dries without cracking or damage and is an extremely durable wood, even resisting termite attack. Teak is strong, making it a valued wood in shipbuilding and for high-quality furniture. One problem with harvesting teak is that the wood is very dense, so that when it is first felled and has not been dried it sinks in water. It cannot be extracted from forests by floating down rivers unless the wood has been dried first. In India, trees are ring-barked while still standing and are allowed to die and dry in that position for two years before being felled. The outcome is that the wood is lighter and will float when it is eventually harvested.

Over the course of time, wild specimens have been extracted from most areas of teak's natural growth, and its increasing rarity and value have meant that it is now usually used as a veneer rather than as a solid wood. Teak can be grown as a forestry tree and is planted in Sri Lanka, Trinidad, and Nigeria.

ment supplies the regenerating forest below, which is in a state of active biomass growth so is able to absorb the extra nutrients, which are thus not lost to the ecosystem. The next strip to be harvested lies immediately above this, and so on until the cycle is complete and the lower slopes are ready for their next harvest. It is wise to avoid harvesting the very summit of the hill because this could lead to permanent loss of soil from the hill tops.

There are several advantages to this system of forestry. The disturbance to the forest by harvesting and extraction is localized and kept to a minimum. Biodiversity is maintained because areas of forest in different stages of secondary succession are present alongside one another so that reinvasion can easily take place. Nutrients are conserved within the forest because all leached elements are available to the rapidly growing forests immediately down slope. The management system thus makes both ecological and economic sense because less fertilizer is needed to encourage the regrowth of trees.

Many tropical trees have proved suitable for forestry growth, including afara (*Terminalia superba*) and meranti (*Shorea* species). These trees grow rapidly, reaching a harvestable size in less than 40 years, and they are reasonably resistant to insect attack even when grown in monocultures. Some of the traditional forest hardwoods, such as mahogany, however, are less suitable because they become infested by shoot-boring insects when grown in pure stands. This demonstrates how the biodiversity of the natural forest and the wide spacing of trees of any given species assist in controlling the intensity of insect predation on plants.

Harvesting tropical forest timber results in great losses of nutrients from the ecosystem. These losses can be reduced by a system of felling trees in strips along the contours of a slope, starting at the lower end of the slope and working progressively uphill. When a strip is harvested, some nutrients are released into the soil and drain downhill, where they can be captured by the growing trees on the lower slopes.

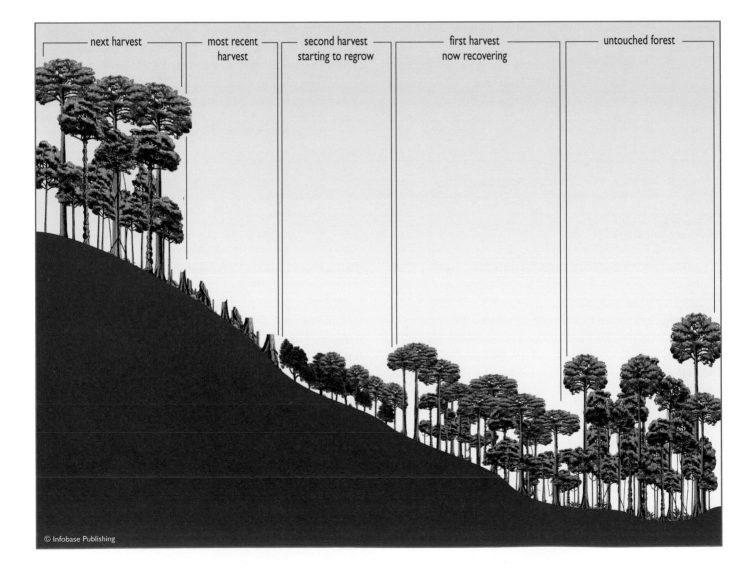

next harvest | most recent harvest | second harvest starting to regrow | first harvest now recovering | untouched forest

■ NON-TIMBER PRODUCTS OF THE FOREST

The tropical forests produce many materials apart from timber that are useful to people. Many tropical fruits are valuable for food manufacture—few more so than cocoa (*Theobroma cacao*). The very word *Theobroma* means "food of the gods." The fruits of the cocoa tree have been harvested for well over 2,000 years and were first taken to Europe by Hernán Cortes (1485–1547) following his travels in southern Mexico. He had been offered cocoa by the Aztec emperor Montezuma in the form of a drink called "chocolatt," which the Aztecs prepared using vanilla and spices. The Spanish replaced the spices with sugar and thus concocted a material closely resembling modern cocoa. The habit of drinking cocoa spread rapidly through Europe and became the first of the New World tropical fruits to attain such popularity. As the fruit became increasingly popular, speculators introduced the tree from which it came to other tropical countries, including Ghana, which is now the world's greatest producer of cocoa.

The cocoa tree is an understory tree, growing to a height of only 15–20 feet (4.5–6 m). It branches profusely and the branches become weighed down with heavy pods that grow directly onto the branches and trunk. A tree can produce up to 70 pods in a year and each pod may contain up to 70 seeds or beans. In the wild, monkeys break open the pods and eat the fleshy material surrounding the seeds, which they drop and disperse. When people harvest the pods, they open them and scoop out both the flesh and the seeds, then ferment the mixture in curing sheds. Yeasts conduct the fermentation and produce ethyl alcohol, which on contact with air subsequently oxidizes to acetic acid. After a week, the whole mix is laid out to dry for an additional two weeks, and then the beans are collected and cleaned. In the final stage of preparation the beans are roasted and ground to produce cocoa powder. It takes between seven and 14 pods to produce one pound (0.45 kg) of cocoa. The massive worldwide demand for cocoa and chocolate ensures that this product of the tropical forest is extremely valuable in economic terms. The removal of cocoa trees to other parts of the tropical world means that the original source areas of the product no longer receive the main revenue from the product, which is unfortunate for the relatively poor countries of Central America.

One of the most widely consumed fruits of the tropical forest regions is the banana (genus *Musa*), shown in the illustration. The inflorescences of this plant hang from the main stem, which results in the whorls of banana fruits pointing skywards as they develop. The plant is as tall as a tree, but is actually herbaceous, and after fruiting the main stem dies but leaves behind a side shoot that develops into the new plant.

The banana plant is a native of the tropical forests of southern Asia, but is now cultivated around the world for its edible fruits. *(Commander John Bortniak, NOAA/Department of Commerce)*

Another valuable tropical product is the breadfruit (*Artocarpus communis*), illustrated on page 204. This plant will always be associated with the infamous mutiny on the *Bounty*, because it was Captain Bligh's commission to take breadfruit trees from the Pacific island of Tahiti and introduce them to the West Indies. Although Captain Bligh was interrupted in his task and failed, the introduction has subsequently been achieved, though breadfruit has not become very popular outside its original Pacific region home.

The spices have been some of the most valuable tropical fruit products over the course of history. Their value was placed alongside gold and silver, and spices were often included among the most precious of gifts given to princes and kings. Nutmeg (*Myristica fragrans*) is derived from the seeds of a tree that grows on the Molucca Islands of Indonesia, and was so precious that the possession of the islands at one time became the cause of war between the Dutch and the British. Cinnamon is derived from the bark

Breadfruit (*Artocarpus communis*) is a plant of the tropical Pacific islands, where its rich starch content provides a staple food for the human inhabitants. *(Blaney Photo)*

linate the plant in its homeland were not present to perform this function in other parts of the world. Spice farmers then devised a method of pollinating the vanilla flowers by hand, and the plant can now be grown in many tropical regions.

Many other forest products that do not supply food and flavoring have proved very useful to people. Before the advent of synthetic fibers and packaging materials, tropical plants were of major significance for these purposes. Jute (*Corchorus capsularis*), for example, produces fibers that have been widely used for coarse cloth, matting, sacking, and rope manufacture. The jute fibers are obtained from the stems of the plant and are derived from the vascular tissues. Farmers harvest the crop after only five months' growth and extract the fiber from the plant stems by a process called retting. The stems are bound together and then left to lie in shallow water in a ditch or pond for up to a month, during which time the softer tissues decompose and the tough fibers remain.

Rattans (*Calamus* species) are climbing palms that clamber through the forest canopy. They can grow up to 600 feet (180 m) in length and hang from the canopy in long ropes, made famous by the movies of Tarzan swinging from tree to tree with their aid. Jungle people harvest them for weaving into furniture and screens. Bamboo (*Bambusa* species) is also used for a range of different construction purposes, such as furniture and the construction of scaffolding. Although bamboo is a grass, it is unusual in producing highly strong yet flexible hollow wooden stems.

Many tropical forest plants produce a variety of chemicals, some of which are proving of great value in industry. Perhaps the most famous of these plants are the producers of latex, including the rubber tree (*Hevea brasiliensis*) of South America (see sidebar on page 205). Latex is a white, sticky substance found in special ducts running through the stems and leaves of a range of different plants and has a protective function. It seeps out of wounds and seals them as it dries, preventing the invasion of fungi. It also deters many insects that find its texture, taste, and toxicity unacceptable in their food. The properties of latex have been exploited for many different purposes by humans ever since prehistoric times (see "Early Forest People," pages 187–188).

Rubber is not the only latex that is commercially important. The zapote tree from Central America (*Manilkara zopota*) is the source of a type of latex that forms the basis of chewing gum manufacture. The tree is grown commercially on the Peninsula of Mexico, where the planted forests serve an additional function, providing a winter habitat for many North American songbirds. The balata tree (*Achris balata*) from the northern parts of South America is the source of a highly elastic type of rubber that is perfect for the construction of golf balls.

Some trees produce gums and resins, rather than latex, as protective materials when they are injured, and these can

of a tree (*Cinnamomum zeylanicum*) that grows in southern India and Sri Lanka and has been used there for at least 4,000 years. During the rainy season the bark of the tree can be removed in strips, but must then be left for a further two years to recover. Vanilla (*Vanilla planifolia*) is derived from a clambering orchid native to Central America, although like so many tropical spices it is now cultivated around the tropical world, especially in Madagascar. The spice is obtained from cured fruit pods, which must be boiled then dried under carefully controlled conditions for six weeks. As in the case of cocoa, it was Hernán Cortes who first took this spice to Europe, where it proved popular as an addition to drinks and tobacco. Unlike cocoa, however, vanilla proved very difficult to cultivate away from its native Central America. It grew perfectly well but refused to fruit. In 1836 botanists finally realized that the problem lay with pollination, because the specific bees and hummingbirds that pol-

Rubber

The rubber tree (*Hevea brasiliensis*) is a member of the spurge family, Euphorbiaceae, many of which contain milky latex in their tissues. It is a tall tree, up to 120 feet (36 m), that grows in the floodplain of the Amazon River, where its seeds are dispersed by water. The elastic properties of the dried latex from a number of tropical plants were known to the Aztecs and fascinated Christopher Columbus when he first came across it. But it was a French explorer, Charles-Marie de la Condamine (1701–75), who first brought the rubber tree to the attention of scientists in Europe. He was traveling in the Amazon Basin and taking measurements of the rate of water flow when he first observed rubber trees in 1751. The raw rubber became a curiosity in Europe because of its softness in warm conditions and relative hardness when cold. The name rubber came from the observation that a hardened piece of latex could erase the marks of a graphite pencil on paper. Rubber remained no more than a curiosity until chemists discovered in 1839 that when it was combined with sulfur, in a process known as vulcanization, the rubber became tough and firm, yet flexible. These properties were to make rubber immensely important when bicycle and motor car tires were invented in the late part of the 19th century. Fortunes were then made by developing industrial methods for harvesting the latex from the trees.

At first, rubber was extracted only from wild trees, which were located in the forest and then cut in spirals around their trunks. The latex that seeped out from beneath the bark was collected in cups strapped to the trunk at the base of each spiral cut. In the 1890s the Brazilian city of Manaus became rich on the proceeds of the rubber trade. The Europeans, however, were anxious to collect samples to grow in other tropical regions where they could control the harvest more effectively. To this end, the Royal Botanical Gardens at Kew, England, arranged for the transport of 70,000 seeds from Brazil to the gardens, where just 2,800 germinated successfully. Some of these were sent out to Sri Lanka (then Ceylon) and Singapore, where the trees were grown in monoculture plantations and the rubber harvested. It proved impossible to grow the trees in monoculture in Brazil

Historical photograph of a rubber tree being tapped in Borneo in 1926. Latex drains from the cuts in the injured bark and collects in a vessel at the base of the tree. *(Leonard Johnson, NOAA/Department of Commerce)*

because pests, particularly a leaf blight fungus, were found to reach epidemic proportions when this was attempted. As a result, Asia became the rubber-producing center of the world (see illustration) and has remained so up to the present day, supplying over 90 percent of the world's rubber. The plantations consist of rows of rubber trees, often with other crops grown between them. They are tapped for rubber when they are five years old, and reach maximum rubber production five years later. At that stage the rubber harvest can be as high as 1,300 pounds per acre (1,460 kg ha⁻¹) each year.

be useful for the production of varnishes in the furniture industry. The Congo copal tree of West Africa (*Copaifera demeusei*) exudes a resin from injured bark that can be tapped in a similar way to rubber, by cutting the bark and collecting the resin in pots strapped to the trunk of the tree. The tree is found in swampy areas of the rain forest so is not easily accessible. The people who tap the trees also dig for lumps of resin buried in the mud beneath the trees, as these materials are harder and more valuable than the freshly harvested resin. Similar products can be obtained from some South American trees, such as *Copaifera reticulata* and *Hymenaea courbaril,* but the materials they produce are softer than the African resins and are usually regarded as inferior as sources of varnish. Foresters collect the gum from *Copaifera reticulata* by drilling a hole into the base of the tree and inserting a hollow tube, usually of bamboo, and collecting the gum as it is exuded. One other member of this genus of trees (*C. langsdorffii*) from Brazil is a potential source of a hydrocarbon fluid that could be used as an alternative fuel to diesel in motor vehicles. The Japanese have experimented with the petroleum nut tree (*Sapium sebiferum*) from the Philippines, which can be planted in a backyard. Each tree will produce up to 15 gallons (57 l) of liquid organic fuel per year. As geological reserves of fuel dwindle and atmospheric carbon dioxide increases, the possibility of using tree-based fuels for providing a carbon-neutral and renewable resource looks increasingly attractive.

Coniferous trees are familiar to most people as sources of resin, and these are also found in the tropical forests and can be exploited commercially. In Southeast Asia, for example, *Agathis alba* is a major source of resin. Like many members of this genus, *A. alba* is a giant of a tree, growing to 200 feet (60 m). Past injuries to branches cause large accumulations of hardened resin on its trunk, which can be collected. Some people, such as the inhabitants of the Celebes Islands, soften the *Agathis* resin by heating it and then burn it to produce light. Fishermen in the area use burning *Agathis* resin to attract fish at night.

The biological function of resin is to protect wood from fungal attack, which is particularly necessary in the hot, wet climate of the tropical forests. Extracts from resins, therefore, can be used for timber protection, and substances like turpentine and creosote are derived from some tropical pines, such as *Pinus caribaea* and *P. merkusii.* Distillation of the extract takes out the low-boiling-point turpentine oil and leaves a residue of resin. The resin, called rosin, is valuable in a variety of industries, including perfumery, insecticides, inks, and adhesives.

The barks of many trees, including tropical forest trees, contain protective compounds collectively called tannins, which contain phenols and other astringent chemicals. These materials have proved useful over many millennia in the treatment of animal hides, resulting in the production of leather. The cutch tree (*Acacia catechu*) of India, for example, is still used for this purpose.

Many tropical plants contain pigments that have been exploited commercially in the dye industry. The sappanwood tree (*Caesalpinia sappan*) of India and Southeast Asia, for example, has a rich orange wood that is the source of a dye used for coloring cotton, wool, and silk fabrics. The logwood (*Haematoxylon campechianum*) of Mexico was being used

as a source of red dye when Cortez first visited the country. It is now used commercially and has proved valuable in the biological sciences as a stain for microscopic preparations.

Perhaps the most extensive and promising of the uses of forest products, however, is in the pharmaceutical industry.

◼ FOREST PHARMACOLOGY

The tropical forests abound with insect and other herbivores, and most tropical plants have developed chemical toxins to deter grazers (see "Plant Defenses," pages 128–130). Chemists have expended a great deal of effort in the study of these toxins because they can be used commercially, both in agriculture and medicine.

The most obvious use of plant toxins is in the defense of crop plants from insect pests. Many natural products have proved useful in agriculture, especially in contrast to some of the synthetic insecticides, such as DDT, which have proved to be so environmentally persistent and harmful. One of the most widely used natural pesticides is derived from the Southeast Asian species of the climbing plant genus *Derris.* The most important species for commercial exploitation are *D. elliptica* and *D. malaccensis,* now cultivated on a large scale in Malaysia, the Philippines, and West Africa. The roots of *Derris* accumulate a poison, rotenone, which protects them from soil pests. As in the case of many forest products, the native people of Asia have long known about the toxic properties of *Derris* roots and have used the extracts as a fish poison and as an arrow-tip poison. Rotenone is extremely potent; even when diluted 300,000 times it can still kill fish. Its wider value, however, lies in its toxicity to insects; it is now widely used as a domestic and horticultural insecticide. It is relatively harmless to mammals and does not persist and accumulate in food chains like DDT does, so it has become a valuable "natural" pesticide worldwide.

All tropical forest societies have adopted the use of herbal medicines for controlling disease, often linking these to rituals and religious ceremonies, and often placing their use in the hands of shamans and witch doctors. An oral tradition, often surviving long periods of time and derived from extensive trial and error experience, has led to the development of a great body of knowledge about the uses of plants for medical purposes. The survey and study of such a great body of knowledge, called ethnobotany, is now an important area for pharmaceutical research. Perhaps the best known and most widely applied outcome of ethnobotanical research relates to the use of quinine in the control of malaria (see sidebar on page 207).

Many medicines are actually poisons. Their value in the treatment of human disease is based on the fact that low doses can prove more damaging to the disease agent, such as a parasite or bacterium, than to the human host. One fine example of this is use of the well-known poison strychnine in the treatment of gut disorders and fevers. Strychnine is a highly toxic alkaloid found in seeds of the snakewood tree (*Strychnos nux-vomica*). Its biological significance is clearly to discourage animals from eating the seeds, and in mammals it causes death by inducing strong contraction of the muscles of the chest and diaphragm. It is still used in Southeast Asia as a medicine, but the scientific basis for its use is rather frail, and the risks involved are generally too great. As a result, it is now mainly used to poison mammal pests, such as moles, but even here the risk to other wildlife is considerable and the material is best avoided.

A less dangerous alkaloid is emetine, produced in the roots of the ipecacuanha plant (*Cephaelis ipecacuanha*) of the rain forests of Brazil. As in the case of most toxic alkaloids, it must be greatly diluted before it is safe to humans, in this case just one part in 100,000. But even at this very low concentration, emetine is fatally poisonous to the dysentery amoeba (*Entamoeba histolytica*). This parasite causes a widespread and debilitating disease throughout the Tropics, often proving fatal in young children, so the drug is extremely valuable as a treatment. The plant is now grown commercially, particularly in Southeast Asia.

Some of the chemicals accumulated by tropical plants for their own protection act resemble animal hormones and act on the grazer's physiology in this way. The yam (*Dioscorea* species), for example, produces an alkaloid called dioscin that can cause respiratory failure if eaten in large quantities. Its chemistry is very similar to that of animal sex hormones, so it is used commercially for the manufacture of contraceptive pills and various other steroids used in the treatment of asthma, skin complaints, and rheumatoid arthritis.

One of the main arguments for the conservation of biodiversity of the tropical forests is the likely existence of many undiscovered sources of medicines and drugs. A prime example of a species on the verge of extinction that has proved of great pharmaceutical value is the Madagascar periwinkle (*Catharanthus roseus*), which grows in the remnants of the forests of the island of Madagascar, off the East African coast. The plant contains an extensive armory of at least 80 toxic alkaloids to protect it from predators, so it is chemically a well-defended plant. Scientists tested some of these compounds as possible aids in the treatment of diabetes, but in doing so they discovered a much more valuable property: The extract had a strong delaying effect in the progress of acute lymphocytic leukemia. It also proved

Quinine

The cinchona tree (*Cinchona officinalis*) is just one of about 65 species of this genus found in the forest of South America, mainly in the montane forests of the Andes foothills of Columbia, through Ecuador, Bolivia, and Peru. Spanish Jesuit priests who settled in Peru were the first to make written records of the use of the bark of cinchona for the treatment of malaria in 1630 (see sidebar "Mosquitoes and Malaria," page 191). According to a slightly unreliable account, the priests recommended the use of the traditional remedy when the countess of Chinchona became ill, and their proposal resulted in her cure. Whether this is strictly true, it inspired the great botanical taxonomist Carolus Linnaeus (Carl von Linné; 1707–78) to name the plant in her honor. Unfortunately, his spelling was not as acute as his taxonomy, and the name was recorded as *Cinchona* rather than *Chinchona*.

By the end of the 17th century the bark was widely used in Europe in the treatment of malaria, but the imported material was of uneven quality, which led a Portuguese chemist to analyze and extract the active component. He made an alcohol extract and purified a crystalline compound that was eventually given the name quinine. As its reputation rose, the world demand for quinine caused excessive demands to be placed on the wild *Cinchona* trees of South America, which were felled without being replaced. As the trees became scarce, the price of quinine rose further, so increasing pressure fell on the forest. In time, cultivation of the tree proved a partial answer to this problem, and commercial plantations were established, particularly in Southeast Asia. The young trees need to reach an age of four years before harvesting can begin, and maximum production of the toxic alkaloid is not reached until the tree is about 12 years old. During World War II Southeast Asia became inaccessible to the Allied forces, leading to an intensification of research into alternative synthetic antimalarial drugs. Chloroquine and paludrine gradually replaced naturally produced quinine as both cheaper and more effective alternatives. They also had the advantage of producing less unpleasant side effects. But the malarial parasite is a highly efficient parasite that has proved extremely flexible in its evolution, and drug resistance is now quite widespread in the malarial organism *Plasmodium*. The search continues for new treatments, including new plant-based drugs and the possibility of a vaccine.

effective in the treatment of Hodgkin's disease. The extraction and concentration of the alkaloids, however, is not a simple matter, and approximately two tons of the periwinkle are needed to produce just 0.04 ounce (1 g) of the alkaloids. This is sufficient for about six weeks of treatment for a child with leukemia. Large-scale cultivation is evidently called for if these active compounds are to be used extensively, or, alternatively, a synthetic chemical that apes the activity of these alkaloids must be manufactured.

The list of known pharmaceutically valuable tropical forest plants is very long, and the list of unknown and potentially important drug sources is undoubtedly much longer. Among the people of Polynesia, for example, 427 plant species are used medicinally, and of these, 66 percent are used in no other parts of the world so are completely unknown to medical science. Just 17 percent of the Polynesian plants are already used in Western medicine, so there is great scope for examining the other plants that the people of Polynesia find useful, to determine how these plants could be of wider application in world medicine.

◼ FOOD FROM THE FOREST

The tropical forests are thus an enormous resource for the production of timber and a range of industrial and pharmaceutical chemicals. It is possible that they also contain species of plants and animals that could be brought into domestication and used to help supply the world with food. It is surprising how little progress has been made in the area of domestication since the initial development of the idea some 8,000 to 10,000 years ago. About 90 percent of the world's food production is still based on just 20 species of plants, all of which were brought into cultivation many thousands of years ago. It is unlikely that our ancient ancestors hit upon all of the species worth domesticating.

Botanical surveys, many based on the work of ethnobotanists working in remote jungle areas, have shown that there are at least 30,000 plants that can be eaten, at least in part, and many of these are already used by indigenous people, who are aware of their valuable properties. Some of these also contain the poisons that may have medical properties, so need careful treatment before they can be consumed. Manioc, for example, contains compounds that generate highly toxic cyanide, which can only be destroyed by prolonged cooking, so agriculturalists need to learn about the problems as well as the potentialities of plants. Many members of the pea and bean family, Fabaceae, are very rich in proteins. The winged bean (*Psophocarpus tetragonolobus*) of Southeast Asia, for example, has up to 42 percent protein and a wide range of amino acids. But many members of the family also contain highly toxic compounds, including toxic amino acids. The Caribbean legume *Mucuna* species, for example, contains between 6 and 9 percent of an amino acid that is not involved in protein construction, but is highly toxic to insect grazers. It is called L-DOPA and is extremely valuable in medicine for the treatment of Parkinson disease.

The animal life of the tropical forest is extensively used by indigenous people as a source of food. In the Amazon Basin, approximately 50 percent of the protein intake by humans comes from the hunting of wild game, often referred to as "bush meat." While human populations are low in density, the forest can sustain this degree of predation, which in the Amazon is reckoned to account for between 67,000 and 164,000 tons each year. Between 10 and 24 million animals are killed each year in the Amazon—mainly larger mammals that are most easily hunted. One of the problems associated with forestry is that roads are built into the deep jungle to provide access for loggers so that they can extract felled timber, and bush meat hunters then use these roads and tracks for the hunting of game. Instead of using bush meat simply as a protein supplement for small communities and villages, these hunters collect meat on a large scale and transport it to towns where it can be sold in markets. The situation in West Africa is far worse than in South America, with perhaps 20 to 50 times as many animals being killed, including many primates, such as chimpanzees and gorillas. Not only is this a serious threat to the conservation of tropical forest mammals, but it also places the hunters at risk because some of the diseases of the apes are transferable to humans, including HIV.

There are opportunities for domesticating more forest animals for human consumption, which would perhaps moderate the amount of bush meat hunting. Many of the domestic animals currently used by humans as food are temperate in origin and are open-grassland grazers, such as sheep and goats. The savanna grasslands of the Tropics have many grazing species, including gazelle, antelope, and zebra, that are efficient herbivores in tropical conditions because they are more resistant to the insect pests of the low latitudes and are better adapted to grazing the natural vegetation of these regions. The forests of the world are the natural habitats of wild pigs and cattle, and many of these have attributes that could improve the efficiency of domestic stock in tropical environments. An example is banteng cow of Java (*Bos banteng*), an endangered forest species and a possible ancestor of some types of modern cattle. Pigs and peccaries, together with the South American capybara, offer additional possible opportunities for domestication. Not all animals take well to domestication, but selection for a placid temperament can help to develop an appropriate stock for breeding.

The forests are thus a rich source of genetic diversity, and in these days of genetic engineering, used to supplement the breeding programs that agriculturalists have long been using, many of these genes may prove useful. For example, it may be possible to find wild genes that can be used to make corn a perennial crop, provide barley with a better degree of salt tolerance, or increase the productivity of some tropical grasses or the disease resistance of cattle. Biodiversity is more than just a collection of species; it is also a bank of genes lying hidden within wild plants and animals, accumulated over thousands of years of selection by the stresses of nature and honed to a high degree of efficiency. It would be wasteful of humankind to lose this genetic diversity before it has even been discovered.

■ THE FOREST AS A CARBON SINK

Although the diversity of the forest is one aspect of its great value to humankind, the sheer bulk of the tropical forest is equally important. One of the greatest problems currently facing humanity is how to keep the global circulation of carbon under control. The tropical forests are an important element within the carbon cycle because they

represent a massive reservoir of organic matter, rich in carbon, stored mainly in the form of timber, vegetation, and soil organic matter.

Carbon is an essential element for all living organisms. Along with hydrogen and oxygen, it accounts for the bulk of all living and dead organic matter. Carbon is also a component of the atmosphere, where it is present at quite a low concentration, less than 0.04 percent by volume. It is this atmospheric reservoir of carbon that green plants tap when

The global carbon cycle. Deforestation contributes a significant quantity of carbon to the atmosphere, amounting to approximately 25 percent of that contributed by the burning of fossil fuels. Regrowth of lost forests could provide an important drawdown of carbon from the atmosphere. Figures in boxes are quantities expressed in gigatons (billion tonnes) and figures on the arrows are rates of movement in gigatons per year. The metric tonne, which is used by biogeochemists, is roughly equivalent to the imperial ton.

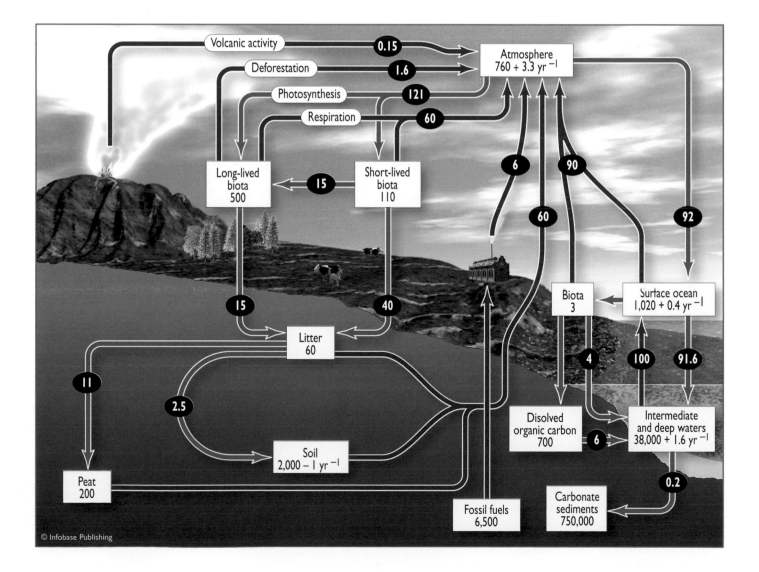

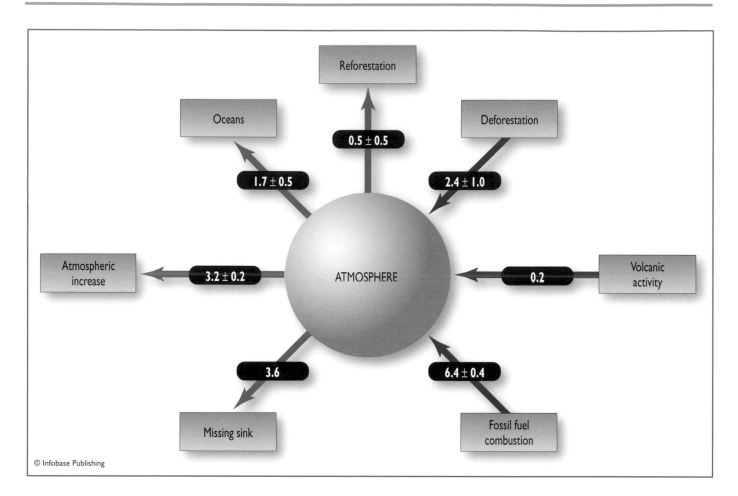

The carbon budget for the atmosphere expressed in gigatons (billion tonnes) per year. The metric tonne is roughly equivalent to the imperial ton. There are both sources and sinks for carbon that either add or remove carbon to or from the atmosphere. Sources of carbon currently exceed the sinks to the extent of 3.2 billion tonnes per year, so the carbon content of the atmosphere is steadily increasing. The "missing sink" consists of the carbon that is evidently leaving the atmosphere but is currently unaccounted for. Perhaps some of this is taken up by the regrowth of forests that had been previously felled.

they photosynthesize and convert the atmospheric carbon dioxide to organic materials (see "Primary Productivity in the Tropical Forests," pages 78–82). When organic matter decays it returns to the atmosphere, completing the carbon cycle. In fact, the carbon cycle, as shown in the diagram on page 209, is more complex because there are other paths along which the element can move. Atmospheric carbon dioxide dissolves in the oceans, interacts with the minerals in rocks, and is released into the atmosphere by volcanic eruptions. Carbon is also stored in ancient organic materials produced by photosynthesis in the distant past—hundreds of millions of years ago—and this store is currently being used as a source of energy by people, thus releasing fossil carbon into the atmosphere.

The accompanying, simplified diagram of the global carbon cycle distinguishes between reservoirs where carbon resides for a period of time and paths along which it moves in its constant cycling. The figures are approximate values for the quantity of carbon stored in reservoirs and the rates at which carbon moves along its various paths. The reservoirs labeled *biota* represent the total biomass of living organisms; it is evident that the total carbon in the world's biomass is only a little less than the total in the atmosphere at any given time. Of this terrestrial (land-based) biomass, about 80 percent resides in the forests of the world, of which

the tropical forests form the greatest proportion. The tropical forests, therefore, are one of the world's biggest reservoirs of carbon, and are also the most vulnerable reservoir. It would be difficult to convert the dissolved carbon in the oceans or the stored carbon in all the limestone rocks of the world into atmospheric carbon dioxide. Converting forest biomass to a gas, however, is relatively easy and is currently proceeding quite rapidly.

Carbon is also being taken up during the process of primary productivity as the trees and vegetation photosynthesize. But carbon is also being returned to the atmosphere as the trees respire and as they drop leaves, lose roots, are eaten by herbivores, or simply die. Just because primary productivity is high in the tropical forest does not mean that the forests are a *carbon sink* for atmospheric carbon dioxide. The word sink implies that a material is constantly disappearing into it, and the forests could only act as a sink for carbon dioxide if they were taking up more carbon than they are releasing. If this is taking place, then the biomass reservoir of carbon should be increasing. If the biomass of the tropical forests is constant, then the forest can be regarded as carbon neutral—in other words, it is neither an overall source nor a sink for carbon. It is very difficult to measure whether the tropical forests globally are growing, shrinking, or staying roughly the same. Atmospheric chemists constantly analyze the carbon dioxide levels of the atmosphere, and from the seasonal changes in concentration they can obtain some idea of how the balance is operating. Atmospheric carbon rises in winter, when productivity is low and respiration remains high, but drops in summer as the plants increase their carbon uptake. There is some slight evidence that more carbon is taken out of the atmosphere by primary producers each year, but is this due to the growth of tropical forests? Probably not. The effect is seen most strongly in the Northern Hemisphere and is likely to be a consequence of the expansion of forests in some of the northern lands rather than changes in the Tropics. Indeed, the Tropics are experiencing net deforestation (see "Forest Clearance," pages 213–216), so the region is more likely to be a source of carbon than a sink.

The carbon budget for the atmosphere (see diagram on page 210) shows that globally deforestation is contributing more carbon to the atmosphere than is being removed as a result of reforestation. Volcanic activity is responsible for only a small input of carbon to the atmosphere; the burning of fossil fuels by humans is a major source. The oceans are a sink, but overall the atmosphere is increasing in its carbon dioxide content with each successive year. Scientists worry about the likelihood of climate change (see "Climate Change and Tropical Forests," pages 217–218), because this gas acts as a thermal blanket around the Earth, allowing the light of the Sun to enter but preventing the radiated heat from escaping. Carbon dioxide is thus a greenhouse gas. But the yearly increase in carbon dioxide is not as great as might be expected on the basis of known sinks and sources, so there must be an additional sink, a missing sink, that is as yet unaccounted for. It is possible that there are several missing sinks and that the current sinks underestimate the uptake of carbon from the atmosphere. The oceans, for example, may be absorbing more carbon than is currently believed, or the biomass of the Earth may be increasing faster than has been estimated. Rising atmospheric carbon dioxide could well cause plants to photosynthesize more rapidly because this gas often limits the rate at which productivity can take place.

There are still some great unknowns in the global carbon equation, but as far as the tropical forests are concerned it is clear that this is a massive carbon reservoir and its destruction could lead to a considerable rise in atmospheric carbon dioxide, possibly by 50 percent or more. So the destruction of forest undoubtedly acts as a source of carbon. When forests enter the process of secondary succession following clearance (see sidebar "Succession," page 63), it acts as a sink for atmospheric carbon for awhile as the biomass grows and finally stabilizes. But whether areas of mature, stable forest act as a sink for carbon dioxide is currently difficult to determine. If their biomass is indeed stable, then they are carbon neutral; only if their biomass is gradually increasing can they be regarded as a sink, and this has not been demonstrated.

Recent studies of the carbon dynamics of tropical forests have revealed an additional problem that could greatly modify current views on the interaction between the forests and the atmosphere. Tropical forests have been found to produce methane, which is a greenhouse gas with about 30 times greater potential for heat absorption than carbon dioxide. The production of methane by tropical forests is particularly associated with the abundance of *methanogenic bacteria* in the soils of *wetland* areas, but even the dry woodlands of the tropical savannas produce significant quantities of methane. It is possible that as much as half of the 125 million tons of methane released globally each year by vegetation comes from the tropical forests. The role of the forests in the global carbon cycle and the impact of tropical forests on the greenhouse gas concentration of the atmosphere thus remain areas of scientific uncertainty.

■ CONCLUSIONS

Although tropical forests are of great value to people, it is not always possible to put an economic price on this value. The value of the forest for timber production is relatively easily measured because a price can be calculated for timber extraction. Many forest trees are of great economic importance, including mahogany and teak, but many of these trees, especially large specimens, have become scarce in the wild and future timber production needs to employ more managed forestry rather than systematic felling. Selective felling of the more valuable trees reduces the general ecological impact of clear felling, but still causes very extensive damage to the forest because of the problems of timber extraction.

Many forest plants are valuable to people for reasons other than timber production. Some are a source of food, including the cacao plant, whose beans are the basis of chocolate manufacture; many others produce spices, including cinnamon and nutmeg, used for flavoring. Fibers, canes, and weaving materials are obtained from jute, rattans, and bamboo. Other plants are important in industry, such as the rubber tree, which produces latex that can be hardened into rubber. Resins and gums, obtained from some tropical trees, are used in a variety of industries, including perfumery, wood preservation, and varnish production.

Many forest plants have developed chemical defenses to withstand the constant grazing of insects and other herbivores, and some of these chemicals have proved to be medicinally useful. Among the most famous of these is quinine, which is an effective agent against the malarial parasite. Some compounds, such as poison from *Derris,* are useful as environmentally friendly insecticides. Others, like the extract of the Madagascar periwinkle, which combats leukemia, and the alkaloid derived from yam, which acts as a contraceptive, are now widely used in medicine. Indigenous forest people have extensive knowledge of the properties of forest plants, based on a wisdom that has been accumulated from the experience of many generations. The study of this great body of knowledge, called ethnobotany, is an increasingly important area of research for scientists involved in pharmacological developments.

Some forest plants and animals could be exploited more extensively if they were cultivated and domesticated, including some tropical beans. Many more contain genetic adaptations that could be incorporated into domestic plants and animals through genetic engineering. One of the great values of biodiversity, including genetic diversity, is the storehouse of attributes contained within the forest biota.

The forest is also valuable as a major reservoir of the Earth's carbon. If the tropical forests were all cleared it would result in a very substantial rise in the atmospheric concentration of carbon dioxide that would undoubtedly have a profound impact on the global climate. Whether the forest is currently acting as a sink for atmospheric carbon is debatable; the mature forest is likely to be carbon neutral, while forest clearance in many areas is converting it into a source of carbon. Although it is not possible to place an economic value on the tropical forests, it is clear that the loss of this global storehouse of carbon and genetic information would be extremely damaging to human life, welfare, and economy.

11

The Future of Tropical Forests

The tropical forests, as has been shown in the last chapter, are valuable for a wide variety of reasons. There is a very real danger, however, that human greed will lead to the immediate exploitation of some of the valuable assets, such as the timber, and in the process destroy other, more important and long-lasting features of the forest, such as its biodiversity. The future of the forests, therefore, depends on rational conservation and the development of sustainable harvesting techniques that will not irreparably damage this very valuable resource. This policy must include the *rehabilitation* of damaged forest areas.

■ FOREST CLEARANCE

North America was once rich in deciduous and coniferous forests, but the development of agriculture, the rise of industry, and the increase in human population have led to the decimation of forest cover. The history of forests in Europe and many other temperate regions is very similar, and now this history threatens the tropical forests. The clearance of tropical forests for timber production and agricultural development is a process that has only just begun when viewed in the long course of human history, but if the history of the temperate world is any guide, the stripping of forest can take place very rapidly. It is already difficult to calculate how much tropical forest was present before the advent of human clearance, but one estimate suggests that there were almost four billion acres (1.6 billion ha). Over the last few decades, something like one third of this area has already been lost. The rate of loss is greatest in Southeast Asia, where 88 percent of the original forest has now been destroyed. Every minute of the day and night 69 acres (28 ha) of tropical forest fall to the ax, which amounts to 56,000 square miles (140,000 km²) each year. In some parts of the tropical world the rate of loss is over 1 percent per year; clearly the harvest cannot be sustained for much longer. The loss of resource and biodiversity incurred can never be fully recovered.

Deforestation is accompanied by a decline in biodiversity, but it has proved difficult to place a precise figure on such losses because of a lack of monitoring data. Some observations imply that species loss is not as fast as one might expect. A case in point is the coastal rain forest of eastern Brazil, where 90 percent of the original forest has been destroyed, but no bird species extinction has been recorded in the area. Perhaps there is a delay in the process of extinction and the habitat is now oversaturated with species, which will gradually fall over the brink into extinction. Singapore, with an area of 208 square miles (540 km²), has been quite well monitored and has lost 95 percent of its forest cover over the last two centuries. For birds and mammals, the recorded losses are close to 40 percent of the original totals of species, although this accounts only for those recorded. Allowing for the poor quality of the baseline survey 200 years ago, the losses could be closer to 60 percent or higher. On the basis of this information, one can extrapolate into the future: The present rate of deforestation could well result in the extinction of half the species in Southeast Asia over the coming century.

Forests are cleared primarily because of the economic value of the timber harvested; the land that has been cleared is rarely used for forestry and the renewal of a timber crop. More frequently the clear felled areas are burned and then planted or grazed (see illustration on page 214). The poor fertility of the tropical forest soils means that cultivation, or even pastoral activities, cannot be sustained for very long before productivity levels become economically marginal. Even with a well-managed system of cattle grazing following clearance and burning, the productivity after five years is poor.

The developed world has a high level of responsibility for driving the process of deforestation. The demand for wood has increased by approximately 25 percent over the course of 30 years, increasing the economic viability of tropical deforestation. The demand for high-protein foods, especially meat, has drastically increased, and one calculation concludes that the production of a single quarterpound hamburger actually consumes 67 square feet (6.25 m²) of

Once the forest has been felled, the understory and ground vegetation are burned prior to agricultural development. In the process, much carbon is taken from the biomass and is transferred to the atmosphere. *(Large-Scale Biosphere-Atmosphere Scientific Conference)*

forest. The control of deforestation in remote areas of poor countries is almost impossible. The legal processes of clearance and harvesting are accompanied by much illegal timber extraction, particularly in Southeast Asia, and the cultivation of ground for drug production, especially in South America. The illicit cultivation of coca (*Erythroxylum coca*), whose leaves are the source of the drug cocaine, accounts for about 2.5 million acres (1 million ha) of forest in Peru alone. The cultivation of the steep slopes of the inaccessible Andes foothills for cocaine production has led to extensive soil erosion and environmental degradation.

From the point of view of timber extraction, clear felling is not often worthwhile because only a small proportion of the tropical trees are valuable for their timber. In the forests of Southeast Asia this may amount to eight trees per acre (20 trees ha^{-1}), in the Amazon Basin just 1.5 trees per acre (6 trees ha^{-1}), and in West Africa only 0.5 trees per acre (2 trees ha^{-1}). But in the course of the felling and extraction of these valuable trees much of the remaining forest is damaged. Selective logging, therefore, is not a viable option for effective conservation. The construction of roads for loggers then gives access to cultivators and agriculturalists, who then clear and burn the remaining forest in preparation for crops or grazing animals. Sustainable forestry is a better option but requires a 35-year cutting cycle, which requires advance investment and a degree of patience. It is much more tempting to make fast money from standing timber, and this wasteful practice is still being conducted by international logging companies to supply the developed world, especially Japan, which consumes over one third of the world's tropical timber production. India leads the world as the country with the highest record of tropical forest plantations.

Plantation forest is less diverse than primary forest, but it is preferable to open grassland or arable land for biodiversity conservation. Plantations can be managed to enhance biodiversity, as in the case of coffee plantations in Central and South America. Traditional coffee cultivation has taken place in the shade of canopy trees, and this structured habitat has provided suitable conditions for overwintering songbirds from North America, including Swainson's thrush and the Northern waterthrush, whose migratory movements are shown in the accompanying maps. In recent years farmers have turned to high-yield coffee varieties that prefer to grow in full sunlight, and this change in vegetation structure over wide landscapes may well account for the recent declines in migrant songbirds in North America, some species having population falls of over 40 percent. As is so often the case, increased agricultural productivity comes at the cost of biological diversity. In the case of coffee, the sun varieties can produce yields three times those of the traditional shade varieties, so farmers will need compensation if they are to be encouraged to use bird-friendly systems of agriculture.

Economics also underlie the farmer's policy of transforming the forest to grassland for meat production, even though the gains are largely short term. Natural forest has a high overall productivity, but it is difficult to channel this into human food supplies. As has been demonstrated (see "Human Populations and Disease in the Tropical Forests," pages 190–194), the human populations in tropical forests are low compared to those of agriculturally developed regions, in part due to the difficulty of sustaining a high population density on hunting and gathering or on slash-and-burn farming. Grassland may have a lower overall primary productivity, but a greater proportion of the productivity can be directed into the human food chain via cattle. In fact, very little of the cattle meat produced as a result of forest clearance goes into the stomachs of local people. Most is exported to the meat-demanding countries of the world, largely in North America and Europe. The economic returns to the farmer certainly encourage the replacement of forest by grassland.

When a forest is converted to grassland, the local climate is altered. Overall humidity falls, temperature at ground level rises because of better penetration of sunlight, the daily variation in temperature is greatly increased, and wind speeds increase close to the ground. The transpiration of water from grassland is less than that from the forest canopy, so the air above the ground is less humid, influencing rainfall. An intact rain forest canopy absorbs more solar energy than grassland, reflecting only 12 percent compared to 20 percent for grassland. This stimulates transpiration and generates convection currents, leading to rainfall. So, the replacement of forest by grassland changes the climate of an area, leading to a reduction of precipitation by as much as 30 percent. If the forest is replaced by open soil or desert conditions, the reduction in precipitation can be as great as 70 percent.

The prospects for forest conservation on a large scale, therefore, are bleak, but there are some hopeful signs. Rates of deforestation in the Amazon Basin, for example, appear to be slowing down in recent years. In 1991, 4,258 square miles (11,000 km²) of forest were cleared. By 1995 this had

The summer and winter ranges of Swainson's thrush (*Catharus ustulatus*) and the northern waterthrush (*Seiurus noveboracensis*). Both spend their summer breeding seasons in the northern temperate regions of North America and their winters in the tropical forest zones of Central and South America.

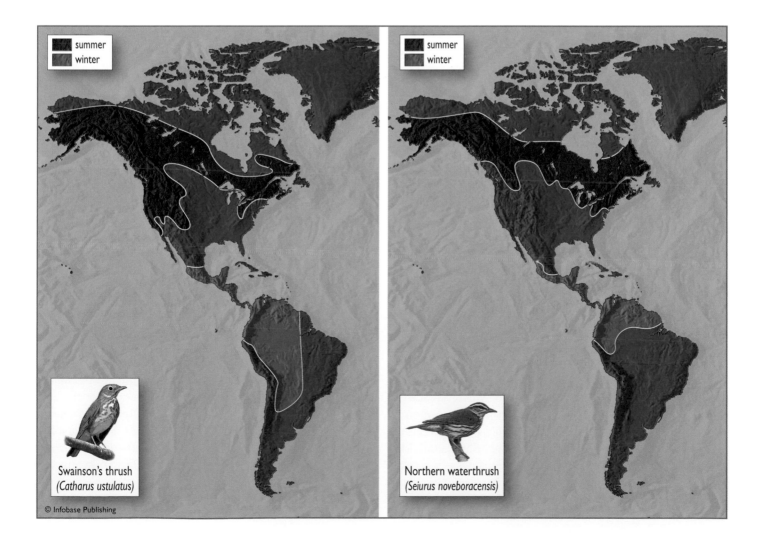

© Infobase Publishing

risen to 11,200 square miles (29,000 km²) in the course of the year, but it had fallen back to 7,250 square miles (18,800 km²) in 2005. The year 2006 saw a further decline to 5,057 square miles (13,100 km²) of forest cleared, so there is some evidence that the rate of deforestation is declining, now resembling that of the early 1990s. Brazil is also steadily increasing its protected forest areas, and added a significant new reserve in late 2006. This forest reserve is situated in the state of Pará, and extends over an area of 63,000 square miles (164,000 km²), approximately the size of Denmark. Despite such advances in conservation effort, and despite the reduction in the annual rate of deforestation in the Amazon Basin, predictions still suggest that only 40 percent of this enormous ecosystem will remain intact by the year 2050. Most of the developed world has already removed its forest and replaced it with agricultural systems designed to feed higher densities of human populations. It seems unreasonable for the developed nations to demand that the developing countries apply a brake to their own population expansion and economic growth. On the other hand, the global consequences of tropical deforestation could be grim.

FIRE IN THE FOREST

Deforestation is often followed by the use of fire to remove unwanted residual vegetation and economically worthless trees prior to cultivation or grazing. Fire is also a means of accelerating nutrient cycles, converting the elements present in biomass to a mineral form, some of which enters the soil and becomes available for plant growth. In practice, however, some of the most valuable elements are lost in smoke. Nitrogen is chief among these; burning biomass can result in over 70 percent of the nitrogen from the plant organic matter being lost into the atmosphere. Over 50 percent of the sulfur content is also lost in this way, but most of the potassium, phosphorus, calcium, and magnesium passes into the soil. The soils of tropical forests, however, are extremely leaky, and elements rapidly pass through as they are leached away by heavy rainfall, moving into streams and rivers and eventually to the sea. Burning does not, therefore, substantially increase the fertility of forest soils, and the rapid losses of elements accounts for the decline in agricultural productivity over the subsequent five years.

One of the main elements to be released to the atmosphere during biomass burning is carbon, which is oxidized to carbon dioxide. The great biomass reservoir of carbon is thus shifted into the atmospheric reservoir, where it adds to the greenhouse effect and global warming. Much emphasis is given to the effects of fossil fuel combustion as a source of atmospheric carbon, but forest clearance and biomass combustion account for a movement of carbon equivalent to approximately 40 percent of that due to fossil fuels. Each

year tropical forest fires consume over 50 million acres (20 million ha), an area roughly half the size of California. Obviously, this is a major process on a global scale.

Fires started deliberately to complete the clearance of an area are also liable to penetrate into the forest interior, especially where the forest is already fragmented. In this way, fires become extensive and, especially in a dry period, can have long-lasting effects on the rain forest wildlife. The intensity of fire is important—that is, the temperature attained and the time over which high temperature is sustained—but so is the frequency of fire. In fragmented forest areas, the residual patches of forest border onto many clearings, so they are exposed to fires from all sides and may experience regular and frequent burning as a consequence. All tropical forests are occasionally subjected to fire from natural causes, even wet-forests, although these are less flammable because the canopy traps layers of humid air. The dry forests are generally less badly damaged by fire because many of the tree species have thicker, protective bark, but wet-forest trees usually have thin bark and are more seriously damaged by a high fire frequency. Any disturbance of the forest, especially one that results in the removal of parts of the canopy, renders it more prone to the spread of fire because the complex structure of the rain forest conserves humidity and gives it a degree of fire resistance. Lowered humidity in disturbed forest thus greatly increases the risk of fire. Fire is one of the main problems following selective logging of a forest.

When a forest burns, the surface litter of leaves and fruits is often completely destroyed, leading to the loss of up to 85 percent of the viable seeds on the forest floor. As a result, recovery following fire is slower than that following other natural disasters, such as wind damage or flood. A different type of vegetation usually invades following a fire, consisting mainly of grasses and vines, so the secondary succession is deflected along a different path. Repeated fires during the recovery phase lead to the selection of fire-resistant species, so the entire composition of the forest is altered. In regions where there is a dry season, this can eventually result in the conversion of forest to savanna woodland.

Animal life responds to fire in a complex variety of ways. Some mobile animals, such as birds, many mammals, reptiles, and flying insects, can escape the fire and later reinvade from outside. Some survive fires by burrowing into the ground. The temperature gradient in soils during a fire is very steep, so that an animal only a few inches below the surface may avoid the damaging temperatures of the surface. Others find shelter within crevices in bark, or beneath stones, or in rotting logs. But the landscape following a fire is greatly changed, with open conditions and lower humidity. As a result, the survivors are subjected to more stresses, such as predation and desiccation.

In many parts of the world fire is a natural and regular feature of tropical forests, but an artificially increased fire frequency and the penetration of fire into normally very wet forests are undoubtedly harmful to biodiversity. The wet bog forests of Southeast Asia are particularly at risk of damaging fires.

■ CLIMATE CHANGE AND TROPICAL FORESTS

The temperature of the Earth has risen by about 1.1°F (0.6°C) over the past 150 years. This increase, however, has not been steady and uniform. There was a steep rise in temperature between 1900 and 1945, followed by a cooling over the next 25 years. Since 1970 global temperatures have been rising steeply once again. Climate change has always been a part of the history of the Earth, so fluctuations are not unexpected, but the changes observed during the last century or so have been surprisingly fast. Most scientists who have studied these recent changes agree that human activities during this time have contributed substantially to the temperature rise, especially the injection of greenhouse gases (carbon dioxide, methane, nitrogen oxides, chlorofluorocarbons, and ozone) into the atmosphere. On the assumption that human behavior patterns will not alter substantially and that the current trend in rising global temperature continues, one can predict that great changes will take place in the world's climate and its vegetation over the next century or so.

It is very difficult, however, to predict with any degree of precision the likely temperature and precipitation changes that will result from a continuation of current trends. Warmer temperatures may lead to increased evaporation, and this could cause increased cloud cover and thus the reflection of some incoming sunlight. Higher atmospheric carbon dioxide may lead to increased primary productivity, affecting the global balance of vegetation. Raised global temperature will cause increased rates of ice melt in the two major ice sheets of the world in Greenland and Antarctica, and this will raise the sea level and may cause alterations in the pattern of oceanic currents. So it is very difficult to create models of the world under new sets of conditions, and even more difficult to predict the rate at which changes will take place or how these changes may affect the global distribution of biomes.

Each species on Earth has its climatic preferences and tolerance limits. A change in climate, therefore, is likely to induce a change in distribution pattern for individual species, causing its extinction in some areas and an extension of its range in other regions. If all of the climatic requirements of all the world's plants and animals were known, it would be easier to map the outcome of particular climate changes. In practice, however, scientists have documented relatively few

species in this way. Biogeographers have used this approach in Mexico, assembling the climatic information on 1,870 species of animals and predicting the outcome of an increase in temperature of 3.6°F (2°C) and a fall in precipitation of four inches (10 cm), which is the likely situation in the year 2055. Most species will change their geographic ranges considerably. For example, the Mexican chachalaca (*Ortalis poliocephala*), an endemic bird of the region, will decline in the inland region but will survive in the coastal areas and move up into the foothills of the mountains. Overall, they estimate that over 40 percent of the animals in any particular location will change as a result of the climatic alterations.

Use of this model approach to the problem of responses to climate change has its difficulties, however. It assumes that the animal or plant is capable of moving into areas that become increasingly suitable as it declines in those areas that become increasingly unsuitable. In a world where habitats, especially forests, are becoming more and more fragmented, this free movement of organisms is unduly optimistic. Plants and animals have survived past climate changes because they could alter their ranges by spreading into new regions, but a lack of corridors for movement may well contribute to substantial extinctions as a result of the coming climatic alterations. Species with very limited range of distribution will also be at risk in changing conditions because there are no alternative sources for reinvasion in case of calamity.

Studies of past changes in species distributions as a response to climatic change show that individual species migrate according to their own specific needs. There is no evidence that whole communities, ecosystems, and biomes migrate as units. The outcome of individual shifts in range, however, is that new communities are created, as in the Amazon forests, which contained more montane tree species during the height of the last glacial episode (see "Tropical Forests in the Last Two Million Years," pages 180–182). At present the information about the climatic requirements of most tropical organisms is too poor to permit detailed predictions, and the climate models are not entirely consistent in their predictions about the future conditions in the Tropics. The indications are that the changes in climate are likely to be much more extreme in the high latitudes than in the low latitudes, so the boreal forest and tundra biomes will be placed under greater climatic stress than the tropical forests.

Sea level rise will become a serious problem for many coastal regions in the Tropics and will affect many regions of coastal forest and many of the small tropical islands. As in the case of climate change, it is difficult to predict precisely how rapidly the world's oceans will rise, but increasing temperature will lead to an expansion in ocean volumes, and melting ice will add to the problem. It is uncertain how rapidly the ice sheets will melt because warmer seas and higher levels of evaporation and precipitation could lead to faster ice accumulation in some areas, but the general trend is likely to be one of melting. Current predictions suggest that by the year 2100, sea levels globally will probably be 1.1 feet (0.35 m) higher than at present and could be as much as 2.6 feet (0.8 m) higher.

This may not seem like a great increase, but some parts of the world would be badly affected by such changes, including many tropical forest regions. In Bangladesh, for example, a rise in sea level of just 20 inches (50 cm) would flood 10 percent of the inhabitable part of the country, including the Sundarban mangrove swamps, resulting in the displacement of about six million people. The land surface in this part of the world is slowly subsiding, so this will make the problem more urgent. It will leave the region even more subject to devastating cyclones, storm surges, and tsunamis. Low-lying small islands (see "Tropical Volcanoes and Islands," pages 41–43) will also be at risk, such as the Maldives in the Indian Ocean and the Marshall Islands in the Pacific. In the case of the Marshall Islands, almost all of their land surface lies within 10 feet (3 m) of sea level, so a rise in the ocean would substantially reduce their surface area. Perhaps even more serious would be the contamination of groundwater by seepage from the ocean. As sea level rises, the lens of freshwater that underlies tropical islands would be reduced and could become incapable of sustaining human populations and agriculture.

Coral atolls, when in a healthy condition, are constantly growing upward, so they may be able to keep pace with the rise in sea level. They can cope with a rise of about 20 inches (50 cm) in a century. Many coral regions of the world, however, are suffering from a process of bleaching, in which the algal component of the coral dies. This may itself be related to increased water temperature and means that the coral growth will not be able to keep pace with sea level rise.

◼ MINING AND POLLUTION

Some areas of tropical forests contain mineral resources that can be harvested and are economically attractive. Unfortunately, such harvesting almost invariably involves habitat destruction and often leaves areas in such a devastated condition that recovery is virtually impossible.

The coastal bog forests of Indonesia (see "Lowland Swamps and Bog Forests," pages 68–71) sit on a very large mass of peat, which is very attractive to those who seek a source of energy in a world that is rapidly becoming depleted of its fossil fuels. Indonesia contains over 100,000 square miles (27 million ha) of tropical peat forest, of which about 30,000 square miles (8 million ha) have peat over six feet (2 m) deep. These bog forests have already suffered from illegal logging and are currently at great risk from nonregulated peat extraction. Much peatland has already been drained

and converted to agricultural use in Indonesia, but this type of land use is mainly restricted to shallow peat areas. The deep peat deposits become poor in plant nutrients because of their elevated surfaces (see sidebar "Peat," page 169), which are leached by percolating water. These deep peat sites are now the most attractive for peat harvesting so are now threatened by industrial development, and the policy of establishing protected areas needs to be fully implemented and enforced.

Precious metals are sometimes present in the rocks beneath the rain forest, including gold in Amazonia. Guiana has suffered through the use of hydraulic mining, where the sediments of rivers are extracted and washed on site. Small nuggets of gold are extracted, and the large quantities of waste sediment are a source of pollution to the river systems. Not only does the waste produce silting downstream, it also contains high levels of toxic elements, especially mercury and cyanide used in the extraction processing. The mercury contaminates fish in the river and finds its way into the human food chain. In Malaysia, the "tailings" from hydraulic mining of tin now cover considerable areas of ground around Kuala Lumpur and have proved very difficult to revegetate because of toxicity. Similar problems exist in Papua, New Guinea, as a result of copper mining.

Obtaining the energy required for mineral extraction also creates problems in the forest. In Amazonia, hydroelectricity has become an important source of energy for mining, but the construction of dams has flooded large areas of forest and destroyed an ecosystem that survived on periodic flood and dry seasons. The flood regime of regions downriver of the dams is also strongly affected. When energy is available, surface mines of considerable depth can be created, such as the Sungai Besi tin mine at Serdang, Malaysia, which is over half a mile deep. Surface mines often flood during heavy rain, carrying toxic waste into waterways. One side effect of the development of the mining industry is that migrant workers are attracted to the mining area, and extensive and often poorly planned settlement takes place. Deforestation accelerates and the felled timber provides an additional energy resource to power the mining. In the Grande Carajas Project in the eastern Amazon region of Brazil, for example, 347,000 square miles (900,000 km²) have been exploited for iron ore, copper, manganese, aluminum ore, and nickel. Each year 2,700 tons (2,743 tonnes) of pig iron are extracted, using charcoal for smelting and demanding up to 600 square miles (1,500 km²) of forest clearance.

Industrial and agricultural development in the Tropics is leading to an increase in air pollution, some of which finds its ways into the tropical forests. Fossil fuel burning and the use of agricultural fertilizers result in the release of nitrogen compounds into the atmosphere that may be washed out over the forests. In temperate regions, forests are important sinks for aerial nitrogen compounds, readily trapping them and using them for enhanced growth. In the Tropics, some forests absorb nitrogenous pollutants in the same way, but many forests are limited by phosphorus rather than nitrates for their growth, so they simply recycle any additional nitrogen back into the atmosphere.

Atmospheric pollution, particularly by chlorofluorocarbons, has resulted in reduced quantities of the gas ozone in the stratosphere, mainly in the polar regions. High-altitude ozone absorbs harmful ultraviolet radiation (UV) from the Sun, so it plays a vital role in shielding terrestrial plants and animals from its damaging effects. The regular creation of an "ozone hole" over the Antarctic as a result of accumulating pollutants is a cause for concern in the high latitudes, but its effects may even reach the tropical forests. Atmospheric scientists have recently been monitoring the levels of UV over Central and South America and have found that the frequency of high-intensity episodes increased from 5 percent to 15 percent of days over a 20-year period. They claim that this higher incidence of damaging radiation could account for an observed decline in amphibians in South America and particularly in Central America, where the amphibian decline is most marked and the UV exposure is greatest. Pollution often has far-reaching impacts so distant from the source that links between cause and effect can be difficult to establish firmly. Monitoring the consequences of pollution is hard to achieve even in temperate regions where research personnel and equipment are relatively abundant. It is even more difficult in the remoteness of the tropical forests.

■ ECOTOURISM

Tropical forests, along with all the great wilderness regions of the world, are proving increasingly attractive for recreational activities. Ecotourism is a particular type of tourism in which the participant seeks to observe nature unspoiled by destructive human activities, so this tourism is particularly concentrated in wilderness regions. Promoters of ecotourism also try to avoid damaging the places being visited and are especially concerned not to leave pollution and destruction in its wake. The activity can be summarized by the ideal that not even a footprint should be left behind to indicate that a visit has taken place. The ecotourist is keen to support additional conservation efforts to sustain the environment and to encourage local human communities in activities conducive to conservation, often by the maintenance of traditional ways of life.

Tourism in the Tropics has often been closely associated with the ocean, consequently islands have long proved popular for general recreational activities. Such tourist developments often have the advantage that beaches will be kept clean and pollution-free, but the disadvantages include the construction of access, such as airstrips or harbors, the latter often involving building on coral reefs or even harvesting the limestone of the reef for construction work. Residential facilities need to be built, often meaning that coastal habitats, such as mangroves, need to be drained and felled. Provision needs to be made for sewage treatment, and the discharge even of treated sewage into lagoons causes algal growth and ecosystem disturbance. Trash and waste materials may be disposed of locally, often as part of landfill operations for the reclamation of mangrove habitats. Food and other facilities have to be supplied, which either puts a strain on local resources or, more often, results in the import of materials that adversely affects local traders. Water is usually the scarcest resource on tropical islands that is in much demand by tourists, who insist on frequent showers. Among the most popular recreational activities are water sports, often involving the use of fast boats, jet-skis, and other vehicles that are both disturbing and damaging to local marine and terrestrial ecosystems. So the impact of tourism on a small island may bring some economic advantages to the residents, but more often it results in severe environmental degradation, with the bulk of the profits going to investors living elsewhere. The profits can, however, be considerable. The island of Guam attracts over a million visitors each year, while its resident population is only one tenth that number. Hawaii has an even bigger tourist industry with five million visitors each year.

Whereas the mass tourists gather at beach resorts and gradually erode them, ecotourists often wish to penetrate into the interior of islands and experience the undisturbed natural peace of the tropical forest. However well intentioned the visitor, there remains the danger that the very presence of ecotourists, especially in large numbers, can begin the process of destruction of the very habitat that is regarded so highly. Penetrating a forest requires the construction of trails, otherwise unregulated passage of many visitors will result in general degradation. Boardwalks may be required if the ground is wet and particularly liable to erosion. The trail must be robustly constructed and well maintained, and the number of people passing along it each day needs to be regulated to avoid wildlife disturbance. A long trail may need refreshment facilities, rest rooms, and receptacles for trash disposal. Very soon, the ecotourist's demands begin to damage the environment.

It is not difficult to encourage the ecotourist to be quiet, to take trash home, to avoid picking flowers or disturbing animals, and to refrain from feeding wildlife unsuitable materials. The ecotourist should take nothing into an ecosystem and remove nothing apart from photographs and memories. Those who organize ecotourist activities often wish to educate and interest their clients in the natural world, though the people who indulge in this type of tourism are usually relatively well educated and committed to conservation attitudes. The construction of nature trails, in which items of interest are described in a trail leaflet or on information boards, is often a high priority, particularly for young people.

Tropical forests raise particular challenges for the ecotourist and those who seek to construct an ecotourism industry. The complexity of the forest canopy is difficult to appreciate from the ground; indeed, the general lack of vegetation at ground level can make a walk through a tropical forest a dark and dull experience. Ideally, the forest needs to be viewed from high in the canopy, where most of the activity takes place. This can be achieved by constructing aerial boardwalks, but it does entail considerable expense and can be regarded as an undue interference with the natural habitat. An alternative is to lay trails along hillsides where natural rock outcrops occasionally present the visitor with views over the forest, or where streams and waterfalls create open areas and the penetrating light leads to greater variety of ground vegetation.

The problems of giving the ecotourist a full visual picture of the structure and composition of tropical forests without disturbing them may well limit the demand for this type of habitat as a potential tourist development. The wide landscapes and the ease of viewing animals in biomes such as the savanna grasslands and woodland, and the tundra cannot be fully matched in the forest. But the experience of dark, dank habitats filled with exotic smells and sounds can compensate for the visual problems, and the demand for tropical forest ecotourism is likely to increase as a consequence. Such developments may well concentrate on specific goals, such as the viewing of mountain gorilla families in the Virunga and Ruwenzori regions of Uganda.

■ HUNTING AND DISEASE

As ecotourism has become more popular, hunting has become more common through the viewfinder of a camera rather than along the sights of a gun. There are situations where culling of animals is a necessary part of the management of a region—for example, where landscapes become overgrazed or where food scarcity leads to starvation and disease. But culling usually concentrates on the weaker members of a population rather than animals in their prime. Trophy hunters, however, are usually interested in obtaining the largest and most impressive animals in a herd. The selection of large male animals might be considered least harmful, as breeding demands a greater input from females than from males. For many species, relatively few males are needed in a population to maintain stability, and the nonbreeding males could be harvested without causing serious harm. Trophy hunters are more likely to target the largest and most successful males, which are pre-

cisely the animals needed to sustain high genetic quality in the population. Recent studies on the genetics of animal populations in which large males are selectively removed indicate that this hunting practice results in a weakening of the gene pool. Tourist hunting is generally harmful and needs to be carefully controlled. On the other hand, if the hunting is regulated it can bring considerable revenue, which can be used to stimulate the interest of local people in the conservation of habitats.

Hunting by local people for bush meat is likely to increase because of the general scarcity of protein among many tropical forest cultures. This is now a major conservation problem in the tropical forests of the world (see "Food from the Forest," pages 208–209). Large mammals, particularly primates, including the great apes, are most at risk. Approximately 80 percent of the world's gorillas and most of the chimpanzees are restricted to Gabon and the Republic of Congo in western Africa. Between 1993 and 2000 the populations of both these great apes was cut in half, mainly as a result of commercial bush meat hunting. As the remaining areas of forest are fragmented and opened up for hunting, numbers will undoubtedly continue to fall. The decline in apes is accelerating because of the spread of disease among them. Ebola hemorrhagic fever is spreading rapidly among apes in Gabon and Congo and is probably killing as many as the hunters. Contact with humans is undoubtedly assisting in the spread of this disease, which we share with the apes. The densest areas for apes lie farthest from those human settlements where the Ebola virus is present.

Anthrax is a disease usually associated with ruminant mammals, but new evidence suggests that this is also infecting ape populations in West Africa. Anthrax is caused by a bacterium, *Bacillus anthracis,* which infects the respiratory system and the gut. It results in a very high rate of fatalities among infected animals and is an additional cause for concern among great-ape conservationists. As in the case of Ebola, increasing contact with humans and their domestic stock is leading to new epidemic problems in the forest.

The future of the forest, therefore, looks rather bleak. Logging, fire, climate change, mineral extraction, and bush meat hunting are all proceeding at a great pace and may even be out of control. The political will for forest conservation is often present among the forested countries, but economic opportunities and temptations can easily erode good intentions. Having removed so much of their own forest heritage and having extracted minerals for industrial growth, the developed nations are not in a strong moral position to dictate land use and conservation to the developing countries. Conservation on a global scale is nonetheless clearly vital if the human race is to survive for any length of time on this planet, because any other option is equivalent to sawing through the branch on which we are sitting.

■ CONCLUSIONS

There are many pressures on the tropical forests of the world, and it is inevitable that their area of cover will decline in the coming decades. This is of particular concern because they contain about half of the Earth's total biodiversity. Timber is the most valuable resource, and the harvesting of the forests is proceeding at a rate of about 1 percent loss per year. The process of logging involves the construction of access roads for the extraction of timber, then the harvesting of blocks by clear felling or selective felling of the more valuable trees. The outcome for the landscape is fragmentation, which encourages local extinction of species, especially those that need large areas of deep forest or are unable to migrate across open country to the next area of forest. Selective logging takes only the most economically valuable timber, but it still causes very considerable damage to the remaining trees and seriously damages the structure and composition of the forest.

Cleared areas may be allowed to regenerate, but often they are maintained in an open condition so that they can be used for agriculture. Forest soils are poor, however, and crops are unlikely to succeed over any prolonged period without additional fertilization. The removal of forest creates problems not only for the resident birds and mammals but also for those migratory birds that spend the winter in the forest before returning to temperate latitudes for breeding. Some plantation crops, such as cocoa, have a structure sufficiently similar to the forest to support some of these migrant birds, but the number of songbirds in temperate countries such as the United States is declining, in part because of changes in the birds' winter quarters.

The replacement of forest by grassland has many attractions for the agriculturalist because beef cattle can be raised, leading to a profitable harvest. But such change in vegetation causes great alterations to the local climate. Less water moves from the soil into the air as a result of tree removal, and the level of precipitation declines by 30–70 percent. It is possible, therefore, to convert tropical forest to desert.

Fire is a natural event in the tropical forest, even in the wet rain forest, and is relatively common among the forest with a dry season. Burning after felling, however, or frequent burning of forest edges where they border on the savanna woodlands can result in the permanent loss of forest cover or the creation of forest with a changed composition, consisting largely of fire-tolerant trees.

The climate of the Earth is never stable, the last 100 years seeing a sustained rise in temperature that is likely to due in part to human industrial activity. If this upward trend in temperature continues, the forests, along with all other biomes, will be placed under further stress. Each species of plant and animal has its limits of climatic tolerance and its optimum for growth. When the climate changes, species will tend to change their geographical range accordingly, and some model predictions indicate that the kind of increased warmth expected in the next 50 years could result in a 40 percent turnover of species.

Sea levels are currently rising as a consequence of thermal expansion of the oceans, coupled with ice melt from the ice sheets. It is possible that the level of the oceans by the end of this century will be raised by one to two feet (0.3–0.7 m). This rise in sea level will affect mainly the coastal tropical forests, such as the mangroves, and the bog forests of Indonesia. Low-lying islands will also be badly affected, and some of the forested islands may disappear altogether.

Some of the tropical forest regions are rich in mineral resources or peat deposits and mining of these materials is already causing considerable damage to the forests. In Southeast Asia the bog forests are in danger of being exploited for their considerable peat resources, and the Amazon Basin has already suffered from surface mining of iron, copper, manganese, nickel, and even gold. The gold is extracted from river sediments by washing, and the wastewater is often rich in toxins, such as cyanide and mercury.

Atmospheric pollution is not widespread in the forest regions, but areas close to industrial or farming activity may receive additions of nitrogen compounds through the atmosphere and rainfall. The tree growth of most tropical forests, however, is limited by phosphorus rather than nitrogen, and it is likely that excess nitrogen is passed back into the atmosphere.

Ecotourism, in which the visitor seeks to experience the natural world without unduly disturbing it, is a major growth industry and will undoubtedly expand in the tropical forests. These habitats are more difficult to observe than other types of biome, however, and those who organize ecotourism holidays will need to develop imaginative ways of displaying the biodiversity and atmosphere of the tropical forest without disturbing it.

Hunting in the forest is not likely to attract tourists bent on recreation, but it will become an increasing problem as a result of local populations harvesting bush meat. Primates, especially the great apes, are particularly at risk because of their size. Populations of these animals in West Africa have been halved in recent years. Their situation is made even more precarious because they share certain diseases, such as Ebola and anthrax, with humans. The fragmentation of forests together with increasing exposure to people is causing widespread epidemics. The future of the forests, therefore, is far from secure, and the governments of the world need to give immediate attention to their conservation for the good of all humankind.

12

General Conclusions

The tropical forests are the most complex and diverse ecosystem on Earth, yet they are also the least studied and understood. They rank among the world's most precious resources, yet they are rapidly being destroyed. It is important, therefore, that people around the world come to appreciate the value, the beauty, and the irreplaceable nature of the tropical forests before they are needlessly lost.

The tropical forests occupy a broad belt on either side of the equator, covering a range of climatic regimes. The equatorial rain forests lie at or close to the intertropical convergence zone, where warm, moist air masses from the north and the south converge to create convection currents and atmospheric turbulence. Rising air cools and moisture condenses, resulting in a constant, year-round supply of water. When combined with the warm conditions of the Tropics, the climate is ideal for plant growth, leading to high primary productivity and massive quantities of biomass accumulation. Tall trees thrust upward into the light above the dense canopy, and shade-tolerant species of trees form layers beneath these emergents, creating a complex architecture within which lie many different microclimates. The high organic productivity and the complex structure of the forest supports a great variety of animal life, and these two factors are probably the main causes of the exceptionally high biodiversity of the forest. As many as half of all the species of animals, plants, and microbes on the planet live in the tropical forests.

As one moves away from the equator, the climate and the nature of the forest change. The tilt of the Earth's axis creates seasonal variations in climate at higher latitudes; this is felt in the Tropics by the occurrence of a dry season during the "winter," when the intertropical convergence zone has moved into the other hemisphere. The dry season becomes more marked at greater distances from the equator, and this has an impact on ecosystem function and structure. Whereas the equatorial rain forest is largely evergreen, the seasonal forests contain a greater proportion of deciduous trees, and the dry season is accompanied by leaf fall and partially bare canopies. The primary productivity of the seasonal forest is lower than that of the true rain forest, and its biomass and structural complexity are less. As productivity and architectural complexity diminish, biodiversity also falls, but the species of plants and animals present in the seasonal forest are often different from those of the equatorial rain forest, so when the tropical forests as a whole are considered, the total biodiversity is increased as a result of the variations in seasonality.

Some very high mountains occur within the equatorial region, including parts of the Andes in South America, and Mount Kenya and Kilimanjaro in East Africa. The tropical forests change in their structure and composition with altitude, generally becoming less tall and structurally simpler with increasing altitude up the side of a mountain. New types of forest are found at higher levels, including montane forests, cloud forests, bamboo forests, and elfin forests, all of which add further to tropical forest diversity.

The tropical forests thus vary with latitude and altitude, as well as with longitude. The forests form three major blocks around the Tropics, consisting of the New World block (Central and South America and the Caribbean), the African block (including Madagascar), and the Asian block (ranging from India and Sri Lanka, through Burma and Thailand, into Malaysia, Indonesia, northern Australia, and the South Pacific islands). Just as forest biodiversity varies with latitude, so too does it vary with longitude. Overall, the New World Tropics are richest in species, followed by the Asian block, with Africa being the least diverse. But even within the different forest blocks are regions particularly rich in species, the biodiversity hotspots, which often have large numbers of endemic organisms, species that have arisen in the area and are confined to that region. Exactly why hotspots of biodiversity exist is still not fully understood; these regions are clearly priority sites for conservation effort.

The great variety of life in the forests means that there are many complex interactions between species. These interactions include competition, where two organisms demand the same resource; predation, where one organism uses another as a food source; and parasitism, which is a more subtle type of exploitation of one animal by another that does not lead to the death of the host. Some plants use other plants as a means of support, including the climbers or lianes, and the epiphytes that live on the branches and trunks of trees, supported in well-lit positions and having no contact with the ground. In the tropical forest ecosystem there are also many examples of cooperative behavior, or dependency of one species upon another. Many plants rely on animals to pollinate their flowers and disperse their seeds. They allocate some of their photosynthetic resources to the attraction of these cooperative animals, gaining their attention and providing them with food.

The complexity of the forest structure, especially the layering of the tree canopies, leads to a partial separation of different zones of activity above the ground. At ground level, in the permanent shade and humidity of the forest floor, there is little vegetative growth, so most animals rely on what falls from above for their food, almost like the dwellers on an ocean bed. Organic, energy-rich materials, such as leaves, fruits, dead insects, and feces, rain down from above. The ground level thus supports a detritivore food web in which dead organic matter is the main energy input. Invertebrates and microbes are important in this ground-level ecosystem, supporting larger animals, such as ground-dwelling birds and mammals, such as tapirs, which feed on fallen fruit. Large-vertebrate predators, including the jaguar, lie at the top of this ground-based food web. In the seasonally flooded forests of the Amazon, fish are important seed consumers, and reptiles, such as caimans, lie at the top of the food chain.

The lower and middle canopy is a different world from the forest floor. Most of the plants have their roots down in the ground, their leaves adapted to low-light levels. The epiphytes create their own gardens along the branches, forming microhabitats in which animals can carry out their entire lives. Many of the birds, reptiles, and mammals of the canopy will never touch the ground from birth to death, feeding on the plant and animal life that occupies the three-dimensional systems of channels and tunnels among the twigs and leaves. The flowers of the canopy, including trees and orchid epiphytes, create colorful patches that attract the attention of butterflies, bees, hummingbirds, beetles, and bats. The canopy ecosystem is thus largely independent of the ground, except for the mineral elements brought up from the soil by the trees and returned to the ground in excrement.

The high canopy contains its own range of plants and animals, living high above the ground and insulated from all the activity below. Life at the top requires some very special adaptations. The plants need to be capable of raising water to great heights and resisting the occasional drought that inevitably accompanies exposure to high light levels and temperatures. Epiphytes, in particular, need to be very drought resistant to survive in the emergent trees. For animals the power of flight is a great advantage. Birds and many insects have no problems in moving among the open conditions of the upper canopy, seeking out favored locations for feeding or nesting. Among mammals, however, the bats are the only truly flying representatives. Others, such as the flying squirrels, have developed webbing between their limbs that allows them to glide from tree to tree. The sloths survive by being able to cling tightly to branches with strong claws, while primates have developed great agility in using arms and legs for swinging through the open canopy. Cats are not very successful as top predators at great heights, but some snakes are able to take the less agile prey. Flight, however, is a great advantage to upper-canopy predators, and birds such as the harpy eagle lie at the very top of the food chain here.

The tropical forests have a very long history. Vegetation with a similar structure but very different composition from that of the present day is found back in the Carboniferous, coal-forming swamps. Conifers and cycads dominated the tropical forests of Permian, Jurassic, and Cretaceous times, but the flowering plants gradually took over dominance and emerged as the main components of the forests during the last 65 million years. Apart from the high tropical mountains, the tropical forests have not been subjected to glaciation and extreme cold, unlike the high latitudes. Climatic changes have taken place, however, that have caused considerable alterations in forest composition. During the ice ages of the past two million years, the tropical forests have experienced cooler and drier conditions, leading to some contraction and fragmentation, together with increased dominance of montane trees. Some areas of forest may have escaped the most extreme of climatic fluctuations, and these were likely to be refuges in which species survived and from which they expanded when conditions were more favorable. The survival of such refuges during periods of cold and drought has undoubtedly contributed to the overall diversity of tropical forests.

The tropical forests should not be regarded as wilderness areas that lack any sign of human life. On the contrary, they have long supported human populations and continue to do so. The primates, the animal group to which the species *Homo sapiens* belongs, evolved within the tropical forests and is still largely confined to them. Human cultures based on hunting and gathering the products of the forest are still found throughout the Tropics. Agricultural developments encouraged the practice of forest clearance for growing crops, leading to temporary settlement until soils were exhausted and the community moved on. This shifting cultivation, or "slash and burn," continues as the way of life for many forest peoples. But only low population densities of humans can be supported in this way, and populations have increased rapidly with better health and nutritional provision. This increase has led to more pressure on the forests as they are cleared for more economically attractive ways of making a living.

Forest resources include the great timber reserves, for which the world has a seemingly insatiable appetite. The harvesting of timber continues to erode the world's tropical forests at a rate of about 1 percent per year. Roads are created into the depth of the forest so that areas can be clear felled or selectively felled and the massive tree trunks extracted. Even selective felling of the most valuable timber causes great damage to the complex structure of the forest, which is the mainstay of its biodiversity. Fragmentation by clearance leads to a patchwork of standing forest interspersed with cleared areas, resulting in the local extinction of many species. Plants and animals of the deep forest fail to survive in small residual patches of trees, and the rate of loss is greatest in the smallest patches. The distance between patches also affects biodiversity because reinvasion by animals and plants lost from a site becomes less likely when the locations are far apart. Conservationists are thus eager to protect selected areas of forest that must be chosen on the basis of their biodiversity and must be of sufficient size to ensure that extinction is kept to a minimum. It is important that this selection and protection be carried out before casual and patchy exploitation of the forest results in the elimination of all large intact areas that are suitable for conservation.

Why is it important to ensure the survival of at least some of the tropical forests and their biodiversity? There are many arguments that can be used to support this policy. When a forest is cleared, the local climate is altered, usually resulting in hotter and drier conditions at ground level. The forest effectively generates its own rainfall by transpiring water from the soil to the atmosphere, so forest clearance will alter tropical climates and could create deserts or barren scrublands in places. Harvesting of trees can be conducted in a rational and sustainable way, thus ensuring a supply of timber that is renewable, but there is always a temptation to seek short-term economic gain by taking from the forest without giving anything back. One hopes that rational forestry programs will be established throughout the tropical world before this valuable resource is lost.

The forest also contains a wealth of additional resources, many of which remain as yet undiscovered. Some forest products are now vital to human survival, ranging from the industrial uses of rubber, to the recreational uses of chocolate, to the medical applications of quinine. The great diversity and complex interactions of forest organisms has led to the development of a vast array of biochemical compounds that could be of great value to humans. One of the consequences of the struggle for survival in the forest has been the development of chemical compounds for both defense and attack, and many of these can act as powerful drugs when carefully applied to human subjects. Plants have developed a wide range of toxins to protect themselves against herbivores, and some animals, such as some reptiles and spiders, have developed venoms to incapacitate their prey. Many of these protective and offensive compounds attack the nervous or the circulatory systems of their victims; used in moderation as drugs, such chemicals can prove very valuable in the treatment of human diseases. The pharmaceutical opportunities offered by the forest dwellers are enormous, so the conservation of all members of the forest ecosystem is potentially of great importance for medical science.

There are thus strong utilitarian arguments for ensuring that the tropical forests are retained and are used in a carefully planned and sustainable fashion. But there is one additional argument that rings true in the minds of many people, and that is the moral responsibility of humanity for the care of the Earth, ensuring that wild places survive and continue to inspire all that is most admirable in the human spirit. Ecotourism, if it is developed in a true spirit of conservation, will enable more people to experience and enjoy the wildness of the forest first hand. Television brings the wilderness into the homes of those who will never be able to travel to the remote places of the Earth. Viewers can be uplifted by the knowledge that wildness still exists, relatively uncontaminated by the excesses of human greed and exploitative attitudes. The mental and spiritual benefits to be gained from the forest wilderness have inspired many great writers and poets, but it is the American conservationist John Muir who has perhaps most effectively captured the call of the wild in his writings. Describing forests far removed from the Tropics, but equally impressive, he writes, "I beheld the countless hosts of the forests hushed and tranquil, towering above one another on the slopes of the hills like a devout audience. Never before did these noble woods appear so fresh, so joyous, so immortal." It is the responsibility of humanity to ensure that the tropical forests remain fresh, joyous, and, most especially, immortal.

Glossary

aerosol   particles suspended in the atmosphere as a result of their very small size

albedo   the reflectivity of a surface to light

allelopathy   the capacity of a plant to produce and disperse chemical compounds that interfere with the germination and growth of potential competitors

allogenic   forces outside a particular ecosystem that may cause internal changes; for instance, rising sea level can influence water tables in freshwater wetlands further inland. It is therefore considered an allogenic factor

anaerobic   lacking oxygen. *See* ANOXIC

anion   elements or groups of elements carrying a negative charge, e.g., NO_3^-, HPO_3^-

anoxic   lacking oxygen. *See* ANAEROBIC

aspect   the compass direction in which a surface is orientated

ATP   adeonosine triphosphate. The molecule in cells that acts as a temporary storage system for energy

autogenic   forces within an ecosystem that result in changes taking place, e.g., the growth of reeds in a marsh result in increased sediment deposition. *See also* FACILITATION

autotrophic   organisms capable of constructing complex organic molecules from inorganic sources, such as green plants

biodiversity   the full range of living things found in an area, together with the variety of genetic constitutions within the species present and the range of microhabitats available at the site

biological spectrum   the proportions of different LIFEFORMS within an ecosystem

biomass   the quantity of living material within an ecosystem, including those parts of living organisms that are part of them but are, strictly speaking, nonliving (e.g.,

wood, hair, teeth, claws), excluding separate dead materials on the ground or in the soil

biome   a large-scale community defined on the form of its vegetation. Biomes include tundra, boreal forest, desert, tropical rain forest, and others

biosphere   those parts of the Earth and its atmosphere in which living things are able to exist

biota   the sum of living organisms, plants, animals, and microbes

blue-green bacteria (cyanobacteria)   once, wrongly, called blue-green algae. Microscopic, colonial, photosynthetic microbes that are able to fix nitrogen. They play important ecological roles in some wetlands because of their nitrogen-fixing ability

bog   a peat-forming wetland in which the water supply arrives entirely by rainfall rather than from groundwater

bog forests   acidic, rain-fed, domed tropical mires (true BOGS) that accumulate deep peat deposits in some equatorial coastal regions, particularly in Southeast Asia. They are regarded as the closest modern equivalent to the Carboniferous coal-forming swamps

bog mosses   a distinctive group of mosses, all belonging to the genus *Sphagnum*. They have the capacity to hold up to 20 times their own weight in water and are also able to retain CATIONS. Most species are associated with acidic mires, including tropical BOG FORESTS

boreal   northern, usually referring to the northern temperate regions of North America and Eurasia, which are typically vegetated by evergreen coniferous forests and wetlands. Named after Boreas, the Greek god of the north wind

calcareous   rich in calcium carbonate (lime)

capillaries   fine tubes, as in the structure of partially compacted soil

carbon neutral an ecosystem that generates as much atmospheric carbon dioxide through respiration as it captures by photosynthesis

carbon sink an ecosystem that stores carbon, taking in more carbon than it releases to the environment

catchment a term meaning a region drained by a stream or river system (equivalent to WATERSHED)

catena of soils, refers to changes in the profile of the soil along an environmental gradient

cation elements or groups of elements with a positive charge, e.g., Na$^+$, NH$_4$$^+$, Ca^{++}

cation exchange the capacity of certain materials (such as peat and clay) to attract and retain CATIONS, and to exchange them for hydrogen in the process of LEACHING

cavitation a break in the water column within the vessels of a tree trunk

charcoal incompletely burned pieces of organic material (usually plant) that are virtually inert and hence become incorporated into lake sediments and peat deposits, where they provide useful indications of former fires

climax the supposed final, equilibrium stage of an ecological succession. Many question whether real stability is ever achieved

coal ancient peats that have been physically and chemically altered as a consequence of long periods of compression, sometimes at high temperature

community an assemblage of different plant and animal species, all found living and interacting together. Although they may give the appearance of stability, communities are constantly changing as species respond in different ways to such environmental alterations as climate change

competition an interaction between two individuals of the same or different species arising from the need of both for a particular resource that is in short supply. Competition usually results in harm to one or both of the competitors

composition the range of species associated in a community

consumer an organism that relies on other organisms for its food (HETEROTROPHIC). Consumers may be primary, secondary, tertiary, and so on, depending on their position in a FOOD WEB

Coriolis effect the tendency of free-moving objects, such as air masses, to be deflected by the rotation of the Earth on its axis. Deflection is to the right in the Northern Hemisphere and to the left in the Southern Hemisphere

Crassulacean acid metabolism (CAM) a photosynthetic mechanism in which carbon dioxide is temporarily fixed in the form of organic acids (often during the night) and is later released within the plant cells to be fixed again by conventional metabolic processes

cyanobacteria See BLUE-GREEN BACTERIA

cyclone a spinning mass of air created by converging air masses of different temperature, when the warmer, less dense air is pushed upward by the cooler, denser air, resulting in unstable storm conditions

deciduous a plant that loses all its leaves during an unfavorable season, which may be particularly cold or dry

decomposer a MICROBE involved in the process of DECOMPOSITION

decomposition the process by which organic matter is reduced in complexity as microbes avail themselves of its energy content, usually by a process of oxidation. As the organic materials are respired to carbon dioxide, other elements, such as phosphorus and nitrogen, are returned to the environment where they are available to living organisms once more. It is therefore an important aspect of the NUTRIENT CYCLE

deterministic a process in which the outcome is predictable and does not allow for chance (STOCHASTIC) events

detritivore an animal (usually invertebrate) that feeds upon dead organic matter

diameter at breast height (DBH) the diameter of a tree trunk measured at 4.25 feet (1.3m) above the ground

dimension analysis estimating the biomass or the timber volume of a tree on the basis of its DIAMETER AT BREAST HEIGHT

disjunct of populations of an organism that are widely separated geographically

dissociation the separation of two elements from one another in solution, forming charged IONS

diversity a term that includes both the variety of elements in an assemblage and the relative evenness of their representation

drip tip the apex of a leaf extended into a long, descending point and that assists water to drain freely from the leaf

ecological niche See NICHE

ecosystem an ecological unit of study encompassing the living organisms together with the nonliving environment within a particular habitat

emergent when used in a forest context the term refers to extremely tall trees, usually over 130 feet (40 m), which rise above the general forest canopy

endemic a species native to an area and confined to that area

energy sink a region of the world, such as the Arctic, where energy balance is sustained because of constant import of energy from another region, such as the Tropics

epiphyllous of a plant, lichen, or microbe that grows on the surface of the leaf of another plant

epiphyte a plant that grows on the trunk or branches of another plant (usually a tree) to gain a well-lit position. No parasitism is involved

erosion the degradation and removal of materials from one location to another, often by means of water or wind

euphotic zone the upper, well-lit region of the forest canopy. *See also* OLIGOPHOTIC ZONE

eustatic a change in sea level resulting from changing quantities of water being locked up in ice, or the volume of the oceans changing because of alterations in their temperature. *See also* ISOSTATIC

eutrophic rich in plant nutrients. *See also* OLIGOTROPHIC

eutrophication an increase of fertility within a habitat, often resulting from pollution by nitrates or phosphate, or by the leaching of these materials into water bodies from surrounding land. This increase in fertility results in enhanced plant (often algal) growth, followed by death, decay, and oxygen depletion

evaporation the conversion of a liquid to its gaseous phase. Often applied to water being lost from terrestrial and aquatic surfaces

evapotranspiration a combination of evaporation from surfaces and the loss of water vapor from plant leaves

evergreen a leaf or a plant that remains green and potentially photosynthetically active through the year. Evergreen leaves do eventually fall, but may last for several seasons before they do so

exfoliation the erosion of rocks by the flaking off of surface layers, often caused by frost

facilitation one of the forces that drives ecological succession. When a plant grows in a particular location, it may alter its local environment in such a way that it enables other plants to invade. When a water lily grows in a lake, for example, its leaf stalks slow the movement of water and encourage the settlement of suspended sediments. The lake becomes shallower as a consequence and other species of plants are able to invade, eventually supplanting the water lily

ferricrete *See* LATERITE

floodplain the low-lying, alluvial lands running alongside rivers over which the river water expands during time of excessive discharge

food web the complex interaction of animal feeding patterns in an ecosystem

fossil ancient remains, usually applied to the buried remnants of a once-living organism, but the term can be applied to ancient buried soils, or even the organic remains known as fossil fuels

fragmentation the breaking of a habitat into smaller patches, often as a result of human land use

frass dead invertebrates, shed skins, fecal material, and other animal-derived deposits that fall from the forest canopy and form part of the LITTER

frugivore a fruit-eating animal

fundamental niche the potential of an organism to perform certain functions or to live in certain areas. Such potential is not always achieved because of competitive interactions with other organisms. *See also* REALIZED NICHE

gene pool the sum of genetic variation found within a population of an organism

geothermal energy the heat energy released as the core of the Earth cools

girth the circumference of a tree trunk, usually measured at a height of 4.25 feet (1.3 m) above the ground

gleysol a soil that forms under waterlogged conditions, which is therefore ANOXIC

greenhouse effect the warming of the Earth's surface as a result of short-wave radiation passing through the atmosphere, then being converted to long-wave radiation as a result of interception and reflection by the Earth. Long-wave radiation is more likely to be absorbed by the atmosphere after reflection because of the presence of GREENHOUSE GASES

greenhouse gas an atmospheric gas that absorbs long-wave radiation and which therefore contributes to the warming of the Earth's surface by the greenhouse effect. Greenhouse gases include carbon dioxide, water vapor, methane, chlorofluorocarbons (CFCs), ozone, and oxides of nitrogen

groundwater water that soaks through the soils and rocks, as opposed to water derived from precipitation

habitat structure the architecture of vegetation in a habitat

halophyte a plant adapted to life in saline conditions as a result of its physical form, its physiology, or both

hemiparasite a plant that is only partially parasitic on its host and is capable of maintaining some photosynthetic production

heterotrophic organisms that are unable to fix their own energy and are, therefore ultimately dependent upon green plants or other AUTOTROPHIC organisms as an energy source

homiothermic   warm-blooded

hotspot   a region with exceptionally high levels of biodiversity

hydrogen bonding   a bond between two molecules in which one of the components is a hydrogen atom, often linked to oxygen or a halogen. The hydrogen bonding between water molecules contributes to the great cohesive strength of this material, which is vital for the uplift of water in tall plants

hydrology   the study of the movement of water in its cycles around ecosystems and around the planet

igneous rocks   rocks created as a result of volcanic activity

inbreeding   a population in which genetic exchange from outside is severely restricted, resulting in lack of genetic variation. *See also* OUTBREEDING

intertia   the ability of an ecosystem to resist disturbing forces

insectivorous   an organism that feeds on insects and other invertebrates. The term may be applied to certain plants that trap insects and digest them as a source of energy and nutrient elements

interception   the activity of plant canopies in preventing rainwater from reaching the ground directly. Intercepted water may continue on its way to the ground or may be evaporated back into the atmosphere

interspecific   taking place between different species

intraspecific   taking place within a population of a given species

ion   a charged element or group of elements. *See* ANION and CATION

intertropical convergence zone (ITCZ)   the band of the low-latitude zone of the Earth where air masses from the north and the south converge. Its position varies with season, migrating poleward during the summer season in each hemisphere. It is a region of low atmospheric pressure and consequently high precipitation

invertebrate   an animal lacking a backbone, such as insects, mollusks, and crustaceans

isostatic   a change in the relative position of the sea level to land level as a result of crustal warping under the influence of such factors as ice volume during glaciation

jet stream   a rapid movement of air from west to east resulting from rotation of the Earth

kleptoparasitism   a form of parasitism in which theft is involved, as in the removal of insect prey from carnivorous plants by ants

K-selected species   those species that expand their populations relatively slowly but are able to survive long term because of their competitive ability in tapping environmental resources

lapse rate   the rate at which temperature falls with increasing altitude

latent heat of evaporation   the energy needed to convert liquid into vapor at the same temperature

laterite   a tropical soil in which a combination of high temperature and precipitation results in the depletion of silica, leaving high concentrations of iron oxides

latitude   conceptual lines running around the world that are named according to the angle subtended to the equatorial plane. Thus the equator is regarded as 0° and the poles as 90° north or south. The equatorial regions lie in the low latitudes and the polar regions in the high latitudes

leaching   the process of removal of IONS from soils and sediments as water (particularly acidic water) passes through them

leaf area index   the area of leaf cover divided by the area of ground overlain

lek   a mating system in which males display as a group for the attentions of females, which are responsible for mate selection

liane   a climbing plant with a long, winding stem rooted in the ground but using trees for support

life-form   a system of classifying plants according to where they carry their perennating organs

light compensation point   the light intensity at which photosynthesis in a plant is just sufficient to provide for the plant's respiration needs, so there is no net gain in weight

limestone   sedimentary rocks containing a high proportion of calcium carbonate (lime)

litter   the dead remains of plant and animal material that falls from the canopy and accumulates on the surface of the soil

longitude   conceptual lines running from pole to pole and intersecting the equator. They are numbered from 0° at the Greenwich Meridian in southeastern England running east and west to 180° running through the Pacific Ocean

lycopsids   a group of plants related to modern horsetails (*Equisetum*) that once included large wetland species that dominated the coal-forming swamps of Carboniferous times

macrofossils   FOSSILS that are large enough to be examined without the use of a microscope. Sometimes referred to as megafossils

management in the context of wetlands ecology, a term that refers to the process of manipulation by humans to achieve a particular end (e.g., flooding, mowing, burning, harvesting, and so on)

mangal forested coastal ecosystems in the Tropics. The mangrove trees characteristically have upwardly bending roots that extend above the water level and act as respiratory organs

mangrove a term applied to both the MANGAL habitat and the trees that typify this habitat

metamorphic rocks rocks that have been modified in their structure and composition as a result of high temperature and pressure in the vicinity of volcanic or tectonic activity

metapopulation a population of an organism that is naturally patchy or has been broken into separate patches as a result of habitat FRAGMENTATION

methanogenic bacteria bacteria that produce methane gas as a result of their metabolism

microbes bacteria, fungi, and viruses

microclimate the small-scale climate within habitats, such as beneath forest canopies, or in the shade of desert rocks

microfossils FOSSILS that can be observed only with the aid of a microscope, such as POLLEN GRAINS, DIATOM FRUSTULES, plankton remains, and so on

migration the seasonal movements of animal populations, e.g., geese, caribou, or plankton in a lake

mimicry the adoption of form or coloration that causes an organism to resemble a different species, usually as a means of protection

minerals inorganic compounds that in combination make up the composition of rocks. The term is also used for the elements needed for plant and animal nutrition

mire general term for any peat-forming wetland ecosystem

monsoon a wet season in the subtropics caused by a change in wind direction bringing warm, humid air over a landmass

net primary production the observed accumulation of organic matter mainly by green plants after they have used some of the products of photosynthesis in their own respiration and metabolism

niche the role a species plays in an ECOSYSTEM. It consists of both spatial elements (where the species lives) and the way in which it makes its living (feeding requirements, growth patterns, reproductive behavior, etc.)

nidifugous of young birds that leave the nest shortly after hatching

nutrient cycle the cyclic pattern of element movements between different parts of the ECOSYSTEM, together with the balance of input and output to and from the ecosystem

occult precipitation precipitation that is not registered by a standard rain gauge because it arrives as mist, condensing on surfaces, such as vegetation canopies

oceanic climate a climate in which summer temperatures are cool and winter temperatures are mild, and often accompanied by high precipitation. Such conditions are most often encountered in regions close to the oceans

oceanic conveyor belt the movement of the oceanic waters of the world in a pattern that distributes energy from the equatorial regions to the polar regions. Warm, low-density water moves northward into polar regions, where it cools and becomes denser, returning to southern regions as deep water currents moving in the opposite direction to those at the surface. *See* THERMOHALINE CIRCULATION

oligophotic zone the lower, poorly lit region of the forest close to the ground. Contrast with the EUPHOTIC ZONE

oligotrophic poor in plant nutrients. *See also* EUTROPHIC

ombrotrophic fed by rainfall. BOGS are ombrotrophic mires, receiving their water and nutrient input solely from atmospheric precipitation

osmosis the movement of water molecules from low SOLUTE concentration to high solute concentration through a semi-permeable membrane that prevents the diffusion of the solute

outbreeding a population in reproductive contact with other populations of the same species and able to exchange genetic material. Such populations generally contain more variety in their GENE POOL than INBREEDING populations

oxbow lake a crescent-shaped body of water produced from an old river channel as a result of a new route being cut, concluding with the isolation of the old channel

pachycaul of plants that have compressed, stout stems, such as the giant senecios of tropical African mountains

paleoecology the study of the ecology of past communities using a variety of chemical and biological techniques

paleomagnetism the retention of magnetic alignment in a rock as a result of being formed within a magnetic field

palynology the study of pollen grains and spores

peat organic accumulations in wetlands resulting from the incomplete decomposition of vegetation litter

pH an index of acidity and alkalinity. Low pH means high concentrations of hydrogen ions (hence acidity). A pH of 7 indicates neutrality. The pH scale is logarithmic, which means that a pH of 4 is 10 times as acidic as pH 5

phanerophyte the LIFE-FORM of a tree in which buds are held well above the surface of the ground

photosynthetic bacteria bacteria possessing pigments that are able to trap light energy and conduct photosynthesis. Some types are green and others purple

physiognomy the general structure of vegetation

pioneer an initial colonist in a developing habitat

pneumatophore (or **pneumorrhiza**) root structures on MANGROVE trees that project above the mud and act as a means of gaseous exchange with the atmosphere. They are needed because of the anaerobic conditions in waterlogged mud that prevent roots from respiring

podzol a soil in which LEACHING has resulted in the movement of iron, aluminum, clay, minerals, and organic matter down the profile to be deposited in lower horizons

poikilohydric of plants that become desiccated under drought conditions but rapidly recover when water is available once more

poikilothermic cold-blooded

polar front the boundary where tropical air masses encounter cooler polar air masses, resulting in unstable weather conditions and the formation of depressions

pole forest the uniform tree cover found on the elevated peat domes of tropical coastal BOG FORESTS of Southeast Asia

pollen analysis the identification and counting of fossil pollen grains stratified in peat deposits and lake sediments

pollen grains cells containing the male genetic information of flowering plants and conifers. The outer coat is robust and survives well in wetland sediments. The distinctive structure and sculpturing of the coats permits their identification in a fossil form

polyphyletic a trait that has arisen a number of times in the course of evolution

population a collection of individuals of a particular species

precipitation aerial deposition of water as rain, dew, snow, or in an OCCULT form

primary forest an area of undisturbed, pristine forest

primary productivity the rate at which new organic matter is added to an ecosystem, usually as a result of green plant photosynthesis

pyramid of biomass a principle that applies to terrestrial ecosystems stating that the succeeding TROPHIC LEVELS of an ECOSYSTEM have a lower total BIOMASS than preceding levels

quality of light, refers to the spectrum of different wavelengths present

raised mire a mire in which the accumulation of peat results in the formation of a central dome that raises the peat-forming vegetation above the influence of groundwater flow. The surface of the central dome thus receives all its water input from precipitation (OMBROTROPHIC)

realized niche the actual spatial and functional role of a species in an ecosystem when subjected to competition from other species. *See also* FUNDAMENTAL NICHE

rehabilitation the conversion of a damaged ecosystem back to its original condition

relative humidity the quantity of water present in a body of air expressed as a percentage of the amount of water needed to saturate the air under the same conditions of temperature and pressure

relict a species or a population left behind following the fragmentation and loss of a previously extensive range

resilience the ability of an ecosystem to recover rapidly from disturbance

resource allocation the division of the products of photosynthesis among different parts of a plant, such as leaves, stem, and roots

resource partitioning the manner in which different species assume different roles (NICHES) in an ecosystem and thus divide the resources between them

respiration the oxidation of organic materials resulting in the release of energy. Waste products include carbon dioxide, methane, or ethyl alcohol, depending on the availability of oxygen (*see* REDOX POTENTIAL) and the type of organism involved

rheotrophic a wetland that receives its nutrient elements from groundwater flow as well as from precipitation. In rheotrophic mires the groundwater flow is usually responsible for the bulk of the nutrient input

r-selected species those species that expand their populations rapidly but have limited ability to survive intense competition. They therefore cope well in unstable, disturbed conditions

salinity the concentration of salts in a solvent

saturated vapor pressure the total amount of water vapor that a volume of air can hold at a given temperature

secondary dispersal the additional transport of a seed following its initial dispersal, as when a mouse or an ant picks up a wind-dispersed seed and takes it farther from the parent plant

secondary forest   an area of forest that has recovered from disturbance, such as felling

sedimentary rocks   rocks formed by the gradual accumulation of eroded materials, either under water on on land, eventually forming a compressed, stratified mass of material

shifting cultivation   *See* SLASH AND BURN

slash and burn   a primitive system of forest management involving local clearance, burning, cultivation, and then abandonment

solute   a material that dissolves in a SOLVENT

solvent   a liquid in which materials (SOLUTES) can dissolve

species richness   the number of species of organisms within a given area; an important component of BIODIVERSITY

spores   the dispersal propagules of algae, mosses, liverworts, ferns, and fungi

stem flow   precipitation intercepted by a forest canopy that drains down the branches and trunk of a tree, eventually reaching the ground

stochastic   a chance event; one that cannot be predicted. *See also* DETERMINISTIC

stomata (singular stoma)   the pores through which a plant exchanges gases with its environment and through which it loses water by TRANSPIRATION

stratification   the layering of the forest canopy

stratum (plural strata)   layers of the forest canopy, often given letter labels from A to E, such as the *C stratum*

stratosphere   the part of the Earth's atmosphere lying above the TROPOSPHERE, from around nine to 30 miles (150–50 km)

structure   the architecture of a forest canopy

subduction zone   a region where one tectonic plate pushes beneath another

succession   the process of ecosystem development. The stages of succession are often predictable as they follow a directional sequence. The process usually involves an increase in the BIOMASS of the ecosystem, although the development of RAISED BOG from CARR is an exception to this. Succession is driven by immigration of new species, FACILITATION by environmental alteration, competitive struggles, and eventually some degree of equilibration at the CLIMAX stage

sunfleck   a patch of light on the forest floor resulting from the uninterrupted passage of sunlight directly to the ground

swamp   a vegetated wetland in which the summer water table remains above the sediment surface so that there is always a covering of water. In North America, the term is restricted to forested wetlands of this kind, while in Europe the term is normally used only for herbaceous reed beds and cattail marshes

texture   refers to the proportions of different-sized particles in soil. A soil containing a relatively even contribution from sand, silt, and clay is called a *loam*

thermohaline circulation   the movement of water masses around the oceans of the world in a circulatory system that varies in the density of waters, caused by a combination of temperature and salt content. *See* OCEANIC CONVEYOR BELT

throughfall   precipitation that penetrates the canopy of a forest

timberline   the upper limit of forest on a mountain

topography   the form of a landscape

transect   a line along which vegetation is recorded

transpiration   the loss of water vapor from the leaves of TERRESTRIAL plants through the stomata, or pores, in the leaf surface

tropopause   the boundary between the TROPOSPHERE and the STRATOSPHERE

troposphere   the lower layer of the Earth's atmosphere, up to about nine miles (15 km)

variegated   of a plant leaf that contains an incomplete complement of chlorophyll

vertebrate   an animal with a backbone

vessels   specialized cells in wood that are responsible for the transport of water up a stem or trunk

watershed   the region from which water drains into a particular stream or wetland (equivalent to CATCHMENT). The term is also used to describe the ridge separating two catchments, literally the region where water may be shed in either of two directions

wavelength   the distance apart of crests in a wave motion

weathering   the breakdown of a rock into MINERALS and chemical elements as a result of the effects of such factors as frost, solution, and biological activity

wetland   a general term covering all shallow aquatic ecosystems (freshwater and marine) together with marshes, swamps, fens, and bogs

xerophyte   plants with structural adaptations associated with drought resistance

zonation   the banding of vegetation along an environmental gradient, as in the transition from lowland forest, through montane forest and cloud forest, to the timberline on a tropical mountain

Further Reading

GENERAL ENVIRONMENTAL REFERENCE

Archibold, O. W. *Ecology of World Vegetation.* New York: Chapman and Hall, 1995. A broad and useful introduction to all the major biomes of the world.

Bradbury, Ian K. *The Biosphere.* 2nd ed. New York: Wiley, 1998. An introduction to global processes that link the various biomes and the human population of the planet.

Brown, J. H., and M. V. Lomolino. *Biogeography.* 3rd ed. Sunderland, Mass.: Sinauer Associates, 2006. An extensive and exhaustive coverage of the scientific principles that unite biology and geography in the study of the living world.

Cox, C. B., and P. D. Moore. *Biogeography: An Ecological and Evolutionary Approach.* 7th ed. Oxford: Blackwell Publishing, 2005. An introductory text dealing with the historical and modern factors that determine species distributions on Earth.

Gaston, K. J., and J. I. Spicer. *Biodiversity: An Introduction.* 2nd ed. Oxford: Blackwell Publishing, 2004. An explanation of the concept of biodiversity, its meaning, and its importance in conservation.

Houghton, J. *Global Warming: The Complete Briefing.* 3rd ed. Cambridge, U.K.: Cambridge University Press, 2004. The authoritative account of the most recent research by the Intergovernmental Panel on Climate Change.

GENERAL TROPICAL FOREST REFERENCE

Bawa, K. S., and M. Hadley, eds. *Reproductive Ecology of Tropical Forest Plants.* Park Ridge, N.J.: Parthenon, 1990. A study of plant–animal interactions in the breeding systems of tropical vegetation.

Longman, K. A., and J. Jenik. *Tropical Forest and its Environment.* 2nd ed. New York: Longman, 1987. A good introduction to the ecology of tropical forests.

Mabberley, D. J. *Tropical Rain Forest Ecology.* 2nd ed. New York: Chapman and Hall, 1992. An introductory book concentrating on the vegetation of tropical forests.

Richards, P. W. *The Tropical Rain Forest.* 2nd ed. Cambridge: Cambridge University Press, 1996. The standard academic text covering all aspects of tropical forest ecology, but especially its vegetation.

Sutton, S. L., T. C. Whitmore, and A. C. Chadwick, eds. *Tropical Rain Forest: Ecology and Management.* Oxford: Blackwell Scientific Publications, 1983. A collection of reviews of tropical forest ecology and the problems of conservation.

Tomlinson, P. B. *The Botany of Mangroves.* Cambridge: Cambridge University Press, 1986. The special adaptations of mangrove trees.

NEW WORLD TROPICAL FORESTS

Kricher, J. A. *Neotropical Companion.* 2nd ed. Princeton, N.J.: Princeton University Press, 1997. A useful guide to the animals and plants of the tropical forests of Central and South America.

Leigh, E. G., A. S. Rand, and D. M. Windsor, eds. *The Ecology of a Tropical Rain Forest.* 2nd ed. Washington, D.C.: Smithsonian Institute, 1996. A detailed account of the ecology of one area of Central American rain forest.

Lobban, C. S., and M. Schefter, eds. *Tropical Pacific Island Environments.* Mgilao, Guam: Univeristy of Guam Press, 1997. The history, geology, ecology, and human sociology of tropical islands in the Pacific Ocean.

McDade, L. A., K. S. Bawa, H. A. Hespenheide, and G. S. Hartshorn, eds. *La Selva: Ecology and Natural History of a Neotropical Rain Forest.* Chicago: University of Chicago Press, 1994. An intensive study of a forest in Costa Rica.

Reagan, D. P., and R. B. Waide, eds. *The Food Web of a Tropical Rain Forest.* Chicago: University of Chicago Press, 1996. An intensive study of the energy flow patterns in a forest in Central America.

Young, A. M. *Srapiqui Chronicle: A Naturalist in Costa Rica.* Washington, D.C.: Smithsonian Institution, 1991. An account of ecological studies in the forests of Costa Rica.

OLD WORLD TROPICAL FORESTS

Alexander, I. J., M. D. Swaine, and R. Watling, eds. *Essays on the Ecology of the Guinea-Congo Rain Forest.* Edinburgh: Proceedings of the Royal Society of Edinburgh B. Vol. 104, 1996. A collection of scientific papers concerning the ecology of West African forests.

Bowman, D. M. J. S. *Australian Rainforests.* Cambridge: Cambridge University Press, 2000. A detailed review of the ecology and biology of the rain forests of Australia.

Groves, R. H., ed. *Austalian Vegetation.* 2nd ed. Cambridge: Cambridge University Press, 1994. A general account of Australian vegetation, including the tropical forests.

Martin, C. *The Rainforests of West Africa.* Basel: Birkhäuser, 1991. A general and readable account of the ecology of the West African tropical forests.

Poorter, L., F. Bongers, F. N. Kouame, and W. D. Hawthorne, eds. *Biodiversity of West African Forests.* Cambridge, Mass.: CABI International, 2004. A detailed description of tropical forest plant species and their distribution in West Africa.

Rieley, J. O., and S. E. Page, eds. *Biodiversity and Sustainability of Tropical Peatlands.* Cardigan, Wales: Samara Publishing, 1997. A collection of research papers concerning the ecology and exploitation of tropical forest peatlands, mainly in Southeast Asia.

Singh, J. S., and S. P. Singh. *Forests of the Himalaya.* Nainital, India: Gyanodaya Prakashan, 1992. A description of the vegetation of the Himalayan forests, including subtropical monsoon forests.

HISTORY AND ARCHAEOLOGY OF TROPICAL FORESTS

Allen, B., ed. *The Faber Book of Exploration.* London: Faber and Faber, 2002. A fascinating collection of accounts from early travelers in many parts of the world, including the tropical forests.

Araujo-Lima, Carlos, and Michael Goulding. *So Fruitful a Fish: Ecology, Conservation, and Aquaculture of the Amazon's Tambaqui.* New York: Columbia University Press, 1997. The remarkable fishes of the Amazon Basin and their role in seed dispersal during times of flood.

Balée, W., and L. E. Clark, eds. *Time and Complexity in Historical Ecology: Studies in the Neotropical Lowlands.* New York: Columbia University Press, 2006. The impact of early human societies on the landscape and vegetation of Central America.

Goulding, Michael Smith, J. H. Nigel, and Dennis J. Mahar. *Floods of Fortune: Ecology and Economy along the Amazon.* New York: Columbia University Press, 1996. The importance of the regular floods to the forests and people of the Amazon Basin.

Hamilton, A. C. *Environmental History of East Africa.* New York: Academic Press, 1982. The results of paleoecological research into the development of East African vegetation during and since the last ice age.

Hather, J. G., ed. *Tropical Archaeobotany: Applications and New Developments.* New York: Routledge, 1994. The study of human use of plants for food and medicine.

Heiser, C. B. *Of Plants and People.* Norman: University of Oklahoma Press, 1985. The history of the interaction between plants and people in Central America.

TROPICAL FOREST EXPLOITATION

Diamond, J. *Guns, Germs, and Steel: The Fates of Human Societies.* New York: W.W. Norton, 1999. A remarkable and gripping book describing the development of current patterns of human culture on Earth.

Faminow, M. D. *Cattle, Deforestation, and Development in the Amazon.* New York: CABI International, 1998. A study of environmental change in the Amazon Basin resulting from forestry and agricultural intensification.

Gash, J. H. C., C. A. Nobre, J. M. Roberts, and R. L. Victoria, eds. *Amazonian Deforestation and Climate.* New York: Wiley, 1996. A study of forest clearance in the Amazon Basin and its possible effects on global climate.

Web Sites

CONSERVATION INTERNATIONAL
URL: http://www.conservation.org
Accessed November 9, 2006. Particularly concerned with global biological conservation

EARTHWATCH INSTITUTE
URL: http://www.earthwatch.org
Accessed November 9, 2006. General environmental problems worldwide

GENERAL INFORMATION ON TROPICAL FORESTS
URL: http://blueplanetbiomes.org/rainforest.htm
URL: http://rainforests.mongabay.com/
Accessed November 9, 2006. Both sites have much information on the geography, climate, and biology of tropical forests.

INTERNATIONAL COFFEE ORGANIZATION
URL: http://www.ico.org
Accessed November 9, 2006. A site with information on the sustainable use of forests for coffee production.

INTERNATIONAL UNION FOR THE CONSERVATION OF NATURE AND NATURAL RESOURCES
URL: http://www.redlist.org
Accessed November 9, 2006. Many links to other sources of information on particular species, especially those currently endangered.

NATIONAL PARKS SERVICE OF THE UNITED STATES
URL: http://www.nps.gov
Accessed November 9, 2006. Information on specific conservation problems facing the National Parks.

RAINFOREST ALLIANCE
URL: http://www.rainforest-alliance.org
Accessed November 9, 2006. The Rainforest Alliance is an organization concerned with conservation and sustainable use of tropical forests.

RAINFOREST CONSERVATION FUND
URL: http://rainforestconservation.org
Accessed November 9, 2006. Concentrating on the Amazonian region, this site has excellent coverage of individual rain forest species.

UNITED NATIONS ENVIRONMENT PROGRAMME WORLD CONSERVATION MONITORING CENTRE
URL: http://www.unep-wcmc.org
Accessed November 9, 2006. Good for global statistics on environmental problems.

WARRAWONG
URL: http://www.warrawong.com
Accessed November 9, 2006. A site devoted to an Australian rain forest sanctuary.

Index